Mineral Mining in Africa

Africa is endowed with commercially viable quantities of several minerals and metals, and, more than ever before, African countries wish to harness their mineral resources for their economic development. The African mining sector has witnessed a revolution in terms of new mining codes and amendments to extant mining codes, which are designed to achieve a multitude of objectives, including the assertion of greater control over exploitation of mineral resources; optimization of resource royalties and taxes; promotion of equity participation in mining projects; enhancement of indigenization in the form of domestic participation in mineral production and local content requirements; value addition and beneficiation in terms of domestic processing of raw mineral ores and metals in Africa; and the promotion of sustainable practices in the mining sector.

This book analyzes the legal and fiscal frameworks for hard-rock mining in several African countries including Botswana, Democratic Republic of Congo, Ethiopia, Ghana, Guinea, Kenya, Namibia, Nigeria, Liberia, Tanzania, Sierra Leone, South Africa, South Sudan, Zambia, and Zimbabwe, with reference to other resource-rich countries. It engages in a comparative analysis of mining statutes in Africa with regard to topics such as the acquisition of mineral rights; types of mineral rights; the nature of mineral rights; the rights and obligations of mineral right holders; security of mineral tenure; surface rights; fiscal regimes including royalty and tax regimes; resource nationalism in the mining sector; management and utilization of mining revenues including benefit-sharing arrangements between mining companies and host communities; environmental stewardship; and sustainable exploitation of mineral resources.

Evaristus Oshionebo is Professor in the Faculty of Law, University of Calgary, Canada.

Mineral Mining in Africa
Legal and Fiscal Regimes

Evaristus Oshionebo

LONDON AND NEW YORK

First published 2021
by Routledge
2 Park Square, Milton Park, Abingdon, Oxon OX14 4RN

and by Routledge
52 Vanderbilt Avenue, New York, NY 10017

Routledge is an imprint of the Taylor & Francis Group, an informa business

© 2021 Evaristus Oshionebo

The right of Evaristus Oshionebo to be identified as author of this work has been asserted by him in accordance with sections 77 and 78 of the Copyright, Designs and Patents Act 1988.

All rights reserved. No part of this book may be reprinted or reproduced or utilised in any form or by any electronic, mechanical, or other means, now known or hereafter invented, including photocopying and recording, or in any information storage or retrieval system, without permission in writing from the publishers.

Trademark notice: Product or corporate names may be trademarks or registered trademarks, and are used only for identification and explanation without intent to infringe.

British Library Cataloguing in Publication Data
A catalogue record for this book is available from the British Library

Library of Congress Cataloging-in-Publication Data
A catalog record has been requested for this book

ISBN: 978-1-138-48320-0 (hbk)
ISBN: 978-1-351-05554-3 (ebk)

Typeset in Galliard
by Taylor & Francis Books

To the memory of my beloved parents,
Lawrence and Christiana
Oshionebo

Contents

List of tables xiii
Preface xv
Acknowledgements xvi
Abbreviations xvii

1 Overview of mining in Africa 1

  1 Introduction 1
    1.1 History of mining in Africa 2
    1.2 Ownership of minerals 3
  2 Generations of mining statutes in Africa 5
    2.1 'Nationalist' and 'statist' statutes of the post-colonial era 6
    2.2 Economic liberalization statutes 7
    2.3 Development-oriented statutes of the post-liberal era 9
  3 Economic significance of mining in Africa 12
  4 Aim and structure of the book 14

2 Types of mineral rights in Africa 19

  1 Introduction 19
  2 Reconnaissance licence 20
    2.1 Rights conferred by a reconnaissance licence 28
    2.2 Terms and conditions of a reconnaissance licence 29
    2.3 Duration and renewal of a reconnaissance licence 30
  3 Prospecting licence 30
    3.1 Rights conferred by a prospecting licence 31
    3.2 Terms and conditions of a prospecting licence 33
    3.3 Duration and renewal of a prospecting licence 35
    3.4 Relinquishment of land upon renewal of a prospecting licence 36

viii *Contents*

    4 Exploration licence 37
        4.1 Rights conferred by an exploration licence 38
        4.2 Terms and conditions of an exploration licence 40
        4.3 Record-keeping and reporting obligations 41
        4.4 Duration and renewal of an exploration licence 42
        4.5 Relinquishment of land upon renewal of an exploration licence 42
    5 Retention licence 43
        5.1 Rights conferred by a retention licence 43
        5.2 Terms and conditions of a retention licence 44
        5.3 Duration and renewal of a retention licence 45
    6 Mining lease 47
        6.1 Rights conferred by a mining lease 47
        6.2 Standard terms and conditions of a mining lease 49
        6.3 Duration, renewal, and amendment of a mining lease 56
    7 Small-scale mining lease 59
        7.1 Rights conferred by a small-scale mining lease 59
        7.2 Terms and conditions of a small-scale mining lease 60
        7.3 Duration and renewal of a small-scale mining lease 61
    8 Common features of mineral rights 61
        8.1 Transferability of mineral rights 62
        8.2 Surface rights ancillary to mineral rights 67
        8.3 Priority of mining operations over other uses of land 70
    9 Conclusion 71

3 Acquisition of mineral rights   72

    1 Introduction 72
    2 Land available for mining operations 73
    3 Requirements for grant of mineral rights 74
        3.1 Eligibility requirements 74
        3.2 Indigenization and domestic incorporation requirements 77
        3.3 Financial and technical competency 79
    4 Methods of acquisition of mineral rights 81
        4.1 Discretionary grant 81
        4.2 Public auction of mineral rights 91
        4.3 Mergers and takeover transactions 93
    5 Registration of mineral rights 95

    6 *An assessment of the mineral rights acquisition process* 97
        6.1 Paucity of domestic and indigenous participation 97
        6.2 Rarity of public auctions 100
        6.3 Consultation with host communities 102
    7 *Conclusion* 112

4 Security of mineral tenure   114

    1 *Introduction* 114
    2 *Sources of mineral rights* 114
        2.1 Statute-based mineral rights 115
        2.2 Contract-based mineral rights 117
    3 *Time of vesting of mineral rights* 120
    4 *The legal nature of mineral rights* 122
        4.1 Overview of common law and statutory regimes 122
        4.2 Statutory regimes on the nature of mineral rights in Africa 125
    5 *Security of mineral tenure* 130
        5.1 Curtailment of ministerial power and discretion 132
        5.2 Convertibility and transition of mineral rights 136
        5.3 Retention of mineral rights 136
        5.4 Statutory and contractual guarantees regarding security of tenure 137
        5.5 Constitutional guarantees regarding protection of property 140
        5.6 Legal safeguards against arbitrary revocation of mineral rights 141
    6 *Conclusion* 144

5 Fiscal regimes for mineral exploitation   146

    1 *Introduction* 146
    2 *Royalty and tax regimes* 146
        2.1 Royalties 147
        2.2 Taxes 150
    3 *Case studies of mineral royalties and taxes* 151
        3.1 Botswana 151
        3.2 Ghana 156
        3.3 Nigeria 158
        3.4 South Africa 159
        3.5 Zambia 161

x Contents

    4 *Fiscal incentives and allowances 163*
        4.1 Investment allowances, tax credits, and loss deductions 164
        4.2 Tax holidays, deferment of royalty, and other exemptions 168
        4.3 Import duty exemptions and currency transfer guarantees 171
        4.4 Allowances regarding environmental rehabilitation 173
    5 *An appraisal of the fiscal regimes 173*
        5.1 Disconnect between mining revenues and the value of minerals 176
        5.2 Abuse of the fiscal regimes through tax-avoidance schemes 177
        5.3 Revenue costs of the generous fiscal regimes 182
        5.4 Lack of institutional capacity to implement royalty and tax regimes 184
    6 *Anti-tax-avoidance rules 185*
        6.1 Suspicious and fictitious transactions 185
        6.2 The arm's-length standard 187
        6.3 Exclusivity standard 188
        6.4 Ring-fencing rules 188
        6.5 Contractual guarantees regarding payment of royalties and taxes 189
    7 *Ineffectiveness of anti-tax-avoidance rules in Africa 190*
    8 *Conclusion 191*

6 Legal stabilization of mining investments   192

    1 *Introduction 192*
    2 *Stabilization regimes 193*
        2.1 Contract-based stabilization clauses 193
        2.2 Statute-based stabilization clauses 198
        2.3 Treaty-based stabilization regimes 202
        2.4 Consequences of treaty-based stabilization regimes 209
    3 *Drivers of stabilization regimes in Africa 211*
    4 *Deleterious effects of stabilization regimes 213*
    5 *Reformation of stabilization regimes in Africa 220*
        5.1 Limiting the scope and duration of stabilization regimes 220
        5.2 Periodic renegotiation of stabilization regimes 222
        5.3 Independent oversight of stabilization regimes 224
    6 *Why Africa should desist from incessant grant of stabilization clauses 225*
    7 *Conclusion 227*

7 Resource Nationalism in the African Mining Industry   228

   1 Introduction 228
   2 Manifestations of resource nationalism 230
      2.1 Fiscal reforms 231
      2.2 Termination of mineral rights 233
      2.3 Indigenization 236
      2.4 State participation in mining projects 238
      2.5 Declaration of minerals as 'strategic national assets' 241
      2.6 Renegotiation of mining contracts 242
      2.7 Beneficiation 248
      2.8 Local content requirements 250
   3 Drivers of resource nationalism in Africa 252
      3.1 Desire to exert greater control over mineral resources 253
      3.2 The developmental state 254
      3.3 Exploitative relationship between MNCs and African governments 255
      3.4 Redressing historical inequalities 256
      3.5 High commodity prices 258
      3.6 Domestic politics 259
      3.7 Competition for access to mineral resources 260
   4 Effectiveness of resource nationalism in Africa 261
   5 Resource nationalism and foreign direct investment 267
   6 Conclusion 272

8 Management and utilization of mining revenues   273

   1 Introduction 273
   2 Revenue management agencies and institutions 274
      2.1 Public agencies and state-owned mining companies 274
      2.2 Semi-autonomous bodies 284
   3 An appraisal of mineral development funds in Africa 291
   4 Towards the prudent management of mining revenues 292
      4.1 Mandatory transparency, accountability, and oversight regimes 293
      4.2 Capacity-building 296
   5 Benefit-sharing schemes 296
      5.1 Community development agreements 297
      5.2 Company-funded foundations, trusts, and funds 308
      5.3 Equity participation by host communities 314
   6 Conclusion 315

xii *Contents*

9 Mining and the environment 317

   1 Introduction 317
   2 Complicity of mining companies in environmental and human rights violations 317
   3 Environmental regulation of mining operations 320
      3.1 Environmental standards 320
      3.2 Environmental rights as human rights 324
      3.3 Mine closure and rehabilitation standards 326
   4 Non-enforcement of environmental standards 328
      4.1 Incapacity of regulatory agencies 328
   5 Citizen enforcement of environmental standards 331
   6 Liability of parent companies for the wrongful actions of subsidiaries 333
      6.1 Piercing the corporate veil 335
      6.2 Parent company's duty of care 339
   7 Conclusion 343

10 Conclusion 344

   1 Towards pragmatic mining regimes in Africa 344
      1.1 Inventory of mineral resources 345
      1.2 Capacity-building 346
      1.3 Transparency and accountability 347
      1.4 Responsive fiscal and tax reforms 348
      1.5 A clear contractual framework for mineral exploitation 349
   2 A mutually beneficial relationship 351

Index 352

Tables

1.1 Contribution of mining to total exports 14
2.1 Types of mineral rights in selected African countries (excluding mineral processing, artisanal mining and quarrying rights) 21
5.1 Royalty and tax rates in selected African countries 152

Preface

The idea for this book arose while I was preparing materials for my Mining Law course at the University of Calgary, Alberta, Canada. This course covers domestic mining in Canada as well as international mining transactions. While searching for books and other materials on international mining transactions, I realized to my astonishment that there is a dearth of books on the legal regimes governing the mining of minerals in Africa. I also discovered that very few universities in Africa teach mining law as a distinct course. Although there are books and other materials on mining law in some African countries, there is a conspicuous absence of books that examine mining regimes across the continent of Africa on a comparative basis. This book aims primarily to fill the lacuna in mining law jurisprudence in Africa. My hope is that the topics covered in this book will spur the teaching of mining law in universities across Africa.

Acknowledgements

I wish to express my appreciation to Chris M. Boettcher (JD 2020, University of Calgary) for his excellent research assistance. Some of the views expressed in Chapter 6 appear in my earlier work entitled "Stabilization Clauses in Natural Resource Extraction Contracts: Legal, Economic and Social Implications for Developing Countries" (2010) 10 *Asper Review of International Business and Trade Law* 1–33. I thank the editors of the *Asper Review of International Business and Trade Law* for permission to use aspects of my earlier work.

Most of all, I thank my loving wife, Liz, and our beautiful and adorable children, Omozuafo, Omegie and Ainosi, for their unwavering love and support.

Abbreviations

ADB	African Development Bank
AEMFC	African Exploration Mining and Finance Corporation (South Africa)
AMV	Africa Mining Vision
ANC	African National Congress
ANRC	African Natural Resources Center
ASM	artisanal and small-scale mining
BEE	Black Economic Empowerment
BIT	bilateral investment treaty
BSGR	Beny Steinmetz Group Resources
CDA	community development agreement
CDF	Community Development Fund (Burkina Faso)
CPA	Community Property Association
DA	development agreement
DACDF	Diamond Area Community Development Fund (Sierra Leone)
DRC	Democratic Republic of Congo
ECOWAS	Economic Community of West African States
EIA	environmental impact assessment
EITI	Extractive Industries Transparency Initiative
ENRC	Eurasian Natural Resources Corporation
FDI	foreign direct investment
FPIC	free, prior, and informed consent
FTF	foundations, trusts, and funds
GDP	gross domestic product
HDP	historically disadvantaged persons
IANRA	International Alliance on Natural Resources in Africa
ICMM	International Council on Mining and Metals
ILM	International Legal Materials
ILO	International Labour Organization
IMF	International Monetary Fund
IPILRA	Interim Protection of Informal Land Rights Act 31 of 1996 (South Africa)
KCM	Konkola Copper Mines Plc

LEITI	Liberia Extractive Industries Transparency Initiative
LFN	Laws of the Federation of Nigeria
LMC	Local Management Committee (Ghana)
LSA	legal stability agreement
MCDS	Mining Community Development Scheme (Ghana)
MCO	Mining Cadastre Office (Nigeria)
MDA	mineral development agreement
MDF	mineral development fund
MIGA	Multilateral Investment Guarantee Agency
MIT	multilateral investment treaty
MLC	Mining Licensing Committee (Zambia)
MNC	multinational corporation
MPRDA	Mineral and Petroleum Resources Development Act 28 of 2002 (South Africa)
MPRRA	Mineral and Petroleum Resources Royalty Act, 2008 (South Africa)
NEEEF	New Equitable Economic Empowerment Framework (Namibia)
NEITI	Nigeria Extractive Industries Transparency Initiative
NGO	non-governmental organization
OECD	Organisation for Economic Co-operation and Development
SAIMM	Southern Africa Institute of Mining and Metallurgy
SWF	sovereign wealth fund
TWNA	Third World Network Africa
UN	United Nations
UNCTAD	United Nations Conference on Trade and Development
UKSC	United Kingdom Supreme Court
UNECA	United Nations Economic Commission for Africa
VRHL	Vedanta Resources Holdings Limited

1 Overview of mining in Africa

1 Introduction

Africa is endowed with commercially viable quantities of several minerals and metals, including gold, diamond, nickel, platinum, copper, cobalt, chromium, manganese, uranium, iron ore, tantalum, titanium, and silver. In fact, Africa accounts for 30% of the world's total mineral reserves.[1] Africa holds the world's largest reserves of several minerals, including platinum, gold, diamond, chromite, manganese, and vanadium.[2] In reality, the current statistical data on mineral deposits in Africa understates Africa's mineral wealth given that African lands have yet to be prospected fully for mineral deposits.

Africa accounts for a significant proportion of the minerals and metals produced across the globe. South Africa alone produces "three-quarters of the world's platinum, 40 per cent of chromium and over 15 per cent of gold and manganese".[3] Two of the top three producers of platinum are African countries.[4] South Africa is the world's largest producer of chromium, manganese, and platinum, while the Democratic Republic of Congo (DRC) leads the world in cobalt and tantalum production.[5] In fact, ten of the world's top 29 producers of manganese are African countries, while six of the top 18 producers of cobalt are African countries.[6] As far back as 2005, Africa not only led the world in the production of diamonds, vanadium, and platinum group minerals but also held more than 60% of the world's known reserves of these minerals.[7]

---

1 ADB, *Mining Industry Prospects in Africa*, https://blogs.afdb.org/fr/afdb-championing-inclusive-growth-across-africa/post/mining-industry-prospects-in-africa-10177
2 UNECA, *Africa Review Report on Mining (Executive Summary)*, at 2, www.uneca.org/sites/default/files/PublicationFiles/aficanreviewreport-on-miningsummary2008.pdf
3 Africa Progress Panel, *Equity in Extractives: Stewarding Africa's Natural Resources for All*, at 44, https://reliefweb.int/sites/reliefweb.int/files/resources/relatorio-africa-progress-report-2013-pdf-20130511-125153.pdf
4 Federal Ministry of Sustainability and Tourism, Austria, *World Mining Data 2019* at 169, www.world-mining-data.info/wmd/downloads/PDF/WMD2019.pdf
5 *Ibid*. at 152, 153, & 169.
6 *Ibid*. at 152 & 155.
7 African Union, *Africa Mining Vision*, at 3, www.africaminingvision.org/amv_resources/AMV/Africa_Mining_Vision_English.pdf

2 *Overview of mining in Africa*

In terms of the share of global production, South Africa produces 48.26% of the world's chromium, 29.07% of the world's manganese, 15.23% of the world's titanium, 72.07% of the world's platinum, and 80.70% of the world's rhodium.[8] Likewise, the DRC accounts for 60.85% of the world's cobalt production and 43.04% of the world's tantalum production, while Botswana accounts for 17.45% of the world's diamond gem production.[9] Africa equally does well with regard to the quality and production value of minerals. Three of the world's top-20 mineral-producing countries based on the value of minerals are African countries. Of these top-20 countries, South Africa, the DRC, and Mozambique occupy the 5th, 17th, and 20th positions, respectively.[10]

1.1 History of mining in Africa

Mining in Africa predates the arrival of European colonialists to the continent. Mining of minerals such as iron, gold, copper, lime, and salt was undertaken by indigenous Africans long before the advent of colonial rule in Africa. Africans mined, processed, and traded copper for centuries prior to colonial rule and, in fact, the major copper-producing areas in modern Africa, including the Copper belt, "were sites of indigenous production for many years before the takeover by colonial foreign mining companies".[11] For example, the Ingwenya mine in Swaziland was exploited by Africans over 2000 years ago for iron ochres for rock paintings.[12] There is evidence that metallurgy with copper and iron began in West, Central and East Africa around 800 BC.[13] By the 8th century AD, the West African region was already famous for its gold industry which was one of the main pillars on which ancient African empires, including the Ghana, Mali and Songhai empires, were built.[14] In the 10th century AD, gold was "a regular article of commerce in the overland trade between the territories bordering the Gulf of Guinea and the Mediterranean countries".[15] In fact, "in the millennia preceding colonialism, west and southern Africa were major exporters of gold to the rest of the world".[16] When Portuguese explorers arrived in the Gold Coast (now Ghana) in

---

8 *World Mining Data 2019, supra* note 4 at 152–70.
9 *Ibid.* at 152–74.
10 International Council on Mining and Metals (ICMM), *Role of Mining in National Economies*, 3rd Edition, at 22, www.icmm.com/website/publications/pdfs/social-and-economic-development/161026_icmm_romine_3rd-edition.pdf
11 UNECA, *Minerals and Africa's Development: The International Study Group Report on Africa's Mineral Regimes* (UNECA: Addis Ababa, 2011) at 11.
12 Africa Union, *Africa Mining Vision, supra* note 7 at 10.
13 Shadreck Chirikure, "Precolonial Metallurgy and Mining across Africa", in Thomas Spear *et al.*, eds, *Oxford Research Encyclopedia of African History* (Oxford University Press USA), https://oxfordre.com/africanhistory
14 Francis N.N. Botchway, "Pre-Colonial Methods of Gold Mining and Environmental Protection in Ghana" (1995) 13 *Journal of Energy and Natural Resources Law* 299.
15 G. Keith Allen, "Gold Mining in Ghana" (1958) 57(228) *African Affairs* 221.
16 UNECA, *Minerals and Africa's Development, supra* note 11 at 11.

1471, they met Africans already trading in gold.[17] During the pre-colonial era, Africans used various mining techniques including alluvial mining, panning, shallow and deep pit mining, underground mining, and underwater sand dredging.[18] Modified versions of these ancient mining techniques and practices are still used by artisanal miners in modern Africa.

However, the mining of mineral resources intensified during the colonial era because the primary objective of European colonial rule in Africa was the exploitation of the continent's rich natural and human resources. In Tanzania, for example, the discovery of gold in the 1890s by European colonialists was followed by the gold rush in the Lupa region in the 1920s.[19] In Ghana, the European colonizers began large-scale mining in the late 19th century, and by 1941 bauxite was being produced in Ghana to satisfy the demands of the allied forces during World War II.[20]

Colonial governments formalized the mining industry in Africa through the enactment of mining Ordinances. For example, in colonial Nigeria, mining was governed by the *Minerals Ordinance, 1916*, while in the Gold Coast, mining rights could only be acquired based on the *Concessions Ordinance, 1990* and the *Mining Rights Regulation Ordinance, 1905*. These mining statutes were aided by land Ordinances (such as Nigeria's *Public Lands Acquisition Ordinance, 1917*) which empowered colonial governments to compulsorily acquire land for public purposes. Interestingly, under these land Ordinances mining was deemed a public purpose, thus entitling colonial governments to seize land from native Africans for mining-related purposes.[21] Worse yet, colonial governments also enacted laws that compelled native Africans to work for foreign miners in deplorable conditions.[22] In addition, colonial governments subjected Africans to a poll tax which, in times of labour shortage, was arbitrarily increased simply to force Africans "from their villages to seek work in the very poor conditions of the mines".[23]

1.2 Ownership of minerals

A pre-eminent legacy of colonialism in Africa is the vesting of ownership of natural resources in the central governments of African countries. During the colonial era,

---

17 Allen, *supra* note 15 at 222.
18 Botchway, *supra* note 14 at 304–6.
19 Deborah F. Bryceson, Jesper B. Jonsson, Crispin Kinabo, & Mike Shand, "Unearthing Treasure and Trouble: Mining as an Impetus to Urbanization in Tanzania" (2012) 30 (4) *Journal of Contemporary African Studies* 631 at 633–5; Abel Kinyondo & Christopher Huggins, "Resource Nationalism in Tanzania: Implications for Artisanal and Small-Scale Mining" (2019) 6 *The Extractive Industries and Society* 181 at 183.
20 Thomas Akabzaa & Abdulai Darimani, *Impact of Mining Sector Investment in Ghana: A Study of the Tarkwa Mining Region* (January 20, 2001), at 7–8, https://pdfs.semanticscholar.org/db7d/5a2fe09a7c14bf82dddee1bd5cb6ea32d003.pdf
21 Gilber Stone, "The Mining Laws of the West African Colonies and Protectorates" (1920) 2(3) *Journal of Comparative Legislation and International Law* 259 at 260.
22 UNECA, *Minerals and Africa's Development, supra* note 11 at 13.
23 *Ibid.* at 13.

4 Overview of mining in Africa

the European colonialists enacted laws vesting ownership of mineral resources in colonial governments in an apparent attempt to ensure absolute control over Africa's natural resources. In the colony of Nigeria, for example, the *Minerals Ordinance, 1916* provided that "the entire property in, and control of all minerals and mineral oils in, under, or upon any lands in Nigeria, is and shall be vested in the Crown".[24] Upon attaining political independence, African countries retained colonial laws regarding ownership of minerals apparently because it suited the interest of the new political elites. For example, the Constitution of modern Nigeria retains the language of the *Minerals Ordinance, 1916* by providing that

> the entire property in and control of all minerals, mineral oils and natural gas in, under or upon any land in Nigeria or in, under or upon the territorial waters and the Exclusive Economic Zone of Nigeria shall vest in the Government of the Federation.[25]

Similarly, in Ghana

> [e]very mineral in its natural state in, under or upon land in Ghana, rivers, streams, water-courses throughout the country, the exclusive economic zone and an area covered by the territorial sea or continental shelf is the property of the Republic and is vested in the President in trust for the people of Ghana.[26]

In South Africa, "mineral and petroleum resources are the common heritage of all the people of South Africa and the State is the custodian thereof for the benefit of all South Africans".[27] The Constitutional Court of South Africa has held that the *Mineral and Petroleum Resources Development Act 28 of 2002* (MPRDA) abolished private ownership of minerals in South Africa and instead vests all mineral rights in the state.[28] Similarly, in Zambia "[a]ll rights of ownership in, searching for, mining and disposing of, minerals wheresoever located in the Republic vest in the President on behalf of the Republic".[29]

The vesting of ownership of natural resources in the central governments of African countries is significant in many ways. First, African governments possess the legal authority to grant, revoke, suspend, or renew mineral rights upon the grounds stipulated in the relevant statute. Second, African governments set the legal and fiscal frameworks for mineral exploitation through statutory enactments and contractual arrangements. African governments establish the parameters for the exploitation of minerals including the rate of royalties and taxes payable by mining companies. Third, persons who wish to exploit Africa's natural resources

24 *Minerals Ordinance, 1916*, s. 3.(1).
25 *The Constitution of the Federal Republic of Nigeria 1999*, s. 44.(3).
26 *Minerals and Mining Act, 2006* (Ghana), s. 1.
27 *Mineral and Petroleum Resources Development Act 28 of 2002*, s. 3.(1) [*MPRDA*].
28 *Minister of Mineral Resources and Others v Sishen Iron Ore Company (Pty) Ltd and Another* [2013] ZACC 45 at paras 16, 63, & 65 (CC).
29 *Mines and Minerals Development Act, 2015* (Zambia), s. 3.(1).

must enter into legal or contractual relationships with African governments as a precondition to the exploration and extraction of mineral resources. These relationships usually manifest in the form of concessions, including licences, permits, and leases, granted by governments to resource extraction companies. Even in cases where the state grants a mineral right, the state retains reversion rights such that where a mineral right is abandoned by the holder, the mineral right reverts to the state. Fourth, royalties and taxes for mineral resources are paid to the central governments rather than the local communities in whose land the resources are located. In turn, the central governments usually reserve the power to determine how resource revenues are distributed among the constituent regions of each country.

Fifth, and perhaps more fundamentally, the vesting of ownership of mineral resources in central governments creates a trust-like relationship between the governments and the citizens of African countries.[30] In Ghana, for example, ownership of minerals is vested in the government "in trust for the people of Ghana",[31] while in South Africa, the government is entrusted with the custodianship of mineral resources "for the benefit of all South Africans".[32] In Ethiopia, the vesting of ownership of mineral resources in the government imposes a correlative duty on the government "to hold mineral resources, on behalf of the peoples, and deploy them for the benefit and development of all Ethiopians".[33] The implication of this trust-like relationship is that, as trustees and custodians, African governments owe a duty not only to manage mineral resources prudently for the benefit of citizens but also to render an account to the citizens regarding the management and utilization of resource wealth. As custodians of mineral resources, African governments should manage mineral resources "with the intention to promote equitable access to, and socio-economic development of, mineral and petroleum resources in an ecologically sustainable manner".[34] Regrettably, as discussed in Chapter 8, most African governments have failed to manage mineral resources prudently. On the contrary, Africa's mineral resources have been consistently mismanaged – hence the unbridled poverty afflicting African citizens.

2 Generations of mining statutes in Africa

Mining statutes in Africa have evolved drastically since the attainment of political independence. This book classifies mining statutes in Africa based on the policy orientation of the statutes and it adopts the position that the classification of mining statutes in Africa ought to begin from the post-colonial period – that is,

---

30 See Elmarie Van Der Schyff, "Who 'Owns' the Country's Mineral Resources? The Possible Incorporation of the Public Trust Doctrine through the Mineral and Petroleum Resources Development Act" (2008) *Journal of South African Law* 757.
31 *Minerals and Mining Act, 2006* (Ghana), s. 1.
32 *MPRDA*, s. 3.
33 *Proclamation No. 678/2010, A Proclamation to Promote Sustainable Development of Mineral Resources* (Ethiopia), art. 5 [*Proclamation No. 678/2010*].
34 HM van den Berg, "Ownership of Minerals under the New Legislative Framework for Mineral Resources" (2009) 20 *Stellenbosch Law Review* 139 at 145.

the period that Africa attained political independence.[35] This is the position adopted by the *Africa Mining Vision* (AMV), the pre-eminent pan-African mining policy championed by the African Union.[36] While the AMV does not expressly classify mining statutes, it identifies two distinct categories of mining statutes in Africa based on the policy orientation of the statutes. The first category is composed of 'statist' or state-centred statutes enacted immediately following attainment of political independence, while the second category consists of the liberalization statutes of the 1980s and 1990s that were initiated by the World Bank.[37] However, since the AMV was adopted in 2009, a new generation of mining statutes has emerged in Africa, culminating in what could be described as the third generation of mining statutes. In effect, since independence, Africa has seen three generations of mining statutes. These are the nationalistic or 'statist' statutes enacted immediately following independence (first generation); the liberal and semi-market-oriented statutes of the late 1980s to the 2000s (second generation); and the post-liberal and development-oriented statutes of the current era (third generation) that seek not only the maximization of resource benefits (rents, indigenization, local content, etc.) but also the exploitation of minerals in a socially sustainable manner.

2.1 'Nationalist' and 'statist' statutes of the post-colonial era

Soon after gaining independence, many African countries enacted mining statutes with nationalistic and 'statist' undertones. These post-independence statutes (i.e. the first generation of mining statutes) not only vested ownership of minerals in the government but also enabled African governments to assume control over mineral operations, including the nationalization of mining operations particularly during the period between the 1960s and the 1980s. The prevailing sentiment immediately following the end of colonialism in Africa was that economic development "could be achieved only if the state had significant or, indeed, full ownership of mining enterprises".[38] Hence, in countries such as Ghana, Zambia, and Zimbabwe, the state actively took control of mining operations soon after independence.[39] This era also saw the emergence of state-owned mining companies,

---

35 For contrary positions, see Hany Besada & Philip Martin, "Mining Codes in Africa: Emergence of a 'Fourth' Generation?" (2015) 28(2) *Cambridge Review of International Affairs* 263–82 (which does not identify any significant policy difference between the second and third generations of mining statutes identified by the authors); and Nneoma V. Nwogu, "Mining at the Crossroads of Law and Development: A Comparative Review of Labor-Related Local Content Provisions in Africa's Mining Laws through the Prism of Automation" (2019) 28(1) *Washington International Law Journal* 137 at 142 (which considers colonial Ordinances in classifying mining statutes in Africa).
36 *Africa Mining Vision, supra* note 7.
37 *Ibid.* at 9–11.
38 *Ibid.* at 10.
39 SAIMM, *The Rise of Resource Nationalism: A Resurgence of State Control in an Era of Free Markets or the Legitimate Search for a New Equilibrium?* at 97–126, www.saimm.co.za/Conferences/ResourceNationalism/ResourceNationalism-20120601.pdf

such as Zambia Consolidated Copper Mines Limited and Zimbabwe Mining Development Corporation, which were established soon after the attainment of political independence. In addition, post-independence statutes restricted foreign access to mineral resources and imposed strict regulations on foreign investments, including restrictions on foreign sourcing of materials.

However, post-independence mining statutes in Africa failed to spur growth in mineral production. Mineral production in post-independence Africa lagged behind the production levels of other regions of the world.[40] While these statutes promoted state ownership of mining operations, newly independent African states lacked the capacity and expertise to produce minerals. State-owned mining companies were poorly governed and undercapitalized, and lacked the resources to expand their production capacity. Moreover, the strict regulations imposed on foreign investors dissuaded the inflow of foreign direct investment (FDI) to the African mining industry.

2.2 Economic liberalization statutes

The stagnation in mineral production, coupled with the ineffectiveness of state ownership of mining operations, led to a clamour for the liberalization of the African mining industry by international institutions such as the World Bank and the International Monetary Fund (IMF). For example, the World Bank reasoned that:

> the recovery of the mining sector in Africa will require a shift in government objectives towards a primary objective of maximizing tax revenues from mining over the long term, rather than pursuing other economic or political objectives such as control of resources or enhancement of employment. This objective will be best achieved by a new policy emphasis whereby governments focus on industry regulation and promotion and private companies take the lead in operating, managing and owning mineral enterprises.[41]

The World Bank actively urged African countries to discard state-ownership policies and instead adopt private-sector-oriented mining policies, including the deregulation and liberalization of the mining industry. Viewing such policy reversal as a "necessary condition for revitalizing mining in Africa and returning to strong sustained growth", the World Bank pointedly recommended that Africa should enact mining legislation designed to "reduce risk and uncertainty for potential investors and ensure easy access to exploration permits and mining concessions".[42] The World Bank also urged Africa to ensure that mining permits and concessions are transferrable with minimum of government interference and that

---

40 World Bank, *Strategy for African Mining* at x, http://documents.worldbank.org/curated/en/722101468204567891/pdf/multi-page.pdf
41 *Ibid.* at x.
42 *Ibid.* at xiii.

8 *Overview of mining in Africa*

"[i]nvestment agreements, where required, should provide additional assurances to protect the investor from unwarranted government interference".[43]

In fact, the IMF required African countries to adopt economic liberalization policies as a condition for the grant of loans. Ultimately, Africa succumbed to the economic liberalization gospel of the World Bank and the IMF during the 1990s and 2000s. During this period, many African countries enacted mining statutes that not only opened up the mining industry to foreign investors but also enabled the privatization of state-owned mining companies. For example, in the 1990s Zambia privatized its mining operations, including the sale of the mineral assets of the state-owned company, Zambia Consolidated Copper Mines Limited. As described succinctly by the AMV, the second-generation statutes constituted

> a fundamental paradigm shift and redefinition of the role of state, from 100% ownership and control, to deregulation and almost complete withdrawal. Many African countries embarked on a radical reform process with the aim of attracting foreign direct investment to rehabilitate their moribund minerals and mining sector. To this end, state enterprises were privatized and efforts and resources were deployed to improve the investment climate. New mineral policies, and legal, regulatory and administrative frameworks more favourable to private investors were formulated and established.[44]

The liberalization policy underlying this second generation of mining statutes in Africa manifested in the easing of access to mineral rights, the issuance of government guarantees regarding investment stability, the strengthening of security of tenure for holders of mineral rights, and the grant of generous fiscal incentives to foreign investors. While, as discussed in Chapter 5, the liberalization policies led to an increase in FDI inflow to the African mining industry, Africa did not obtain commensurate benefits from mining operations, due partly to the overly generous fiscal regimes that tended to diminish mining revenues accruing to African governments. Besides, during this period mining companies engaged incessantly in exploitative behaviour, including tax-avoidance practices such as transfer pricing, thus depriving Africa of a significant proportion of mining revenues. Moreover, the second generation of mining statutes failed to create any significant linkages between the mining industry and other sectors of the economy, particularly because they did not actively encourage beneficiation and processing of mineral ores within Africa.[45] Also, these statutes neither ameliorated the adverse social and environmental effects of mining operations nor provided adequate compensation for local communities that suffered such adverse impacts.

43 *Ibid.* at xiii.
44 *Africa Mining Vision, supra* note 7 at 11.
45 *Ibid.* at 9.

2.3 Development-oriented statutes of the post-liberal era

The commodity boom of the early to mid-2000s apparently forced Africa to realize that the liberalization policies underpinning the second generation of mining statutes in the continent had failed to produce the desired outcomes, particularly in regard to revenue generation. Hence, in the course of the last decade Africa has yet again embarked on policy revision with regard to mining. Such policy revision has culminated in the enactment of a third generation of mining statutes in countries across Africa. This third generation of mining statutes contain elements of both the first generation (i.e. post-independence mining statutes) and the second generation. Like the first-generation statutes, the third generation of mining statutes in Africa are nationalistic in nature and, as discussed in Chapter 7, have ushered in the current era of resource nationalism in Africa. The third generation of mining statutes actively promote indigenization of the mining industry through local private ownership of mining operations, as well as the beneficiation, processing and refining of mineral ores within Africa. And, like the second-generation statutes, the third generation of mining statutes in Africa ease access to mineral rights and provide clear rules for the acquisition of mineral rights. However, unlike the first and second generations of mining statutes, some third-generation statutes in Africa attempt to enhance the management of mining revenues through transparency and accountability frameworks.

The third generation of mining statutes in Africa aim to achieve a multitude of objectives, including the assertion of greater control over exploitation of mineral resources; optimization of resource royalties and taxes; promotion of equity participation in mining projects; enhancement and promotion of indigenization in the ownership of mining projects; the facilitation of community participation in mining through revenue sharing schemes such as community development agreements; the creation of linkages with, and spill-over effects for, other sectors of the economy through local content requirements; the creation of value addition and beneficiation in terms of domestic processing of raw mineral ores and metals in Africa; and the promotion of sustainable practices in the mining industry. The assertion of greater control over mineral resources through new mining statutes in Africa is indicative of a paradigm shift toward value maximization, as well as the leveraging of resource endowment as a catalyst for economic development.[46] Taken together, these policy objectives appear to be informed by the nationalistic zeal of African governments which, as discussed in Chapter 7, has itself led to the phenomenon of resource nationalism in Africa.

In addition, many new mining statutes in Africa are designed to address the apparent defects in predecessor statutes in the hope of redressing the economic and social imbalances arising from such defective statutes. In South Africa, for example, the express purposes of the MPRDA include the promotion of equitable access to mineral resources by all of the peoples of South Africa; the substantial

---

46 Chris W.J. Roberts, "The Other Resource Curse: Extractives as Development Panacea" (2015) 28(2) *Cambridge Review of International Affairs* 283 at 287.

and meaningful expansion of opportunities for historically disadvantaged persons, including women and local communities, to enter into and actively participate in and benefit from the exploitation of minerals; the promotion of employment as well as the advancement of the social and economic welfare of all South Africans; and the promotion of social responsibility on the part of mineral right holders.[47] Likewise, the general principles underlying Zambia's *Mines and Minerals Development Act, 2015* include the promotion of the socio-economic development of Zambia in a manner that ensures that citizens have "equitable access to mineral resources and benefit from mineral resources development".[48]

While the third generation of mining statutes in Africa are spurred primarily by a desire to obtain greater benefit from mining operations, many of these statutes also appear to have been enacted in response to a continent-wide initiative on the part of the African Union to harmonize mining regimes across the continent. In 2009, the African Union adopted the AMV, which is premised on a desire to promote a "transparent, equitable and optimal exploitation of mineral resources to underpin broad-based sustainable growth and socio-economic development".[49] The AMV urges African countries to revise their mining regimes in order for "the current boom [in commodity prices] to catalyze sustainable development in resource rich African states".[50] The AMV aspires to fully integrate the African mining industry with other sectors by creating "mutually beneficial partnerships between state, the private sector, civil society, local communities and other stakeholders".[51]

The AMV advocates a resource-based development and industrialization strategy for Africa, a strategy underpinned by the belief that mineral resources can be used to catapult Africa to modernization.[52] Mineral resources can catalyze broad-based growth and development only if Africa maximizes "the concomitant opportunities offered by a mineral resource endowment, particularly the 'deepening' of the resources sector through the optimization of linkages into the local economy".[53] Hence, among others, the AMV recommends that African countries should:

> Promote local beneficiation and value addition of minerals to provide manufacturing feedstock; promote the development of mineral resources (especially industrial minerals) for the local production of consumer and industrial goods; establish an industrial base through backward and forward linkages; encourage and support small and medium-scale enterprises to enter the supply chain; improve the quality of the business environment, increase private sector confidence and participation, and reduce entry barriers and operating costs to achieve external economies of scale; ensure compliance of industry players

47 *MPRDA*, s. 2.
48 *Mines and Minerals Development Act, 2015* (Zambia), s. 4.
49 *Africa Mining Vision, supra* note 7 at v.
50 *Ibid.* at 9.
51 *Ibid.* at v.
52 *Ibid.* at 3.
53 *Ibid.* at 13.

with the highest standards of corporate governance, and environmental, social and material stewardship; harness the potential of mid-tier resources that may not necessarily attract major international companies but high net worth individuals, including local entrepreneurs; establish the requisite enabling markets and common platforms for services (raising capital, commodity exchanges, legal and regulatory support, marketing support and know-how); harness the potential of Public Private Partnerships (PPPs); and promote regional integration and harmonization to facilitate factor flows.[54]

The third generation of mining statutes in Africa were enacted partly in response to the AMV's call for the revision of Africa's mining regimes. More than ever before, African countries are desirous of harnessing their resource endowment for economic development. In fact, since the AMV was adopted by the African Union in 2009, no less than 23 African countries have amended their mining statutes or enacted new statutes.[55]

In keeping with the policy orientation of the AMV, the third-generation statutes adopt an economic development paradigm while also enhancing the business environment through the strengthening of security of mineral tenure. In addition, these statutes attempt to promote sustainable exploitation of mineral resources, including environmental stewardship and corporate social responsibility. Perhaps more significantly, unlike predecessor statutes that overlooked and neglected artisanal and small-scale mining (ASM), some third-generation mining statutes in Africa formalize and enhance ASM by establishing rules for the grant, renewal, and revocation of artisanal mining permits.[56] Some of these statutes actively encourage artisanal miners to form mining cooperatives and other associations. In Kenya and Sierra Leone, for example, artisanal mining permits may be granted to artisanal mining cooperatives, associations, or partnerships, thus enabling artisans to pool resources.[57]

The formalization of ASM could potentially create jobs and increase income generation in Africa, thus broadening the tax base for governments across the continent. Artisanal mining also provides employment for African citizens, particularly in rural areas where most mining operations occur. As of 2009, about 3.7 million Africans were directly engaged in ASM, while about 30 million depended on ASM for sustenance.[58] Because ASM employs more Africans than large-scale mining, ASM has the potential to catalyze economic development in

---

54 *Ibid.* at 4.
55 Nwogu, *supra* note 35 at 139. For example, new mining codes have been enacted in Central Africa Republic (2009), Democratic Republic of Congo (2002, amended in 2018), Guinea (2011), Kenya (2016), Mali (2012), Mozambique (2014), Rwanda (2014 & 2018), South Sudan (2012), Tanzania (2010, amended in 2017), and Zambia (2015).
56 See, for example, *Mining Act, 2016* (Kenya), ss. 92–100.
57 See *Mining Act, 2016* (Kenya), s. 95 and *Mines and Minerals Act, 2009* (Sierra Leone), s. 84.
58 *Africa Mining Vision, supra* note 7 at 26.

rural areas by increasing local purchasing power and fostering local economic multipliers.[59] The formalization of ASM could also facilitate access to capital and credit facilities for artisanal miners.[60] Banks and other financial institutions may be more willing to provide credit to artisanal miners in a formalized system, given that the miners are be able to use their mining permits and licences as security for loans.

3 Economic significance of mining in Africa

Mining plays a significant role in the economies of African countries. Mining operations lead to job creation in Africa, thus contributing to employment in the continent. Mining-related employment could be direct employment by mining companies and indirect employment through the procurement of domestic goods and services. It has been suggested that "mining typically contributes only around 1–2 per cent of total employment in a country – but when indirect and induced employment is included, this can jump to 3–15 per cent".[61] For example, the mining industry is estimated to have created 453,543 direct jobs in South Africa as of 2018,[62] while in Ghana direct employment by mining companies stood at 10,503 in 2017.[63]

More significantly, mining is the primary revenue generator for many African countries. In low- and middle-income countries, mining contributes anywhere between 2% and 20% of total revenues.[64] Depending on the royalty regimes in a country, the contribution of mining to government revenues could increase drastically with increases in production capacity as well as during periods of high commodity prices. For example, between 1985 and 1994, mining accounted for 50.9% of government revenues in Botswana, and the number increased slightly to 52% between 1995 and 2004.[65] However, the percentage of contribution declined to 39.9% between 2005 and 2014 due partly to a decline in production and partly because of the slump in commodity prices during that period.[66] Similarly, between 2000 and 2013, mining accounted for 45% and 23% of government revenues in Botswana and Guinea, respectively.[67] Mining accounted for 28% of

---

59 *Ibid.* at 26–7.
60 Agatha Siwale & Twivwe Siwale, "Has the Promise of Formalizing Artisanal and Small-Scale Mining (ASM) Failed? The Case of Zambia" (2017) 4 *The Extractive Industries and Society* 191 at 197
61 ICMM, *supra* note 10 at 3.
62 Minerals Council South Africa, "Facts and Figures 2018" at 8, www.mineralscouncil.org.za/industry-news/publications/facts-and-figures
63 Ghana Chamber of Commerce, *Performance of the Mining Industry in 2017* at 14, https://ghanachamberofmines.org/wp-content/uploads/2016/11/Performance-of-the-Industry-2017.pdf
64 ICMM, *supra* note 10 at 3.
65 ANRC & ADB, *Botswana's Mineral Revenues, Expenditure and Savings Policy* at 8, www.europarl.europa.eu/intcoop/acp/2016_botswana/pdf/study-en.pdf
66 *Ibid.* at 8.
67 ICMM, *supra* note 10 at 35.

revenues accruing to the government of Zambia in 2014,[68] while Ghana generated 15.8% and 16.3% of its revenues from mining operations in 2016 and 2017, respectively.[69]

In countries where mining is the dominant sector, mining usually accounts for a significant proportion of the gross domestic product (GDP) and export earnings. In Botswana, for example, mining accounted for 42.2% of the GDP between 1985 and 1994, 30.9% between 1995 and 2004, and 22% between 2005 and 2014.[70] In 2014, mining accounted for 12% of Zambia's GDP.[71] Likewise, the mining sector contributed 22% to the DRC's GDP in 2015, while mining accounted for 5.6% of Ethiopia's GDP in 2016.[72] In South Africa, the mining sector's contribution to the GDP was 7% in 2016, 6.8% in 2017,[73] and 7.3% in 2018.[74] As of 2015, mining contributed about 15% to the GDP in Zimbabwe but that number has decreased in recent years apparently due to political instability and a decline in production of minerals.[75]

Moreover, as indicated in Table 1.1 below, mining is the dominant source of export in many African countries, and hence mining is the primary source of foreign exchange earning in these countries. In fact, as of 2014 mining accounted for approximately 92% of Botswana's total exports.

As is evident from Table 1.1 below, the export dependency ratio in many African countries increased steadily between 1996 and 2014 due to the exponential rise in commodity prices, which itself spurred an increase in mineral production.[76] In addition, mining contributes to economic and social development of host communities. As discussed in Chapter 8, mining companies often provide development assistance to host communities by signing community development agreements and by establishing foundations and trusts for the benefit of host communities. Through these schemes, mining companies provide infrastructural facilities such as roads, schools, hospitals, and electricity in host communities.

Although mining contributes significantly to revenues in Africa, the economic significance of mining appears undermined by the generous fiscal incentives offered to investors by African countries. As discussed in Chapter 5, these generous fiscal

---

68 World Bank, "Zambia Mining Investment and Government Review" (April 2016) at 1–2, http://documents.worldbank.org/curated/en/305921468198529463/pdf/105820-REVISED-PUBLIC-Zambia-Report-ONLINE.pdf
69 Ghana Chamber of Commerce, *supra* note 63 at 10.
70 ANRC & ADB, *supra* note 65 at 8.
71 World Bank, "Zambia Mining Investment and Government Review", *supra* note 68 at 1–2.
72 MACIG, *The Official Mining in Africa Country Investment Guide 2018*, at 40 & 64, www.gbreports.com/wp-content/uploads/2018/02/MACIG-2018-Web-Version-1.pdf
73 Minerals Council South Africa, "Facts and Figures 2017" at 6, https://www.mineralscouncil.org.za/industry-news/publications/facts-and-figures
74 Minerals Council South Africa, "Facts and Figures 2018", *supra* note 62 at 9.
75 Wayne Malinga, "From an Agro-Based to a Mineral Resources-Dependent Economy: A Critical Review of the Contribution of Mineral Resources to the Economic Development of Zimbabwe" (2018) 45(1) *Forum Development Studies* 71 at 86.
76 ICMM, *supra* note 10 at 23–4.

14 *Overview of mining in Africa*

Table 1.1 Contribution of mining to total exports

Country	1996	2012	2014
Botswana	80.90%	91.60%	91.92%
DRC	72.40%	81.50%	78.92%
Zambia	76.10%	69.20%	69.05%
Mauritania	35.90%	62.90%	58.82%
Guinea	76.30%	60.10%	53.15%
Burkina Faso	8.20%	46.30%	49.65%
Mali	8.40%	42.30%	47.12%
Sierra Leone	27.80%	50.60%	45.91%
Rwanda	2.80%	39.10%	45.18%
Central African Republic	56.00%	44.60%	44.75%
Liberia	49.30%	24.00%	43.43%
Mozambique	6.10%	35.90%	41.55%
Lesotho	3.50%	44.50%	37.83%
Namibia	38.00%	53.40%	37.65%
Eritrea	62.50%	60.50%	35.68%
Madagascar	8.00%	17.80%	33.58%
Tanzania	4.00%	35.20%	32.91%
South Africa	29.60%	33.00%	32.50%

Source: International Council on Mining and Metals (ICMM), *Role of Mining in National Economies*, 3rd Edition, at 23.

incentives diminish mining revenues, thus negatively impacting the economic transformation of Africa through mining operations. Moreover, African countries often grant mineral concession agreements containing stabilization clauses, some of which freeze the fiscal regimes for mining for a considerable length of time.

4 Aim and structure of the book

This book analyzes the legal and fiscal frameworks for hard-rock mining in several African countries, including Botswana, Ghana, DRC, Ethiopia, Guinea, Kenya, Liberia, Mozambique, Namibia, Nigeria, Sierra Leone, South Africa, Tanzania, Zambia, and Zimbabwe. As discussed above, mining is a key component of many economies in Africa – hence Africa's desire to establish continent-wide linkages in the mining industry to catalyze economic development. Regrettably, there is currently lacking a text on mining law in Africa. Although there are books on country-specific mining regimes in Africa, and while there are a few monographs on environmental and social aspects of mining in Africa,[77] a comprehensive and

77 See, for example, Bonnie Campbell, ed., *Regulating Mining in Africa: For Whose Benefit?* (Uppsala: Nordiska Afrikainstitutet, 2004); Victoria R. Nalule, *Mining and the Law in Africa: Exploring the Social and Environmental Impacts* (London: Palgrave Macmillan, 2020).

comparative analysis of the legal and fiscal regimes governing the mining of hard-rock minerals in Africa has yet to be undertaken. This book is a modest attempt to fill the apparent void in the literature.

This introductory chapter provides a brief overview of mining in Africa, including the history of mining in Africa as well as the regulatory implication of state ownership of mineral resources. This chapter also traces the evolution of mining statutes in Africa, culminating in the current development-oriented statutes that view mineral resources as catalysts for the economic transformation of African countries. While recognizing the economic significance of mining in Africa, this chapter observes that the enactment of new mining statutes across Africa in recent years is itself a manifestation of the growing phenomenon of resource nationalism in the continent.

Chapter 2 discusses the types of mineral rights in Africa, the terms and conditions of mineral rights, and the rights and obligations of mineral right holders. Chapter 3 analyzes the acquisition of mineral rights in Africa, including the parameters for the exercise of the discretionary powers vested in the Minister to grant mineral rights. This chapter analyzes the statutory criteria for the grant of mineral rights, including eligibility requirements, indigenization requirements, and financial and technical competency requirements. It discusses the methods and procedures for acquiring mineral rights, as well as the registration of mineral rights. It identifies certain deficiencies in the mineral right acquisition process across Africa that may ultimately undermine Africa's desire to use mineral wealth as a catalyst for economic transformation. These defects include the paucity of indigenous participation in mineral exploitation, the rarity of competitive auction of mineral rights, and the utter lack of consultation with local communities prior to the grant of mineral rights. The chapter argues that, in view of the adverse impacts of mining on local host communities, African governments ought to engage in deep consultation with these communities prior to the grant of mineral rights.

Mining operations occur in phases and they usually have long gestation periods from prospecting to production of minerals. Thus, it is imperative that mineral right holders are assured of the continuity of their rights. In view of this, Chapter 4 analyzes security of mineral tenure in Africa, particularly in relation to the degree to which mineral right holders are protected against interference by governments. As a prelude to the analysis, this chapter examines two preliminary threshold issues – namely, the source of mineral rights and the legal nature of mineral rights. The source and legal nature of mineral rights are significant factors in determining the degree of security afforded to mineral right holders in a country. For example, security of mineral tenure is enhanced in countries where mineral rights are legislatively characterized as 'property'. The chapter notes that, unlike predecessor statutes, modern mining statutes in Africa promote security of mineral tenure by curtailing Ministerial discretion in relation to the grant and renewal of mineral rights, by providing clear rules for transition from prospecting licence to mining lease, as well as by enabling the transferability of mineral rights.

16 *Overview of mining in Africa*

Chapter 5 analyzes the fiscal regimes governing mineral mining in Africa, including taxes, royalties, rents, fees, and financial incentives offered by African countries to investors in the mining industry. This chapter assesses the underlying features of the fiscal regimes against the backdrop of the policy objectives informing the regimes. The overarching argument in this chapter is that while the primary objective of the fiscal regimes in Africa is to attract FDI and boost government revenues, the fiscal regimes have failed to enhance economic development due largely to the generous nature of the regimes. Foreign mining companies engage in unethical and illegal practices in order to avoid or diminish their tax liability to African governments. Mining multinational corporations (MNCs) in Africa often engage in tax-avoidance schemes, and hence there is a disconnect between the value of the minerals produced in Africa and the amount of mining-related revenues generated by African governments.

The political instability in some African countries, particularly countries in the sub-Saharan region, poses significant risks to foreign investment. Instability in political governance could lead to abrupt changes in the laws and regulations governing investment projects as successor governments may want to amend existing laws or enact new laws to suit their political agenda. Such abrupt changes in the law could adversely impact investment projects, particularly mining projects with a long gestation period. Thus, foreign investors in the African mining industry often prevail on weak African governments to grant contractual and statutory guarantees regarding the stability of the laws governing their investment projects. Such guarantees are prevalent in the African mining industry and manifest in the form of stabilization clauses in mining contracts or stand-alone stability agreements. Chapter 6 analyzes contractual and statutory provisions guaranteeing stability of the fiscal regimes for mineral mining in Africa. The chapter accounts for the prevalence of stabilization clauses in Africa's mining industry by situating these clauses within wider pan-African developments, including the incapacity of African governments to negotiate investment contracts, dependence on foreign companies for mineral extraction, corruption, and Africa's excessive reliance on extractive industries for revenue generation. It highlights the deleterious effects of 'freezing' stabilization clauses on the economic and social wellbeing of Africans. For example, the broad scope and long duration of stabilization clauses diminish mining revenues and hinder the ability of African governments to promote human and environmental rights. Thus, the chapter proposes strategies for ameliorating these ill-effects, including the shortening of the duration of stabilization clauses and the periodic review and renegotiation of stabilization clauses in extant mining contracts.

Chapter 7 examines the legal and fiscal regimes for mining in Africa in the context of resource nationalism. The introductory chapter has asserted that the enactment of new mining statutes in Africa is a manifestation of resource nationalism in the continent. Through these new mining statutes, Africa hopes to assert greater control over the exploitation of mineral resources, thereby optimizing resource royalties and taxes. Chapter 7 assesses the degree to which the policy objectives informing Africa's resource nationalism are being achieved.

In particular, it attempts to determine whether the fiscal regimes for mineral exploitation in Africa advance the policy objective of economic development championed by the AMV. In doing so, the chapter analyzes the manifestations of resource nationalism in Africa, including fiscal reforms, indigenization, renegotiation of mining contracts, beneficiation and processing of minerals, and local content requirements. It argues that although resource nationalism appears entrenched in the African mining industry, and while evidence exists to suggest an upswing in mining revenues, overall the policy objectives informing adoption of resource nationalism in the African mining industry have yet to be fulfilled. For example, despite recent increases in royalty rates for minerals, African governments have failed to optimize mining revenues due to corruption and inefficiency. Likewise, local content provisions have failed to produce the desired spill-over effects primarily due to lack of institutional capacity to enforce the provisions. And, despite the indigenization requirements in mining statutes, the African mining industry remains dominated by foreign companies.

Chapter 8 analyzes the management and utilization of mining revenues, including the distribution of mining revenues among various levels of governments in Africa. It examines revenue management institutions, including public agencies and semi-autonomous institutions such as Sovereign Wealth Funds and Mineral Development Funds. The overarching observation in this chapter is that mining revenues in Africa are poorly managed due to multiple factors, including institutional incapacity of revenue management institutions, lack of independence, weak oversight, and, of course, corruption. Given the deplorable state of revenue management in Africa, this chapter argues that Africa can promote the prudent management of mining revenues through enactment of mandatory transparency and accountability regimes encompassing the disclosure of mining-related information. In this regard, the chapter analyzes transparency statutes in Liberia, Nigeria, and Tanzania. Finally, the chapter analyzes private-sector schemes designed to share mining revenues with local host communities in Africa focusing on community development agreements between mining companies and host communities; company-funded foundations, trusts, and funds; and equity participation in mining projects by host communities in Africa.

Mining operations often cause adverse impacts in host communities, particularly in relation to human rights and the environment. This is particularly the case in Africa where laws and regulations are rarely enforced against mining companies. Chapter 9 focuses on these adverse impacts and examines recent transnational tort litigation that attempts to hold parent companies (based in the economically advanced countries) liable for the wrongs committed by their subsidiaries in Africa. In particular, this chapter analyzes the 'new' duty of care imposed on parent companies in *Chandler v Cape PLC*[78] and *Vedanta Resources PLC and another v Lungowe and others*.[79] This duty of care obliges parent companies to provide proper supervision for their subsidiary companies, the failure of which

78 [2012] EWCA Civ. 525.
79 [2019] UKSC 20.

could lead to liability for the violations of the rights of third parties by subsidiary companies. In *Vedanta Resources*, for example, the United Kingdom Supreme Court held that, in appropriate cases, a parent company owes a duty of care to third parties (such as host communities in developing countries) that are adversely impacted by the operations of subsidiary companies.[80] This chapter concludes that although parent companies have yet to be held liable for human rights and environmental rights violations committed by their subsidiaries in Africa, the day may not be far off when such liability will be imputed to parent companies.

Chapter 10, the concluding chapter, draws lessons from the discussion and analysis in previous chapters. Overall, this book demonstrates that although Africa desires to utilize mining revenues as catalyst for economic development, this objective has yet to materialize. This is due primarily to the lack of capacity to manage mining revenues, the non-optimization of mining revenues resulting from the overly generous fiscal incentives in the African mining industry coupled with the tax-evading practices of mining MNCs, and the corruption and unaccountability of African governments. Hence, not only does the book advocate for capacity building within Africa countries, but it also urges the institutionalization of transparency and accountability in the management of mining revenues. If nothing else, transparency could dissuade corruption and misuse of mining revenues in Africa.

Finally, a brief note on citation of the statutes analyzed in this book. Many of the mining and tax statutes analyzed in this book bear similar titles and are thus likely to confuse readers. For example, mining statutes in Africa bear similar titles such as 'Mining Act', 'Mines and Minerals Act', or 'Mines and Minerals Development Act', while tax statutes are often titled 'Income Tax Act'. To avoid confusion, I have added the name of the appropriate country to the titles of the statutes. For example, the *Mining Act No. 12, 2016* of Kenya is cited as 'Mining Act, 2016 (Kenya)'; Ghana's *Minerals and Mining Act (Act 703), 2006* is cited as 'Minerals and Mining Act, 2006 (Ghana)'; Zambia's *Mines and Minerals Development Act, 2015* is cited as 'Mines and Minerals Development Act, 2015 (Zambia)'; and Botswana's *Income Tax Act* is cited as 'Income Tax Act (Botswana)'.

---

80 *Vedanta Resources PLC and another v Lungowe and others* [2019] UKSC 20 at para. 52.

2 Types of mineral rights in Africa

1 Introduction

Mining statutes in Africa define 'mineral' as any substance in solid, liquid, or gaseous form that occurs naturally in or on the earth or under the seabed and formed by or subject to a geological process.[1] In some countries, such as Kenya, Nigeria, South Africa, and Zambia, the word 'mineral' is expressly defined to include residue substances, stockpiles, and deposits in tailing ponds.[2] The statutory definitions of 'mineral' reinforce the three primary characteristics of mineral substances: minerals occur naturally; they are situated in or under the surface of the land and seabed; and they require excavation and extraction through mining operations. However, petroleum and water are expressly excluded from the definition of 'mineral' in Africa.

The authorization of the government is required for the mining of mineral resources in Africa. Government authorization manifests in the form of mineral rights classified as reconnaissance licence; prospecting licence; exploration licence or exploration permit; retention licence; mining lease (also referred to as mining licence or mining right); and a small-scale mining lease. These mineral rights are required for the prospecting, exploration, and production phases of mining operations. As is apparent from Table 2.1 below, some African countries (Angola, Botswana, DRC, Ethiopia, Guinea, Mozambique, Tanzania, Zambia, and Zimbabwe) conflate the prospecting and exploration phases by requiring a single licence for both phases, while other countries separate the prospecting and exploration phases. Countries in the latter category require either a reconnaissance licence and an exploration licence (as in Nigeria, Sierra Leone, and South Sudan) or a prospecting licence and an exploration licence (as in Equatorial Guinea and Liberia), or a reconnaissance licence and a prospecting licence (as in Ghana, Kenya, Namibia, and South Africa). These mineral rights are granted over a specified area of land; they cover the specific minerals enumerated in the mineral

---

1 *Minerals and Mining Act, 2006* (Ghana), s. 111.(1); *Mines and Minerals Development Act, 2015* (Zambia), s. 2; *MPRDA* (South Africa), s. 1; *Mining Act, 2016* (Kenya), s. 4; *Nigerian Minerals and Mining Act, 2007*, s. 164.
2 *Mining Act, 2016* (Kenya), s. 4; *Nigerian Minerals and Mining Act, 2007*, s. 164; *MPRDA*, s. 1; *Mines and Minerals Development Act, 2015* (Zambia), s. 2.

permit, licence, or lease; and they are granted subject to conditions such as minimum work and minimum expenditure requirements. The conditions regarding minimum work and minimum expenditure capture what is commonly referred to as 'use it or lose it' because they ensure that mineral right holders 'use' their mineral rights or 'lose' the mineral rights where the conditions are not fulfilled.

2 Reconnaissance licence

In the context of hard-rock mining, reconnaissance means any operation undertaken in connection with the search for minerals by geological, geophysical, and photogeological surveys, including any remote sensing techniques, but does not include any prospecting or exploration operation, such as drilling and excavation.[3] More specifically, reconnaissance means

> the operations and works to carry out the search for minerals through physical observation, rock sampling, geological surface analysis, geophysical surveys, geochemical surveys, photogeological surveys by other non-obstructive surveys or studies of surface geology or by other remote sensing techniques, laboratory testing and assays.[4]

Thus, a reconnaissance licence (also referred to as a reconnaissance permit or reconnaissance permission) enables the licensee to undertake preliminary scientific and non-intrusive investigation in order to determine whether it is worthwhile to explore the land. A reconnaissance licence does not permit the holder to drill, excavate land, or engage in other subsurface activities.[5] Neither does it entitle the holder to "conduct any prospecting or mining operations for any mineral in or on the land in question".[6]

A reconnaissance licence may cover a defined area or it may be applicable to all lands available for mining in a country. For example, Ghana's *Minerals and Mining Act, 2006* requires that the area of land covered by a reconnaissance licence shall "be a block or any number not more than five thousand contiguous blocks each having a side in common with at least one other block the subject of the" licence.[7] Unlike Ghana, a reconnaissance permit in Nigeria covers "any land within the territory of Nigeria available for mining operations", but does not include land that is the subject of a subsisting mineral right.[8]

---

3 MPRDA, s. 1; *Minerals and Mining Act, 2006* (Ghana), s. 111.(1).
4 *Nigerian Minerals and Mining Act, 2007*, s. 164.
5 See *Minerals and Mining Act, 2006* (Ghana), s. 32.(3); *Nigerian Minerals and Mining Act, 2007*, s. 56.(1)(b); *Proclamation No. 678/2010* (Ethiopia) Art. 16; *Mines and Minerals Act, 2009* (Sierra Leone), s. 65(1)(c).
6 MPRDA, s. 15.(2)(a).
7 *Minerals and Mining Act, 2006* (Ghana), s. 31.(3).
8 *Nigerian Minerals and Mining Act*, s. 58.

Table 2.1 Types of mineral rights in selected African countries (excluding mineral processing, artisanal mining and quarrying rights)

Country	Types of mineral rights	Duration of mining lease	Conditions for the renewal of mining lease
Angola	1 Prospecting title 2 Mining title 3 Mining permit (for civil construction and appliances) 4 Mining pass (for small-scale mining)	30 years	Renewal granted if the holder complies with the conditions of the mining title, terms of their contract, and tax and social security conditions.
Botswana	1 Prospecting licence 2 Retention licence 3 Mining licence 4 Mineral permit (for small-scale operations)	25 years	Renewal granted if: – the holder is not in default of statutory obligations and terms of the mining licence; – mining operations have proceeded with reasonable diligence; – proposed programme of mining operations will ensure the most efficient and beneficial use of mineral resources in the mining area; and – in relation to diamond mining, the mining licence holder has reached an agreement with the government regarding state participation.
Democratic Republic of Congo	1. Exploration licence 2. Exploitation licence 3. Small-scale mining exploitation licence	30 years 10 years (small-scale mining)	Exploitation licence is renewed if the holder: – meets the conditions of the licence, including the obligation to commence operations within the stipulated period and make all requisite payments to the government; – demonstrates existence of ore reserves; – demonstrates that they possess financial resources; – obtains the necessary environmental approvals; – demonstrates ability to engage in continuous exploitation of minerals; – *etc.*

(*continued*)

Table 2.1 (continued)

Country	Types of mineral rights	Duration of mining lease	Conditions for the renewal of mining lease
Ethiopia	1 Reconnaissance licence 2 Exploration licence 3 Retention licence 4 Small-scale mining licence 5 Large-scale mining licence	20 years (large-scale mining) 10 years (small-scale mining)	Mining licence is renewed if the holder: – demonstrates the continued economic viability of mining the deposit; – has fulfilled the obligations specified in the licence; and – is not in breach of any provisions of the Mining Proclamation, regulations, or directives which constitute grounds for the suspension or revocation of the licence.
Ghana	1 Reconnaissance licence 2 Prospecting licence 3 Mining lease	30 years	Mining lease is renewed if the holder has materially complied with their obligations under the *Minerals and Mining Act, 2006*.
Guinea	1 Exploration permit 2 Industrial or semi-industrial operation permit 3 Mining concession	25 years (mining concession)	A mining concession is renewed if the holder: – files a new feasibility study; – meets the obligations stipulated in the original concession instrument and the mining agreement; and – meets the obligations under the Mining Code and regulations.

Country	Types of mineral rights	Duration of mining lease	Conditions for the renewal of mining lease
Kenya	*Large-scale operations*: 1 Reconnaissance licence 2 Prospecting licence 3 Retention licence 4 Mining licence *Small-scale operations*: 1 Prospecting permit 2 Mining permit	25 years or the forecast life of the mine, whichever is shorter	A mining licence is renewed if the holder submits satisfactory documents and information to the Cabinet Secretary, including: – a proposed programme of mining operations to be carried out during the term of the renewal; – a plan of the area in respect of which renewal is sought, including all of the contiguous blocks in the mining licence area; – an approved EIA licence, social heritage assessment, environmental management plan, and community development agreement; and – any additional information as may be required by the Cabinet Secretary.
Liberia	1 Reconnaissance licence 2 Prospecting licence 3 Exploration licence 4 Mining licence (Class A, Class B, and Class C)	25 years	Conditions for renewal not stated in the mining statute.
Mozambique	1 Prospecting and research licence 2 Mining concession 3 Mining certificate (for small-scale mining)	25 years (mining concession) 10 years (small-scale mining)	A mining concession may be extended for 25 years based on the economic life of the mine, and the concession holder's compliance with legal duties.
Namibia	1 Reconnaissance licence 2 Prospecting licence 3 Mineral deposit retention licence 4 Mining licence	25 years	Mining licence is renewable for periods not exceeding 15 years at a time. However, the conditions for renewal are not specified in *Mineral Act, 1992*.

(*continued*)

Table 2.1 (continued)

Country	Types of mineral rights	Duration of mining lease	Conditions for the renewal of mining lease
Nigeria	1 Reconnaissance permit 2 Exploration licence 3 Mining lease 4 Small-scale mining lease	25 years	A mining lease is renewable if the holder demonstrates: – compliance with the conditions of the lease; – they are not in default under the Act; – the existence of mineral reserves justifying renewal of the lease, and – that all other requirements under the Act and its regulations are met.
Rwanda	1 Exploration licence 2 Mining licence (covering small-scale, medium-scale, or large-scale mining)	15 years	A mining licence is renewed if the holder has successfully implemented the programme of mining operations and they submit a statement containing the following details: – proved, estimated, or inferred mineral reserves verified by an independent consultant approved by the competent authority; – capital investment to be made, the mining operations costs, and revenue forecasts for the period of renewal; – mining operations to be undertaken during the period of renewal; – any expected changes in methods of extraction; and – the likely social and environmental impacts and the proposed mitigation and compensation measures.

Country	Types of mineral rights	Duration of mining lease	Conditions for the renewal of mining lease
Sierra Leone	1. Reconnaissance licence 2. Exploration licence 3. Small-scale mining licence 4. Large-scale mining licence	25 years or the estimated life of the ore body, whichever is shorter	A large-scale mining licence is renewed if the holder: – has met all requirements under the Mining Act for the renewal of the licence; – is not in default of any of the provisions of the Act or any of the conditions of the licence; and – has developed the mining licence area with reasonable diligence. In addition, the Minister must be satisfied that the proposed programme of mining operations will ensure the most efficient and beneficial use of the mineral resources in the licence area.
South Africa	1. Reconnaissance permission 2. Prospecting right 3. Mining right 4. Mining permit 5. Retention permit	30 years	A mining right is renewed if the holder has complied with the: – terms and conditions of the mining right and is not in contravention of any relevant provisions of the MPRDA or any other law; – mining work programme; – requirements of the prescribed social and labour plan; and – conditions of the environmental authorization.
South Sudan	1. Reconnaissance licence 2. Exploration licence 3. Small-scale mining licence 4. Large-scale mining licence 5. Retention licence	25 years (large-scale mining licence) 10 years (small-scale mining licence)	Large-scale mining licence is renewed if the holder meets the prescribed conditions. However, the conditions are not specified in the *Mining Act, 2012*.

(*continued*)

Table 2.1 (continued)

Country	Types of mineral rights	Duration of mining lease	Conditions for the renewal of mining lease
Tanzania	1 Prospecting licence 2 Retention licence 3 Special mining licence (for large-scale mining operations) 4 Mining licence (for medium-scale mining operations) 5 Primary mining licence (for small-scale mining operations)	Duration of a special mining licence is the estimated life of the ore body, or such period as the holder may request, whichever period is shorter	A special mining licence is renewed if the holder has complied with statutory obligations, including the terms and conditions of the licence and they submit, to the satisfaction of the Minister: – a statement on the implementation of the conditions of the licence; – a statement of the period for which the renewal is sought; – details of the latest proved, estimated, and inferred ore reserves; – the capital investment, production costs, and revenue forecast for the period of renewal; – any expected changes in mining methods; – any expected increase or reduction in mining activities and the estimated life of the mine; – a proposed programme of mining operations for the period of renewal; and – an environmental certificate issued in respect of the proposed mining operations during the renewal period.

Country	Types of mineral rights	Duration of mining lease	Conditions for the renewal of mining lease
Uganda	1 Prospecting licence 2 Exploration licence 3 Retention licence 4 Mining lease 5 Location licence (for small-scale mining)	21 years or the estimated life of the ore body, whichever is shorter	Mining lease is renewed if the holder has complied with statutory and contractual obligations; and they submit satisfactory statements regarding: – proved, probable, and possible ore reserves; – the capital investment, production costs, and revenue forecasts in respect of the renewed period; – any expected changes in mining methods; and – any likely environmental effects and the measures to mitigate such effects.
Zambia	1 Exploration licence 2 Mining licence	25 years (large-scale mining) 10 years (small-scale mining)	Conditions for renewal of a mining licence not stated in the *Mines and Minerals Development Act, 2015*.
Zimbabwe	1 Prospecting licence 2 Mining lease 3 Special mining lease	25 years (special mining lease)	A special mining lease may be renewed, but the conditions for renewal are not specified in the *Mines and Minerals Act*.

Sources: Mining Statutes and Regulations

2.1 Rights conferred by a reconnaissance licence

A reconnaissance licence confers the right to conduct preliminary survey or research regarding potential mineral deposits by means of geochemical, geophysical, and remote sensing equipment. It empowers the holder to enter on or fly over the land covered by the licence for the specific purpose of conducting reconnaissance operations on the land.[9] In some African countries, the holder of a reconnaissance licence has the "exclusive right to carry on reconnaissance in the reconnaissance area for the minerals to which the reconnaissance licence relates and to conduct other ancillary or incidental activity" in the area.[10] However, given the preliminary nature of reconnaissance operations, a reconnaissance licence may not always confer exclusive rights on the holder. For example, in Ethiopia and Sierra Leone a reconnaissance licence grants a non-exclusive right to conduct operations in the reconnaissance area.[11] Likewise, in Nigeria the holder of a reconnaissance permit has the right to "obtain access into, enter on or fly over any land within the territory of Nigeria available for mining purposes to search for mineral resources on a non-exclusive basis".[12]

Although a reconnaissance licence entitles the holder to conduct reconnaissance operations, it does not convey title to the minerals discovered during such reconnaissance operations. Rather, minerals obtained in the course of reconnaissance operations remain the property of the government.[13] The holder of a reconnaissance licence can, with the prior written permission of the government, take and remove specimens and samples from the area for assaying and testing.[14] However, the holder of a reconnaissance licence cannot dispose of the minerals without the written consent of the government.[15] The holder of a reconnaissance licence may take timber and water from lakes, rivers, and watercourses for purposes connected with their reconnaissance operations, and erect camps and temporary buildings for reconnaissance purposes.[16]

---

9 *MPRDA*, s. 15.(1); *Nigerian Minerals and Mining Act, 2007*, s. 58.(1); *Mines and Minerals Act, 2009* (Sierra Leone), s. 64.(2); *Mining Act, 2012* (South Sudan), s. 37. South Africa requires that, prior to entering the land, the holder of a reconnaissance permit must give a written notice of at least 14 days to the landowner or the lawful occupier of the land. See *MPRDA*, s. 15.(1).
10 *Minerals and Mining Act, 2006* (Ghana), s. 32.(1); *Minerals (Prospecting and Mining) Act, 1992* (Namibia), s. 59.
11 *Proclamation No. 678/2010* (Ethiopia), art. 16; *Mines and Minerals Act, 2009* (Sierra Leone), s. 64.(1).
12 *Nigerian Minerals and Mining Act, 2007*, s. 58.(1)
13 See, for example, Ghana's *Minerals and Mining (General) Regulations, 2012 (L.I. 2173)*, reg. 11.
14 *Mines and Minerals Act, 2009* (Sierra Leone), s. 64.(2); *Mining Act, 2012* (South Sudan), s. 37; *Minerals and Mining (General) Regulations, 2012 (L.I. 2173)* (Ghana), reg. 11; *Nigerian Minerals and Mining Act, 2007*, s. 58.(1).
15 *Minerals and Mining (General) Regulations, 2012 (L.I. 2173)* (Ghana), reg. 11; *Nigerian Minerals and Mining Act, 2007*, s. 58.(1).
16 *Mines and Minerals Act, 2009* (Sierra Leone), s. 64.(2); *Mining Act, 2012* (South Sudan), s. 37; *Minerals and Mining Act, 2006* (Ghana), s. 32.(2).

The holder of a reconnaissance licence has a right to apply for a prospecting licence over all or part of the land covered by the reconnaissance licence, provided that the holder complies with applicable statutory requirements, as well as the terms and conditions stipulated in the licence.[17] However, a reconnaissance licence may not confer exclusive right on the holder to obtain a subsequent mineral right over the land covered by the reconnaissance licence. In South Africa, a reconnaissance licence does not entitle the holder to "any exclusive right to apply for or be granted a prospecting right, mining right or mining permit in respect of the land to which the reconnaissance permission relates".[18] An additional circumscription placed on a reconnaissance licence is that, unlike other types of mineral rights, a reconnaissance licence in some African countries is personal to the holder and "may not be transferred, ceded, let, sublet, alienated, disposed of or encumbered by mortgage".[19]

2.2 Terms and conditions of a reconnaissance licence

A reconnaissance licence is granted subject to the conditions that the holder shall conduct reconnaissance activities in accordance with the approved reconnaissance programme, conduct their operations in an environmentally and socially responsible manner, pay compensation to the owners and lawful occupiers of land for any damage to land and property, and submit geological samples, information, and periodic reports as prescribed by the government.[20] In Ghana, for example, a reconnaissance licence obliges the holder to submit to the Minerals Commission all geological and financial reports and other information relating to the reconnaissance operations on a quarterly basis; comply with the terms and conditions of any Environmental Permit relating to the operations carried out under the licence; and report any mineral discovery to the Minerals Commission within 30 days after discovery.[21] The holder of a reconnaissance licence is equally required to expend on reconnaissance operations the minimum amount prescribed in the programme of reconnaissance operations, employ citizens, and procure goods and services from domestic suppliers.[22] Some countries require the holder to commence reconnaissance operations within a stipulated timeframe. In Sierra Leone, for example, the holder of a reconnaissance licence must commence reconnaissance operations within 90 days of the grant of the licence.[23]

---

17 *Minerals and Mining Act, 2006* (Ghana), s. 34.(4).
18 *MPRDA*, s. 15.(2)(b).
19 *MPRDA*, s. 14.(5); *Nigerian Minerals and Mining Act, 2007*, s. 56.(4).
20 *Nigerian Minerals and Mining Act, 2007*, s. 56.(1); *Mines and Minerals Act, 2009* (Sierra Leone), s. 65.(1); *Mining Act, 2012* (South Sudan), s. 39.
21 *Minerals and Mining (General) Regulations, 2012 (L.I. 2173)* (Ghana), reg. 10.
22 *Mines and Minerals Act, 2009* (Sierra Leone), s. 65.(1).
23 *Mines and Minerals Act, 2009* (Sierra Leone), s. 65.(1)(a).

2.3 Duration and renewal of a reconnaissance licence

A reconnaissance licence has a short duration, usually one year, as in Ghana, Nigeria, Sierra Leone, and South Africa.[24] The shortness of the duration ensures that the licence holder conducts their reconnaissance operations in a timely manner. In countries where a reconnaissance licence confers exclusive rights, the shortness of its duration prevents the holder from warehousing the land and denying other prospective miners access to the land. A reconnaissance licence may be extended or renewed in some countries. In Nigeria, a reconnaissance permit is renewable annually if the holder demonstrates that they have complied with statutory and regulatory requirements, including the covenants and conditions attached to the permit.[25] Unlike Nigeria where there is no limit on the number of times a reconnaissance permit can be renewed,[26] in Ghana a reconnaissance licence may be extended once for a period not exceeding 12 months.[27] To be entitled to such extension, the holder of a reconnaissance licence must demonstrate to the Minister that they have "materially complied with the obligations imposed by this Act with respect to, (a) the holding of the licence, and (b) the activities to be conducted under the licence".[28] However, unlike the position in Nigeria and Ghana, a reconnaissance permission is not renewable in South Africa and Ethiopia.[29]

3 Prospecting licence

The word 'prospecting' has a narrow or broad meaning depending on the legal scheme in a country. In the narrow sense, prospecting means preliminary investigations and search for minerals in a minimally intrusive manner. Quite often, prospecting involves the geological analysis of an area and may be conducted through conventional techniques such as panning and modern scientific techniques such as magnetic surveys and airborne surveys. The purpose of prospecting is to gather geological information relating to the possible mineralization in an area so that the information can be processed to determine whether it is worthwhile to explore the area. Some minerals such as magnetite "produce easily detectable anomalies in the earth's magnetic field because the rocks containing them become magnetized"; hence, through magnetic surveys, geologists are able to generate magnetic maps that generally record "the distribution of magnetic material in the underlying crystalline basement".[30] The interpretation of the geological

---

24 *MPRDA*, s. 14.(4); *Minerals and Mining Act, 2006* (Ghana), s. 31.(2); *Nigerian Minerals and Mining Act, 2007*, s. 57; *Mines and Minerals Act, 2009* (Sierra Leone), s. 63.(1).
25 *Nigerian Minerals and Mining Act, 2007*, s. 57.
26 *Nigerian Minerals and Mining Regulations 2011*, reg. 33.(5).
27 *Minerals and Mining Act, 2006* (Ghana), s. 33.(4).
28 *Ibid*. s. 33.(3).
29 *MPRDA*, s. 14.(4); *Proclamation No. 678/2010* (Ethiopia), Art. 17.(2).
30 John Milsom, "Geophysical Methods", in Charles J. Moon, Michael K.G. Whateley, & Anthony M. Evans, eds, *Introduction to Mineral Exploration*, 2nd Edition (Malden, MA: Blackwell Publishing, 2006) 127 at 130.

information gathered during the prospecting phase enables mining companies to make decisions regarding exploration, including where to drill or excavate.

However, in many African countries, the word "prospecting" is given a broad meaning to include mining activities that would normally constitute exploration. For example, in South Africa, "prospecting" is defined as

> intentionally searching for any mineral by means of any method – (a) which disturbs the surface or subsurface of the earth, including any portion of the earth that is under the sea or under other water; or (b) in or on any residue stockpile or residue deposit, in order to establish the existence of any mineral and to determine the extent and economic value thereof; or (c) in the sea or other water on land.[31]

Ghana defines the word "prospect" as meaning "to intentionally search for minerals and includes reconnaissance and operations to determine the extent and economic value of a mineral deposit".[32] Similarly, in Botswana, "prospect" means "intentionally to search for minerals and includes determining their extent and economic value",[33] while Tanzania defines "prospect" as the "search for any mineral by any means and to carry out any such works and remove such samples as may be necessary to test the mineral bearing qualities of land, and includes the conduct of reconnaissance operations".[34]

3.1 Rights conferred by a prospecting licence

A prospecting licence grants to the holder an exclusive right to conduct prospecting operations in a designated area for the minerals specified in the licence.[35] It enables the holder to search for minerals by means of geochemical and geophysical equipment in a manner that disturbs the surface or subsurface of the earth, including excavation, drilling, and dredging. In exercising this exclusive right, the licence holder may enter upon the land to which the licence relates, drill boreholes, erect camps and construct temporary buildings for prospecting purposes, and conduct other activity ancillary or incidental to the prospecting of minerals.[36] Although the holder of a prospecting licence has the exclusive right to undertake prospecting operations, they cannot remove minerals from the prospecting area except for the specific purpose of analyzing the minerals or determining the value of the minerals or conducting tests on the minerals.[37] That said,

---

31 *MPRDA*, s. 1.
32 *Minerals and Mining Act, 2006* (Ghana), s. 111.(1).
33 *Mines and Minerals Act, 1999* (Botswana), s. 2.(1).
34 *Mining Act, 2010* (Tanzania), s. 4.(1).
35 *Mines and Minerals Act, 1999* (Botswana), s. 20; *Minerals (Prospecting and Mining) Act, 1992* (Namibia), s. 67; *Mining Act, 2010* (Tanzania), s. 35; *MPRDA*, ss. 5 & 19.
36 *Mines and Minerals Act, 1999* (Botswana), s. 20; *Minerals and Mining Act, 2006* (Ghana), s. 37.(1); *Mining Act, 2010* (Tanzania), s. 35.(2); *MPRDA*, s. 5.
37 *Mining Act, 2010* (Tanzania), s. 98.

the government may allow the holder of a prospecting licence to remove minerals from the prospecting area on such terms and conditions as it deems appropriate.[38]

In South Africa, the holder of a prospecting right has additional exclusive rights designed to ensure the security of their mineral tenure. These tenure-protecting rights include the exclusive right to apply for and be granted a renewal of the prospecting right in respect of the mineral and prospecting area; and the exclusive right to apply for and be granted a mining right in respect of the mineral and prospecting area in question.[39] However, the MPRDA requires the holder of a prospecting right to fulfil the conditions of the prospecting right before they can validly exercise the exclusive right to apply for and obtain a mining right in respect of the prospecting area.[40]

There are certain restrictions on the exercise of the exclusive rights of the licence holder. For example, except as permitted by the government, a licence holder cannot enter upon, prospect for minerals in, excavate, or build on sacred land such as land designated as a place of burial, or land containing an ancient or national monument.[41] In addition, a prospecting licence does not authorize the holder to enter upon or to prospect on land close to residential dwellings or any land used for agricultural purposes, or close to a dam or private water sources.[42] Furthermore, the holder of a prospecting licence cannot exercise any right granted under the licence upon any land designated as public land or land set aside for the purposes of the government.[43] The exclusive right under a prospecting licence may be curtailed in another sense. Where the licence covers land that is used for agricultural purposes, the owner or lawful occupier of the land retains the right to graze stock upon the land or to cultivate the surface of the land, provided that such grazing or cultivation does not interfere with the licence holder's prospecting activities.[44]

Moreover, the exclusive rights under a prospecting licence must be exercised reasonably. In particular, the licence holder must conduct their prospecting activities in a reasonable and proper manner so as to affect as little as possible the owner or occupier of the land on which the mining activities are conducted.[45] Hence, licence holders are prohibited from creating unprotected pits, hazardous waste dumps, or other hazards which are likely to endanger the stock, crops, or other lawful activity of the owner or lawful occupier of the land.[46]

---

38 *Ibid.* s. 98; *MPRDA*, s. 20.(1).
39 *MPRDA*, s. 19.
40 *Ibid.* s. 19.(2).
41 *Mines and Minerals Act, 1999* (Botswana), s. 60.(1).
42 *Ibid.* s. 60.(1). Note, however, that the owner or lawful occupier of such land may in writing permit the licence holder to enter upon, excavate, and prospect for minerals, on the land. See *ibid.* s. 60.(1)(b).
43 *Mines and Minerals Act, 1999* (Botswana), s. 60.(1).
44 *Ibid.* s. 61.(1).
45 *Ibid.* s. 61.(3).
46 *Ibid.* s. 61.(4).

3.2 Terms and conditions of a prospecting licence

African countries impose similar statutory terms and conditions on a prospecting licence, including conditions relating to prospecting operations, annual work and expenditure requirements, reporting of information, and environmental protection. The holder of a prospecting licence is obliged to conduct prospecting operations in accordance with the approved programme of mineral prospecting operations, including the minimum work and minimum expenditure obligations as specified in the licence.[47] They must commence prospecting operations within the statutory timeline (for example, within 120 days from the date on which the prospecting right becomes effective in South Africa and within three months of the date of issue of the licence in Botswana and Ghana).[48] Failure to commence prospecting operations is an offence and constitutes a ground for suspending or revoking the prospecting licence.[49] Moreover, the holder of a prospecting licence must notify the Minister of the discovery of minerals within the specified period (within 30 days of such discovery in Botswana and Ghana).[50]

The statutory regimes in Ghana and South Africa encapsulate the full gamut of the conditions imposed on the holder of a prospecting licence. In Ghana, a prospecting licence enjoins the holder to:

(a) commence prospecting operation within three months after the date of the issue of the licence, or at a time specified by the Minister,
(b) demarcate and keep demarcated the prospecting area in the prescribed manner,
(c) carry on prospecting operation in accordance with the programme of prospecting operations,
(d) notify the Minister through the Commission, of any discovery of minerals to which the prospecting licence relates within a period of thirty days from the date of the discovery,
(e) notify the Minister through the Commission of the discovery of a mineral deposit which is of possible economic value within a period of thirty days from the date of the discovery,
(f) fill back or otherwise make safe to the satisfaction of the Commission a borehole or excavation made during the course of prospecting operations,
(g) unless the Commission otherwise stipulates, remove within sixty days from the date of the expiration of the prospecting licence a camp, temporary building or machinery erected or installed and make good to the satisfaction of the Commission damage to the surface of the ground occasioned by the removal,

---

47 *Minerals and Mining Act, 2006* (Ghana), s. 37.(2); *MPRDA*, s. 19.(2); *Mines and Minerals Act, 1999* (Botswana), s. 21.
48 *MPRDA*, s. 19.(2); *Mines and Minerals Act, 1999* (Botswana), s. 21.(1); *Minerals and Mining Act, 2006* (Ghana), s. 37.(2); *Mining Act, 2010* (Tanzania), s. 36.(1).
49 *MPRDA*, s. 47.
50 *Mines and Minerals Act, 1999* (Botswana), s. 21.(1); *Minerals and Mining Act, 2006* (Ghana), s. 37.(2); *Mining Act, 2010* (Tanzania), s. 36.(1).

(h) subject to the condition of the prospecting licence, expend on prospecting not less than the amount specified in the prospecting licence, and
(i) submit reports of other documents to persons at prescribed intervals and supporting documents containing required information.[51]

Similarly, in South Africa, the holder of a prospecting right must:

(a) lodge such right for registration at the Minerals and Petroleum Titles Registration Office within 60 days after the right has become effective;
(b) commence with prospecting activities within 120 days from the date on which the prospecting right becomes effective in terms of section 17 (5) or such an extended period as the Minister may authorise;
(c) continuously and actively conduct prospecting operations in accordance with the prospecting work programme;
(d) comply with the terms and conditions of the prospecting right, relevant provisions of this Act and any other relevant law;
(e) comply with the conditions of the environmental authorization;
(f) pay the prescribed prospecting fees to the State; and
(g) subject to section 20 and in terms of any relevant law, pay the State royalties in respect of any mineral removed and disposed of during the course of prospecting operations;
(h) submit progress reports and data of prospecting operations to the Regional Manager within 30 days from the date of submission thereof to the Council of Geoscience.[52]

Holders of prospecting licences in Africa are required to keep full and accurate records and data regarding their prospecting operations, including information regarding the boreholes drilled; strata penetrated; minerals discovered; the results of any geochemical or geophysical analysis; the result of any analysis of minerals removed under the licence; the geological interpretation of the records maintained under the licence; the number of persons employed; any other prospecting work undertaken pursuant to the licence;[53] and the expenditure connected with such operations.[54] Failure to keep the requisite information or record and the provision of misleading record or information are offences.[55]

African countries require prospecting licence holders to submit periodic reports regarding their prospecting activities. For example, Botswana requires the holder of a prospecting licence to submit to the government an annual audited statement

---

51 *Minerals and Mining Act, 2006* (Ghana), s. 37.(2). See also *Mining Act, 2010*, (Tanzania), s. 36.(1).
52 *MPRDA*, s. 19.(2).
53 *Mines and Minerals Act, 1999* (Botswana), s. 21.(3); *Minerals and Mining Act, 2006* (Ghana), s. 37.(4); *MPRDA*, s. 21.(1)(a).
54 *MPRDA*, s. 21.(1)(a).
55 *Mines and Minerals Act, 1999* (Botswana), s. 21.(4).

of expenditure directly incurred under the licence.[56] For its part, South Africa requires the holder of a prospecting right to "submit progress reports and data, in the prescribed manner and at the prescribed intervals, to the Regional Manager regarding the prospecting operations".[57] In fact, South Africa prohibits the disposition or destruction of prospecting records, except in accordance with the written directions of the Regional Manager.[58]

3.3 Duration and renewal of a prospecting licence

The initial term of a prospecting licence in Africa varies from country to country. A prospecting licence is valid for a maximum initial period of three years in Botswana, Ghana, and Kenya,[59] four years in Tanzania,[60] and five years in South Africa.[61] However, prospecting licences are renewable across Africa, provided that the holder complies with statutory requirements.[62] For example, in Botswana the holder of a prospecting licence may apply for renewal of the licence for a period of two years, and such renewal is granted if the Minister is satisfied that the applicant is not in default of the terms of the licence and the applicant's proposed programme of prospecting operations is adequate.[63] Holders of prospecting licences in Botswana are entitled to a maximum of two renewals, although each renewal does not exceed two years.[64] However, the Minister may renew a prospecting licence for more than two years where a discovery of mineral resources has been made by the holder of the prospecting licence but evaluation work has not been completed despite the proper efforts of the licence holder.[65]

South Africa permits a single renewal of a prospecting right for a period not exceeding three years.[66] To be entitled to a renewal, the holder of a prospecting right must indicate the reasons and period for which the renewal is required, and submit (1) a detailed report reflecting the prospecting results, the interpretation thereof, and the prospecting expenditure incurred by the applicant; (2) a report reflecting the extent of compliance with the conditions of the environmental authorization; (3) a detailed prospecting work programme for the renewal period;

---

56 *Ibid.* s. 21.(2).
57 *MPRDA*, s. 21.(1)(b).
58 *Ibid.* s. 21.(2).
59 *Mines and Minerals Act, 1999* (Botswana), s. 17.(1); *Minerals and Mining Act, 2006* (Ghana), s. 34.(2); *Mining Act, 2016* (Kenya), s. 74.
60 *Mining Act, 2010* (Tanzania), s. 32.(1)(a) & (b). However, in Tanzania, a prospecting licence for gemstones other than kimberlitic diamonds has a non-renewable duration of one year. See *ibid.* s. 32.(6).
61 *MPRDA*, s. 17.(6).
62 *Mines and Minerals Act, 1999* (Botswana), s. 17.(2) & (3); *MPRDA*, s. 18; *Minerals and Mining Act, 2006* (Ghana), s. 35; *Mining Act, 2010* (Tanzania), s. 32.(2) & (3); *Mining Act, 2016* (Kenya), ss. 81 & 82.
63 *Mines and Minerals Act, 1999* (Botswana), s. 17.(2) & (3).
64 *Ibid.* s.17.(3).
65 *Ibid.* s. 17.(6).
66 *MPRDA*, s. 18.(4).

and (4) a certificate issued by the Council for Geoscience indicating that all prospecting information regarding the project has been submitted by the applicant.[67] The Minister is obliged to grant a renewal of a prospecting right where the applicant has complied with all legal requirements, including the terms and conditions of the prospecting right, the prospecting work programme, and the environmental authorization.[68]

In Ghana, a prospecting licence may, on an application made not later than three months prior to the expiry of the initial term of the licence, be extended for a maximum period of three years.[69] Such extension may be "in respect of all or any number of blocks the subject of the prospecting licence".[70] However, a prospecting licence is extended only if the Minister is satisfied that the applicant has complied with all statutory and contractual obligations.[71] Furthermore, the term of a prospecting licence in Ghana is deemed to be extended, albeit temporarily, in limited situations such as "[w]here a holder of a prospecting licence has made an application for an extension of the term of the licence and the term of the prospecting licence would, but for this subsection expire".[72] In that case, the prospecting licence "shall continue in force with respect to the land and minerals the subject of the application until the application is determined".[73] This provision caters to potential delays in the administrative process and preserves prospecting licences which would otherwise expire where an application for extension is not determined expeditiously by the Minister.

3.4 Relinquishment of land upon renewal of a prospecting licence

In some African countries, the holder of a prospecting licence is required to surrender to the government some of the lands covered by the licence as a condition for renewal of the licence. A prospecting licence usually covers a vast expanse of land (in Botswana, an area not exceeding 1000 square kilometres[74] and, in Ghana, "a block or a number not exceeding 750 contiguous blocks").[75] Thus, the licence holder may be unable to complete their prospecting operations over the entire land within the initial term of the licence. Hence, the government may desire to retrieve a portion of the land covered by the licence so as to make it available to other miners who may be better able to prospect for minerals on the land. In Botswana, Ghana, Kenya, and Tanzania, the holder of a prospecting licence is required, either prior to or at the expiration of the initial term of the licence, to

---

67 Ibid. s. 18.(2).
68 Ibid. s. 18.(3).
69 *Minerals and Mining Act, 2006* (Ghana), s. 35.(1).
70 Ibid. s. 35.(1).
71 Ibid. s. 35.(2).
72 Ibid. s. 35.(4).
73 Ibid. s. 35.(4). See also *Mining Act, 2016* (Kenya), s. 81.(5).
74 *Mines and Minerals Act, 1999* (Botswana), s. 19.(1).
75 *Minerals and Mining Act, 2006* (Ghana), s. 34.(3).

surrender compulsorily "not less than half" of the land covered by the licence.[76] In fact, half of the land covered by a prospecting licence is relinquished each time the licence is renewed, provided that the relinquishment does not reduce the size of the land below the statutory minimum.[77] In Kenya, the size of the area specified in a prospecting licence is reduced upon renewal "by not less than one half the number of blocks", provided that "a minimum of one hundred and twenty five blocks shall [be] subject to the licence".[78] The statutory provision regarding compulsory relinquishment of land incentivizes mining companies to conduct their prospecting operations expeditiously in order to avoid relinquishing part of the land covered by their licence. However, the holder of a prospecting licence is not entitled to compensation for the area that is compulsorily relinquished.[79]

4 Exploration licence

Exploration entails in-depth scientific investigations regarding the presence and value of minerals. As Willard Lacy opines, "mineral exploration is the search for, and evaluation of mineral deposits which have the POTENTIAL of becoming orebodies under expected conditions at some favorable date in the future".[80] In this regard, South Africa's MPRDA defines 'exploration operation' as

> the re-processing of existing seismic data, acquisition and processing of new seismic data or any other related activity to define a trap to be tested by drilling, logging and testing, including extended well testing, of a well with the intention of locating a discovery.[81]

In Nigeria, the word 'explore' means "operations and works aimed at the discovery, the determination of the characteristics and the evaluation of the economic value of mineral resources".[82] Likewise, Zambia defines 'exploration' as

> the search for a mineral by any means and carrying out of such works, and removal of such samples, as may be necessary to test the mineral bearing qualities of any land and define the extent and determine the economic value of a mineral deposit.[83]

---

76 *Mines and Minerals Act, 1999* (Botswana), s. 19.(2); *Minerals and Mining Act, 2006* (Ghana), s. 38.(1); *Mining Act, 2016* (Kenya), s. 84.(1).
77 *Mines and Minerals Act, 1999* (Botswana), s. 19.(2).
78 *Mining Act, 2016* (Kenya), s. 84.(1)(a).
79 *Ibid.* s. 84.(3).
80 Willard Lacy, *An Introduction to Geology and Hard Rock Mining* (Rocky Mountain Mineral Law Foundation, 2015) at 44 [emphasis in the original text].
81 *MPRDA*, s. 1.
82 *Nigerian Minerals and Mining Act, 2007*, s. 164.
83 *Mines and Minerals Development Act, 2015* (Zambia), s. 2.(1).

38 Types of mineral rights in Africa

Exploration may begin with initial conjectures "based largely on either historical evidence or geological and geophysical consideration", but ultimately it involves field searches, aerial geophysical examination of potential sites, and geochemical tests.[84] Through exploration operations, mining companies may "discover evidence of a mineral occurrence and outline its size and character", but often exploration activities are a never-ending exercise involving "multiple stages of rejection and recommendation, discovery and development, decline and abandonment, rediscovery and development, etc., as economic, technological or political conditions change or geological understanding is improved".[85]

The overarching goal of mining companies during the exploration stage is the discovery of orebody, described as "that part of a mineral deposit which can be mined and marketed at a profit under contemporary technological, economic and legal conditions".[86] As Willard Lacy posits:

> The principal objective of mineral exploration is to find economic mineral deposits that will appreciably increase the value of a mining company's stock to the shareholders on a continuing basis, or to yield a profit to the explorer. For an established mining company this may entail discovery or acquisition of new ore reserves and mineral resources to prolong or increase production or life of the company, to create new assets and profit centers by product and/or geographic diversification. Or, in the case of individuals or exploration companies, an objective may be to seek a deposit for sale to, or joint venture with, a major operating company, or to serve as a basis for stock issue and formation of a new company.[87]

4.1 Rights conferred by an exploration licence

An exploration licence grants the exclusive right to explore for minerals in a defined area, including the assaying, testing and determination of the characteristics of minerals, and the evaluation of the economic value of minerals. The holder of an exploration licence has the exclusive right to conduct exploration in the area covered by the licence, including the right to do all such other acts as may be necessary for, or incidental to, the carrying on of the exploration operations.[88] Such incidental rights include the right to enter the land covered by the licence, drill boreholes, make excavations, erect camps and temporary buildings in the exploration area, and take timber and water from rivers and watercourses for exploration purposes.[89]

---

84 Dwight Newman, *Mining Law of Canada* (Toronto: LexisNexis, 2018) at 52.
85 Lacy, *supra* note 80 at 44.
86 *Ibid.* at 44.
87 *Ibid.* at 45.
88 *Mines and Minerals Development Act, 2015* (Zambia), s. 23.(2); *Nigerian Minerals and Mining Act, 2007*, s. 60.(1); *Mining Act, 2012* (South Sudan), s. 47; *Mines and Minerals Act*, 2009 (Sierra Leone), s. 77.(1).
89 *Mines and Minerals Development Act, 2015* (Zambia), s. 23.(2); *Nigerian Minerals and Mining Act, 2007*, s. 60.(1); *Mines and Minerals Act*, 2009 (Sierra Leone), s. 77.(2).

The holder of an exploration licence also has the consequential right to take, remove, and export specimens as is reasonably required for testing and analysis, and conduct bulk sampling and trial processing of mineral resources not exceeding such limit as is reasonably required for determining mining potential.[90] An interesting difference between the mining regimes in Africa is that while some countries (such as Nigeria and Sierra Leone) allow exploration licence holders to sell specimens and samples obtained from exploration activities or from bulk sampling and trial processing,[91] other countries (such as Zambia) permit licence holders to remove mineral samples from the exploration area for the specific purpose of analyzing or testing the minerals.[92] Although Nigeria allows holders of exploration licences to sell specimens and samples of minerals, they are obliged to pay royalty on the specimens and samples "as if the mineral resources sold were obtained under a mining lease".[93]

The holder of an exploration licence has the exclusive right to apply for and be granted a mining lease in relation to the area covered by the exploration licence, provided the terms and conditions of the licence are satisfied.[94] In Nigeria, for example, the holder of an exploration licence who has complied with the terms of the licence is entitled to the grant of a mining lease in respect of the area covered by the licence.[95] Similarly, in South Sudan the holder of an exploration licence "shall have the exclusive right to apply for a small-scale mining licence or large-scale mining licence for the concerned mineral resource over any part of the exploration area".[96]

However, there are certain restrictions on the exercise of the licence holder's exclusive right to undertake exploration activities. These restrictions are designed to protect the environment as well as third parties such as the owners and lawful occupiers of land. Thus, in Zambia the holder of an exploration licence cannot commence exploration activities unless they obtain environmental approval from the Zambian Environmental Management Agency; or written consent from the appropriate authority where the land is subject to restrictions with regard to entry by the licence holder.[97] Land subject to restrictions includes land consisting of a cemetery, land containing ancient and national monuments, land containing a building or dam owned by the government, and land constituting an aerodrome.[98] In addition, the holder of an exploration licence must obtain the consent

---

90 *Nigerian Minerals and Mining Act, 2007*, s. 60.(1); *Mines and Minerals Act*, 2009 (Sierra Leone), s. 77.(2).
91 *Nigerian Minerals and Mining Act, 2007*, s. 60.(1); *Mines and Minerals Act*, 2009 (Sierra Leone), s. 77.(2d).
92 *Mines and Minerals Development Act, 2015* (Zambia), s. 26.(1).
93 *Nigerian Minerals and Mining Act, 2007*, s. 63.
94 *Proclamation No. 678/2010* (Ethiopia), art. 20; *Nigerian Minerals and Mining Act, 2007*, s. 60.(2); *Mines and Minerals Act*, 2009 (Sierra Leone), s. 79.(1); *Mining Act, 2012* (South Sudan), s. 50.
95 *Nigerian Minerals and Mining Act, 2007*, s. 60.(2).
96 *Mining Act, 2012* (South Sudan), s. 50.(1).
97 *Mines and Minerals Development Act, 2015* (Zambia), s. 25.(1)(a).
98 *Ibid.* s. 52.(1)(a).

of the owner and lawful occupier of land prior to conducting exploration activities and pay compensation for any damage arising from their exploration activities.

4.2 Terms and conditions of an exploration licence

The terms and conditions of an exploration licence are standard in the sense that they are statutorily prescribed, but in some instances the terms may be tailored to the specific circumstances of an applicant. As with other types of mineral rights, the terms and conditions of an exploration licence are aimed at ensuring the expeditious exploration of the area covered by the licence. The standard terms of an exploration licence are that the holder shall commence exploration activities within the stipulated time and comply with the minimum work and minimum expenditure and reporting obligations.[99] The condition regarding minimum work requires the holder to conduct exploration activities that meet the minimum threshold specified in the grant instrument, usually a contract. Likewise, the minimum expenditure requirement specifies the minimum amount of money the holder must spend on each block or hectare of land over a specified period. Such expenditure must be directly related to exploration activities, such as expenditure on exploration equipment, cost of geological surveys, and cost of testing and assaying of samples.

These terms and conditions ensure that mining companies do not deliberately neglect to conduct exploration activities with a view to holding on to the land for speculative purposes. An exploration licence can be terminated by the government if the holder fails to commence exploration operations within the prescribed period, or where they fail to meet the minimum work and expenditure requirements. The fear of losing an exploration licence is thus an incentive for the holder to commence and maintain exploration activities over the duration of the licence. In addition, the holder must conduct exploration activities in a safe, efficient, workmanlike manner and in an environmentally and socially responsible fashion.[100] Other conditions imposed on holders of exploration licences include the duty to restore and reclaim land damaged as a result exploration activities, as well as the duty to compensate users or occupiers of land for damage to land and property resulting from exploration activities.[101]

The standard terms and conditions of exploration licences in Africa are epitomized by Zambia's mining statute, which obliges the holder of an exploration licence to:

(a) only commence exploration operations if the holder submits to the Mining Cadastre Office [a letter issued by the Zambian Environmental Management Agency approving the environmental project brief and, in the case of land subject to restrictions, a written consent from the appropriate authority];

---

99 *Nigerian Minerals and Mining Regulations 2011*, regs 43–4 & Schedule 5; *Mines and Minerals Act*, 2009 (Sierra Leone), s. 78.(1).
100 *Nigerian Minerals and Mining Act, 2007*, s. 61.(1)(a) & (b).
101 *Mines and Minerals Development Act, 2015* (Zambia), s. 25.

Types of mineral rights in Africa 41

(b) within one hundred and eighty days of the grant of the exploration licence, register a pegging certificate at the Mining Cadastre Office;
(c) give notice to the Director of Mining Cadastre of the discovery of any mineral deposit of possible commercial value within thirty days of the discovery;
(d) expend on exploration operations not less than the amount prescribed or required by the terms and conditions of an exploration licence;
(e) carry on explorations in accordance with the programme of exploration;
(f) backfill or otherwise make safe any excavation made during the course of the exploration, as the Director of Mining Cadastre may specify;
(g) permanently preserve or otherwise make safe any borehole in the manner directed by the Director of Mining Cadastre and surrender to Government, on termination, without compensation, the drill cores, other mineral samples and boreholes and any water rights in respect of the boreholes;
(h) unless the Director of Mining Cadastre otherwise stipulates, remove, within sixty days of the expiry or revocation of the exploration licence, any camp, temporary buildings or machinery erected or installed and repair or otherwise make good any damage to the surface of the ground occasioned by the removal, in the manner specified by the Director of Mining Cadastre; and
(i) keep and preserve such records as the Director of Mines Safety may determine relating to the protection of the environment.[102]

4.3 Record-keeping and reporting obligations

Holders of exploration licences bear record-keeping and reporting obligations, including the filing of technical reports regarding their exploration activities. The holder of an exploration licence must keep accurate records of their exploration operations and submit to the government certain technical reports detailing the technical work undertaken by the licence holder. The reports must include details regarding geological, geochemical, and geophysical surveys; drilling operations including boreholes, the strata penetrated, and detailed logs of such strata; mineralogical and metallurgical analysis; the results of any seismic survey or geochemical, geophysical, and remote sensing data analysis; and interpretation of geological investigation.[103] In addition, the report must include details of the expenditure incurred by the holder of the exploration licence in the course of exploration, including field charges for surveys, drilling, and analysis; overhead charges for secretarial services, drafting, purchase of equipment, legal expenses and other fees; compensation paid to landowners or occupiers; and transportation

---

102 *Mines and Minerals Development Act, 2015* (Zambia), s. 25.(1).
103 *Nigerian Minerals and Mining Regulations 2011*, reg. 34 & Schedule 5; *Mines and Minerals Development Act, 2015* (Zambia), s. 25.(2); *Mines and Minerals Act, 2009* (Sierra Leone), s. 78.(2).

42 *Types of mineral rights in Africa*

charges.[104] An exploration licence may be revoked if the holder fails to comply with their record-keeping and reporting obligations.[105]

4.4 Duration and renewal of an exploration licence

The maximum initial duration of an exploration licence is relatively short, such as three years in Ethiopia and Nigeria,[106] four years in Sierra Leone and Zambia,[107] and five years in South Sudan.[108] The short duration of an exploration licence is designed to ensure that mineral right holders expedite their exploration activities. A longer duration could incentivize licence holders to delay their exploration operations, thus defeating the host state's desire to accelerate mineral production. An exploration licence is renewable, provided the holder complies with the terms and conditions of the licence, including the minimum work and minimum expenditure obligations.[109] In Nigeria, an exploration licence is renewable for two further periods of two years each,[110] while Zambia permits the renewal of an exploration licence for two further periods not exceeding three years each.[111] Thus, in Zambia the maximum term of an exploration licence, including the initial term and the renewal period, does not exceed ten years.[112] Interestingly, Zambia's mining statute does not stipulate specific conditions for the renewal of an exploration licence, but it is reasonable to speculate that a renewal would be granted where the holder of the licence is not in breach of the conditions of the licence or any provision of the Act, and where the prospecting operations proposed to be undertaken by the licence holder are adjudged by the licensing authority to be adequate. However, in Zambia exploration licences for small-scale exploration as well as exploration licences for gemstones other than diamonds are not renewable.[113]

4.5 Relinquishment of land upon renewal of an exploration licence

Some African countries require exploration licence holders to relinquish part of the land covered by the licence upon expiry of the initial term of the licence. An

---

104 *Nigerian Minerals and Mining Regulations 2011*, reg. 34 & Schedule 5; *Mines and Minerals Development Act, 2015* (Zambia), s. 25.(2).
105 *Nigerian Minerals and Mining Regulations 2011*, regs 34.(2), 43.(2) & 44.(2).
106 *Proclamation No. 678/2010* (Ethiopia), art. 19.(1); *Nigerian Minerals and Mining Act, 2007*, s. 62.
107 *Mines and Minerals Act, 2009* (Sierra Leone), s. 76; *Mines and Minerals Development Act, 2015* (Zambia), s. 24.(1) & (2).
108 *Mining Act, 2012* (South Sudan), s. 44.
109 *Proclamation No. 678/2010* (Ethiopia), art. 19.(3); *Nigerian Minerals and Mining Act, 2007*, s. 62; *Mines and Minerals Act, 2009* (Sierra Leone), s. 76.(2); *Mines and Minerals Development Act, 2015* (Zambia), s. 24.(2); *Mining Act, 2012* (South Sudan), s. 44.
110 *Nigerian Minerals and Mining Act, 2007*, s. 62.
111 *Mines and Minerals Development Act, 2015* (Zambia), s. 24.(2).
112 *Ibid.* s. 24.(2).
113 *Ibid.* s. 24.(4).

exploration licence covers a large expense of land, such as 200 square kilometres in Nigeria[114] and up to 59,880 cadastre units in Zambia.[115] Hence, African governments desire to regain access to portions of the land, particularly where the land is not fully explored on the expiry of the exploration licence. In this regard, Zambia requires the holder of an exploration licence to "relinquish at least fifty percent of the initial exploration area on the first renewal and at least fifty percent of the balance on the second renewal",[116] while in Ethiopia the holder of an exploration licence must relinquish no less than one-fourth of the licence area during each renewal of the licence.[117] The land relinquishment requirement, which is in keeping with industry practice, is designed to incentivize mineral right holders to diligently utilize their mineral right during its lifespan. The knowledge that portions of the land covered by an exploration licence will be relinquished to the government could prompt licence holders to expedite their exploration operations.

5 Retention licence

Of the many countries covered in this book, only a few (Botswana, Ethiopia, Kenya, Namibia, South Africa, South Sudan, and Tanzania) provide for a retention licence or permit. A retention licence or permit enables the holder of a prospecting or exploration licence to retain their mineral title in situations where it is temporarily impracticable to produce minerals. A retention licence is premised on two factors – namely, the discovery of mineral deposit within the prospecting or exploration area, and the temporary inability of the licence holder to produce the minerals. The holder of a prospecting or exploration licence who discovers mineral deposits which, due to economic, social, or political factors, cannot be mined on a profitable basis can apply for a retention licence with regard to the area and the minerals covered by their prospecting or exploration licence.[118] It may be temporarily impracticable to mine minerals for several reasons, including unfavourable market conditions such as low commodity prices, civil unrest in the area covered by the licence, technical constraints, temporary lack of capital, or other factors beyond the control of the holder of the licence.[119] A retention licence thus allows the holder to retain their mineral right during a prescribed period in which the holder is unable to produce minerals from the area covered by their licence.

5.1 Rights conferred by a retention licence

A retention licence preserves the right of the holder of a prospecting or exploration licence, including the right to apply for and obtain a mining lease in relation

---

114 *Nigerian Minerals and Mining Regulations 2011*, reg. 35.(3)(b).
115 *Mines and Minerals Development Act, 2015* (Zambia), ss. 21.(2) & 76.
116 *Ibid.* s. 76.(2).
117 *Proclamation No. 678/2010* (Ethiopia), art. 22.
118 *Mines and Minerals Act, 1999* (Botswana), s. 25; *Mining Act, 2016* (Kenya), s. 85; MPRDA, ss. 31–3.
119 *Mining Act, 2016* (Kenya), s. 85.

to the area covered by the retention licence.[120] It confers on the holder the exclusive right to retain the area covered by the licence for mining purposes, as well as other ancillary rights. More specifically, a retention licence confers on the holder the rights to carry on prospecting operations in the retention area in order to determine from time to time the prospects of mining any mineral to which the retention licence relates on a profitable basis; remove any mineral or sample of a mineral for any purpose other than sale or disposal; and carry on such other investigations and operations as would be necessary to determine the prospect of mining any mineral to which the licence relates on a profitable basis.[121]

The holder of a retention licence has the exclusive right to be granted a mining lease or mining right in respect of the retention area and the minerals covered by the licence.[122] However, in order to enjoy this exclusive right, the holder of a retention licence must satisfy the terms of their existing mineral right. For example, in South Africa the holder of a retention permit is required to show proof that they complied with the conditions of the environmental authorization issued in relation to the permit.[123] They must also pay the prescribed retention fees and "submit a six monthly progress report to the Regional Manager indicating – (i) the prevailing market conditions, the effect thereof and the need to hold such retention permit in respect of the mineral and land in question; and (ii) efforts undertaken by such holder to ensure that mining operations commence before the expiry" of the three-year period specified in the permit or the two-year renewal period, as the case may be.[124]

That said, the holder's exclusive right is confined to the specific area and minerals covered by the retention licence. Thus, the holder of a retention licence does not have an exclusive right to obtain a lease for minerals not covered by the retention licence. In Namibia, for example, a retention licence does not confer on the holder "any preferential right to any other licence in relation to any mineral or group of minerals, other than a mineral or group of minerals to which such mineral deposit retention licence relates".[125] In effect, the area covered by a retention licence may be the subject of a separate mineral right granted to a third party in relation to minerals not covered by the retention licence.

5.2 Terms and conditions of a retention licence

A retention licence is granted subject to certain economic, financial, and environmental conditions that must be fulfilled by the licence holder during the currency of the licence. These conditions, which are similar to the conditions imposed on holders of prospecting and exploration licences, include the duties to demarcate

---

120 *MPRDA*, ss. 32 & 35; *Mines and Minerals Act, 1999* (Botswana), s. 31; *Mining Act, 2016* (Kenya), s. 88.(1); *Mining Act, 2012* (South Sudan), s. 73.(1).
121 *Mines and Minerals Act, 1999* (Botswana), s. 31; *Mining Act, 2016* (Kenya), s. 88.(2).
122 *MPRDA*, s. 35.(1); *Mining Act, 2016* (Kenya), s. 88.(1).
123 *MPRDA*, s. 35.(2)(a).
124 *MPRDA*, s. 35.(2).
125 *Minerals (Prospecting and Mining) Act, 1992* (Namibia), s. 77.(2).

the retention area in the prescribed manner; reclaim and make safe any excavations and boreholes made in the course of prospecting operations; remove any camp, equipment, plant, or building erected in the retention area; and repair or otherwise make good to the satisfaction of the government any damage to surface area caused by the removal of any camp, equipment, plant, or building.[126] The holder of a retention licence is equally obliged to keep accurate records, including details of all minerals discovered and the results of studies, surveys, tests, and other works undertaken in the area covered by the licence.[127] In addition, the holder of a retention licence must keep records relating to imaging, geological mapping, geochemical sampling, drilling, pitting, and trenching conducted by the holder.[128] However, in South Africa, once a retention permit is issued, the terms and conditions of the prospecting right previously granted in respect of the land to which the retention permits relates are automatically suspended, but the conditions of the environmental authorization issued in regard to the prospecting right remain in force.[129]

5.3 Duration and renewal of a retention licence

A retention licence or retention permit has a maximum life span of three years in Botswana and South Africa,[130] two years in Kenya,[131] and five years in Namibia and Tanzania.[132] In South Sudan, the duration of a retention licence is five years when issued under an exploration licence and six years when issued under a mining licence.[133] In South Africa, a retention permit runs concurrently with the duration of a prospecting right where the prospecting right has not expired.[134] Some African countries permit the renewal of a retention licence while others prohibit such renewal. Among the countries that permit the renewal of a retention licence, some allow a one-time renewal while others permit multiple renewals. For example, a retention permit is renewable once for a period not exceeding two years in South Africa, but to be entitled to a renewal, the holder of the permit must show that they have complied with the relevant provisions of the MPRDA and any other relevant law.[135] They must also show that they complied with the terms and conditions of the retention permit and that the prevailing market conditions make it uneconomical to mine the mineral(s) in question.[136] Likewise, a retention licence may be renewed

126 *Mines and Minerals Act, 1999* (Botswana), s. 32; *Mining Act, 2016* (Kenya), s. 89.
127 *Mining Act, 2016* (Kenya), s. 90; *Mines and Minerals Act, 1999* (Botswana), s. 32.(2); *Minerals (Prospecting and Mining) Act, 1992* (Namibia), s. 89.
128 *Minerals (Prospecting and Mining) Act, 1992* (Namibia), s. 89.
129 MPRDA, s. 32.(2) & (3).
130 *Mines and Minerals Act, 1999* (Botswana), s. 30.(1); MPRDA, s. 32.(4).
131 *Mining Act, 2016* (Kenya), s. 87.
132 *Minerals (Prospecting and Mining) Act, 1992* (Namibia), s. 82.
133 *Mining Act, 2012* (South Sudan), s. 69.
134 MPRDA, s. 32.(2).
135 *Ibid*. s. 34.(2) & (3).
136 *Ibid*. s. 34.(2).

once for three years in Botswana.[137] Unlike the situation in South Africa and Botswana, Namibia allows multiple renewals of a retention licence "for such period, not exceeding two years, as may from time to time be determined by the Minister at the time of the granting of any application for the renewal of such licence".[138] However, South Sudan does not permit the renewal or extension of a retention licence.[139]

The government may conduct investigation to determine whether the market conditions that made it temporarily impracticable to produce minerals in the retention area still exist. In South Sudan, a retention licence is reviewed periodically and if, upon such review, the government "reasonably determines that the relevant market conditions and/or other economic factors have changed such that the reasons for the Retention Licence are no longer present", the government may cancel the retention licence.[140] The government may compel the holder of a retention licence to commence mineral production or apply for a mining lease in relation to the area covered by the retention licence where it determines that the market conditions on which the retention licence is predicated no longer exist. In Kenya and Namibia, the government may issue a written notice to the holder of a retention licence requiring them to apply for a mining licence in respect of the area and minerals covered by the retention licence or to undertake further prospecting operations in the retention area.[141] The government may issue such notice if, based on an independent report, it is satisfied "that it has become technically possible and commercially viable for a mineral deposit that is the subject of a retention licence to be mined during the term of the licence".[142] In the case of Namibia, the government may issue such notice if it believes that the minerals to which the retention licence relates may be won or mined on a profitable basis or that further prospecting operations may indicate the existence of any such minerals which may be won or mined on a profitable basis.[143] The failure of a retention licence holder to comply with the written notice ordering them to apply for a mining licence is grounds for revoking the retention licence and, in such cases, the holder is not entitled to compensation.[144] In countries such as Namibia, the failure of a retention licence holder to comply with the written notice is an offence punishable with a fine or imprisonment or both.[145]

---

137 *Mines and Minerals Act, 1999* (Botswana), s. 30.(2).
138 *Minerals (Prospecting and Mining) Act, 1992* (Namibia), s. 82.(1)(b).
139 *Mining Act, 2012* (South Sudan), s. 69.(3).
140 *Ibid.* s. 70.(2).
141 *Mining Act, 2016* (Kenya), s. 91; *Minerals (Prospecting and Mining) Act, 1992* (Namibia), s. 88.
142 *Mining Act, 2016* (Kenya), s. 91.(2).
143 *Minerals (Prospecting and Mining) Act, 1992* (Namibia), s. 88.(1).
144 *Mining Act, 2016* (Kenya), s. 91.(3–5).
145 *Minerals (Prospecting and Mining) Act, 1992* (Namibia), s. 88.(3).

6 Mining lease

A mining lease is required for large-scale mining in Africa. As discussed in Chapter 4, a mining lease (referred to in Botswana and Zambia as mining licence and in South Africa as mining right) is an interest in land or a *profit à prendre* in the sense that it enables the holder to enter land owned by the government and mine the minerals situated in the land. A lease grants the holder the exclusive right to work, win, recover, and dispose of minerals situated in the land covered by the lease. Generally speaking, a mineral lease manifests in the form of a contract that describes the mineral substances and interest conveyed by the government, the area covered by the lease, the duration of the lease, the terms and conditions on which the mineral interest is conveyed, including minimum work and minimum expenditure requirements, the legal rights and obligations of the lessee, the fiscal regimes governing the lease, and the circumstances under which the lease may be terminated.[146] In Africa, a mineral lease is designed as a state-investor contract and it contains additional features such as fiscal stabilization guarantees by the host government (discussed in Chapter 6); arbitration and choice of law clauses; environmental standards and practices; reporting requirements; equity participation by the host government; local content provisions, including transfer of technology; and the conditions for the renewal of the lease.

Although a mining lease allows the lessee to recover minerals, the lease does not pass title to minerals that are *in situ* (that is, minerals in their natural and original position in the ground) to the lessee. Rather, in Africa, minerals that are *in situ* are owned by the government, but upon recovery of the minerals by the leaseholder, ownership of the minerals passes to the leaseholder, subject, of course, to the payment of royalties and taxes to the government.

6.1 Rights conferred by a mining lease

A mining lease confers contractual and statutory rights on the lessee, including the exclusive right to conduct mining operations in the area covered by the lease, as well as the right to recover and dispose of the minerals specified in the lease.[147] For example, the holder of a mining licence in Zambia has "exclusive rights to carry on mining, processing and exploration in the mining area and to do all such other acts and things as are necessary for, or incidental to, the carrying on of those operations".[148] A mining lease authorizes the holder to take and remove from the land the minerals specified in the mining lease and dispose of the minerals in accordance with the approved marketing plan.[149] The lessee also has certain

---

146 Robert Rennie, *Minerals and the Law in Scotland* (Hertfordshire: EMIS Professional Publishing, 2001) at 107–8.
147 *Mines and Minerals Act*, 1999 (Botswana), s. 44.(1); *Minerals and Mining Act, 2006* (Ghana), s. 46; *Nigerian Minerals and Mining Act, 2007*, s. 68; *MPRDA*, s. 5; *Mines and Minerals Development Act, 2015* (Zambia), s. 32.(4).
148 *Mines and Minerals Development Act, 2015* (Zambia), s. 32.(3).
149 *Minerals and Mining Act, 2006* (Ghana), s. 46; *Nigerian Minerals and Mining Act, 2007*, s. 68; *MPRDA*, s. 5.(3c); *Mines and Minerals Act*, 1999 (Botswana), s. 44.(1)(c).

ancillary rights that aid their ability to conduct mining operations. Mining leaseholders have ancillary or incidental rights such as the right to enter the land covered by their lease for the purpose of undertaking mining operations; erect all necessary equipment, plant, camps, and buildings on the land for mining-related purposes; and stack or dump any mineral or waste products in a manner approved by the government.[150] Thus, the holder of a mining lease can build temporary housing for their employees in the lease area. In addition, they can cut and use timber to build temporary housing or an access bridge within the lease area, provided they comply with the laws regulating the cutting of timber. Mining leaseholders also have a right to use water from rivers, streams, and watercourses for mining-related purposes subject to compliance with environmental regulations. In South Africa, for example, the holder of a mining right can,

> subject to the *National Water Act, 1998 (Act No. 36 of 1998)*, use water from any natural spring, lake, river or stream, situated on, or flowing through, such land or from any excavation previously made and used for prospecting, mining, exploration or production purposes, or sink a well or borehole required for use relating to prospecting, mining, exploration or production on such land.[151]

The Nigerian mining statute encapsulates the rights conferred on the holder of a mining lease as follows:

> A mining lease confers on the holder the right within the Mining Lease Area to –
> (a) obtain access and to enter the Mining Lease Area;
> (b) exclusively use, occupy and carry out mineral exploration within the Mining Lease Area;
> (c) exclusively carry out exploration within the Mining Lease Area;
> (d) utilize the water and wood and other construction materials as necessary for mineral exploitation in accordance with the Permit and Regulations;
> (e) use such portions as may be required for the purposes of growing such plants and vegetables, or keeping such animals and fish as may be reasonable for use of the employees at the Mine;
> (f) store, remove, transport, submit to treatment and process the mineral resources, and dispose of any waste; and
> (g) market, sell, export or otherwise dispose of the mineral products resulting from the mining operations.[152]

---

150 *Mines and Minerals Act*, 1999 (Botswana), s. 44.(1); *MPRDA*, s. 5.(3); *Mines and Minerals Development Act, 2015* (Zambia), s. 32.(4).
151 *MPRDA*, s. 5.(3)(d).
152 *Nigerian Minerals and Mining Act, 2007*, s. 68.

6.2 Standard terms and conditions of a mining lease

A mining lease is granted subject to certain standard conditions that must be met by the lessee. These conditions relate to the expeditious commencement of mining operations; the continuous production of minerals; minimum expenditure and minimum work; record-keeping and reporting obligations; the undertaking of mining operations in a skilful and efficient manner; domestic beneficiation of mineral ores; state participation in mining projects; employment of citizens and procurement of goods and services from domestic suppliers (discussed in Chapter 7); community development obligations (discussed in Chapter 8); and conditions relating to environmental protection (discussed in Chapter 9). In the ensuing pages, we discuss some of the standard conditions of a mining lease in Africa.

The standard conditions of a mining lease are couched in the form of statutory obligations imposed on the lessee.[153] In Botswana, for example, the standard conditions of a mining lease include the obligations to:

(a) commence production on or before the date referred to in the programme of mining operations as the date by which he intends to work for profit;
(b) develop and mine the mineral covered by his mining licence in accordance with the programme of mining operations as adjusted from time to time in accordance with good mining and environmental practice;
(c) demarcate and keep demarcated the mining area in such manner as may be prescribed and, within three months of the date referred to in paragraph (a), submit to the Minister a diagram of the mining area;
(d) keep and maintain an address in Botswana, full particulars of which shall be registered with the Minister, to which all communications and notices may be addressed; and
(e) notify the Minister as soon as he begins to work his mining area for profit.[154]

Similarly, in South Africa, mining right holders are statutorily obliged to:

(a) lodge such right for registration at the Mineral and Petroleum Titles Registration Office within 60 days and the right has become effective;
(b) commence with mining operations within one year from the date on which the mining right becomes effective in terms of section 23 (5) or such extended period as the Minister may authorise;
(c) actively conduct mining in accordance with the mining work programme;
(d) comply with the relevant provisions of this Act, any other relevant law and the terms and conditions of the mining right;
(e) comply with the conditions of the environmental authorisation;
(f) comply with the requirements of the prescribed social and labour plan;

---

153 See *Mining Act, 2016* (Kenya), s. 109; *Nigerian Minerals and Mining Act, 2007*, s. 70.(1); *Mines and Minerals Development Act, 2015* (Zambia), s. 35.
154 *Mines and Minerals Act*, 1999 (Botswana), s. 45.(1).

50 *Types of mineral rights in Africa*

(g) in terms of relevant law, pay the State royalties; and
(h) submit the prescribed annual report, detailing the extent of the holder's compliance with the provisions of section 2 (d) and (f), the charter contemplated in section 100 and the social and labour plan.[155]

6.2.1 Conditions regarding minimum work and minimum expenditure

The programme of mining operations is usually incorporated in or annexed to the mining lease and contains conditions regarding diligence in mining operations, including minimum expenditure and minimum work requirements. The minimum expenditure condition requires the leaseholder to expend a prescribed minimum amount of money on mineral production in the lease area per year, while the minimum work condition enjoins the leaseholder to undertake the threshold of work prescribed in the lease over the prescribed period. Thus, the holder of a mining licence in Zambia must develop the mining area and undertake mining operations with due diligence in accordance with the programme of mining operations; comply with the forecast of capital investment attached to the licence; and take all measures on or under the surface to mine the mineral to which the mining licence relates.[156] That said, the minimum work and minimum expenditure requirements are not unique to a mining lease as similar conditions are imposed on other mineral right holders.

6.2.2 Condition requiring continuity of mining operations

A prominent condition of a mining lease is that the holder must conduct mining operations in an efficient and effective manner. The holder of a mining lease must commence mineral production within the stipulated timeframe (one year in South Africa and 36 months in Nigeria) unless the period is extended by the Minister.[157] However, in some countries the holder of a mining lease may be allowed to defer the commencement of mineral production in circumstances similar to those predicating the grant of a retention licence. For example, the holder of a mining licence in Zambia may apply to the Director of Mining Cadastre for deferment of commencement of mining operations on grounds that

> (a) the holder has identified a mineral deposit within the exploration area which is potentially of commercial significance; and (b) the mineral deposit cannot be developed immediately due to adverse economic conditions or technological constraints, or both, which are, or may be, of a temporary nature.[158]

---

155 *MPRDA*, s. 25.(2).
156 *Mines and Minerals Development Act, 2015* (Zambia), s. 35.(1)(b–f).
157 *MPRDA*, s. 25.(2); *Nigerian Minerals and Mining Act, 2007*, s. 70.(1); *Mines and Minerals Act*, 1999 (Botswana), s. 45.(1); *Mines and Minerals Development Act, 2015* (Zambia), s. 35.(1).
158 *Mines and Minerals Development Act, 2015* (Zambia), s. 33.(1).

The Director of Mining Cadastre approves the application for deferment of mining operations within 60 days of the receipt of the application if satisfied that commercial development of the mineral deposit is temporarily impossible, "but may be possible within a period of five years".[159] However, any such deferment must not exceed a period of five years.[160] Even then, the Director may order a mining licence holder in favour of whom an order of deferment was previously granted to resume mining operations if, based on independent studies, the Director determines that commercial mineral development of the area is possible during the period covered by the deferment order.[161]

The holder of a mining lease must equally ensure that mining operations are undertaken continuously throughout the duration of the lease. Thus, in many African countries the holder of a mining lease cannot suspend production or otherwise stop mining activities without the approval of the government. In reality, however, economic, political, and social factors may prompt the temporary suspension of mining operations. It is common knowledge that, in Africa, the struggle for control over mineral resources often leads to internecine conflicts. In some instances, these conflicts could become violent. Thus, mining companies operating in conflict zones may need to suspend mineral production in order to ensure the safety of their employees and secure their plant and equipment. In other instances, mining leaseholders may need to suspend production due to the maintenance, installation, or decommissioning of equipment; the occurrence of an accident leading to an unsafe working environment or uncontrollable pollution of the mine site; force majeure; or labour disruption.[162]

Some African countries have enacted statutory provisions enabling mining companies to suspend their operations temporarily, provided requisite conditions are satisfied. In both Ghana and Zambia, for example, the holder of a mining lease may suspend the production of minerals by giving the Minister a written notice of the suspension of production activities and indicating valid reasons for the suspension.[163] However, if, for reasons beyond the control of the holder of a mining lease, they are unable to give such advance notice to the Minister, the mining leaseholder must notify the Minister within 14 days of the suspension of mining operations.[164] Once the Minister receives the leaseholder's written notice, the Minister is obliged to investigate the matter.[165] Upon such investigation, the Minister may do one of two things: approve the suspension of production activities or direct the leaseholder to resume full production at the mine by a specified

---

159 *Ibid.* s. 33.(3).
160 *Ibid.* s. 33.(4).
161 *Ibid.* s. 33.(5)
162 *Ibid.* s. 37.(1).
163 *Minerals and Mining Act, 2006* (Ghana), s. 51.(1); *Mines and Minerals Development Act, 2015* (Zambia), s. 37.(2).
164 *Minerals and Mining Act, 2006* (Ghana), s. 51.(2).
165 *Ibid.* s. 51.(4); *Mines and Minerals Development Act, 2015* (Zambia), s. 37.(3).

52 *Types of mineral rights in Africa*

date.[166] The Minister may approve the suspension of mineral production on such terms as the Minister deems appropriate, but in Ghana the suspension of mineral production must not exceed 12 months, although the period may be extended by the Minister for a further period not exceeding 12 months.[167] Zambia does not prescribe the duration of suspension of mineral production but vests discretion in the Director of Mines to determine the duration.[168]

6.2.3 Record-keeping and reporting conditions

Mining leases in Africa impose certain bookkeeping and reporting obligations on the lessee. For example, the holder of a mining licence in Botswana is obliged to maintain accurate technical and financial records of their mining operations, as well as copies of all maps and geological reports, including interpretations, mineral analysis, aerial photographs, core logs, and test results obtained and compiled by the holder in respect of the mining area.[169] Similarly, the holder of a mining right in South Africa is required to keep proper records of mining activities, including financial records in connection with their mining activities.[170] They must also submit to the government the prescribed monthly returns with accurate information and data; an audited annual financial report reflecting the balance sheet and profit-and-loss account; and an annual report detailing the extent to which the holder has complied with the policy objectives of the MPRDA, including the redressing of historical, social, and economic inequalities and the active participation of historically disadvantaged South Africans in the mining industry.[171]

Mining leaseholders (and other mineral right holders) in Nigeria bear an obligation to keep accurate records of the exploration or mining operations conducted within the mineral title area, as well as records of every mineral and ore reserve found in the area; supply to the Mining Cadastre Office (MCO) copies of the records as may be demanded by the MCO; and provide to the Nigerian Geological Survey Agency for storage and archiving a complete set of all geoscientific data acquired in the course of mining operations, including maps, coring, and samples.[172] Similarly, holders of mining licences in Zambia are required to maintain complete and accurate technical records of mining operations and submit these records to the government. They must maintain at their office copies of all maps and geological reports, including interpretations, mineral analysis, aerial photographs, satellite maps, core logs, analysis, and test results in respect of the mining area.[173] In addition, mining licence holders in Zambia must submit certain

166 *Minerals and Mining Act, 2006* (Ghana), s. 51.(4); *Mines and Minerals Development Act, 2015* (Zambia), s. 37.(3).
167 *Minerals and Mining Act, 2006* (Ghana), s. 51.(3).
168 *Mines and Minerals Development Act, 2015* (Zambia), s. 37.(3).
169 *Mines and Minerals Act, 1999* (Botswana), s. 45.(2).
170 *MPRDA*, s. 28.(1).
171 *Ibid*. s. 28.(2)
172 *Nigerian Minerals and Mining Act, 2007*, s. 43.
173 *Mines and Minerals Development Act, 2015* (Zambia), s. 35.(1)(g).

documents to the government, including annual audited financial statements, as well as records relating to their mining operations, mine plans, primary and secondary developments, ore recovery and treatment, production costs, and records regarding ore resources and reserves.[174]

The mining records and data submitted by mineral title holders to government agencies are kept confidential for a specified period. In Nigeria, for example, mining records

> shall not be disclosed to the general public until the earlier of (a) a period of 5 years after its submission; or (b) a part of the mineral title area is relinquished by the mineral title holder; or (c) when the holder of the mineral title ceases to hold the title either as a result of revocation of the title or relinquishment thereof.[175]

6.2.4 Conditions regarding beneficiation, local content, and community development

Some of the conditions attached to a mining lease are designed to promote spillover effects as well as the integration of the mining industry with other sectors of the economy. For example, a mining lease may be granted subject to a condition regarding mineral beneficiation in the host country. In this context, mineral beneficiation means the process of refining, smelting, or converting mineral products into refined products, including the cutting and polishing of mineral products within the host country.[176] In South Africa, the Minister may prescribe the level of beneficiation as a condition for the grant of a mining right.[177] Likewise, as discussed in Chapter 7, some mining contracts in Botswana require the processing of mineral ores within the country.

In addition, mining leaseholders in Africa are required to comply with local content requirements regarding employment and use of local personnel, goods, and services. In Ghana, an applicant for a mining lease is required to submit to the Minerals Commission a detailed programme for the recruitment and training of Ghanaian personnel.[178] This condition is intended to promote Ghana's 'localization policy' which seeks not only the training of Ghanaians in the area of mining technology but also "the eventual replacement of expatriate personnel by Ghanaian personnel".[179] Similarly, in Zambia certain binding documents are usually attached to a mining licence, the terms and conditions of which form part of the mining licence and are therefore binding on the licence holder. The documents usually cover the lessee's undertaking regarding the employment and training of Zambian citizens; an undertaking regarding the promotion of local business

---

174 *Ibid.* s. 35.
175 *Nigerian Minerals and Mining Act, 2007*, s. 43.(5).
176 *MPRDA*, s. 1.
177 *Ibid.* s. 26.(2A).
178 *Minerals and Mining Act, 2006* (Ghana), s. 50.
179 *Ibid.* s. 50.(3). See also *Mining Act, 2016* (Kenya), ss. 46 & 50.

development; and an undertaking regarding the proper management of the environment in the mining area.[180] Thus, in the case of large-scale mining, the holder of a mining licence must implement the local business development undertaking attached to their licence; employ and train Zambian citizens in accordance with the proposal for employment and training attached to the licence; and comply with the forecast of capital investment attached to the licence.[181]

With regard to community development, a common condition of mining leases in Africa is that the lessee shall negotiate and sign a community development agreement (CDA) with host communities. As discussed in Chapter 8, CDAs are statutorily mandated in Nigeria, Guinea, Kenya, Mozambique, Sierra Leone, and South Sudan. Other African countries such as Ghana, the DRC, South Africa, Equatorial Guinea, Niger, Central African Republic, Ethiopia, and Zimbabwe do not require CDAs, but they impose a statutory obligation on mining companies to provide community development assistance to host communities.[182] The community development obligation imposed on mining companies in these countries often appears as a condition in mining leases.

6.2.5 Condition regarding state participation

As discussed in Chapter 8, state participation in mineral exploitation is rampant in Africa, as mining leases are granted on the condition that the host state would own some equity interest in the mining operations undertaken under the lease. For example, a mining lease in Ghana is granted on the condition that the government of Ghana can acquire a direct stake in the mining operations of the lessee. In this regard, the *Minerals and Mining Act, 2006* provides that "[w]here a mineral right is for mining or exploitation, the Government shall acquire a ten percent free carried interest in the rights and obligations of the mineral operations in respect of which financial contribution shall not be paid by Government".[183] In fact, the government may acquire a greater than 10% stake in mineral operations if it reaches an agreement to that effect with the leaseholder.[184] Where the leaseholder is a corporate entity, the Minister may issue a written notice requiring the corporate entity to issue to the Republic of Ghana a 'special share' for no consideration or at no cost to the government.[185] Although a 'special share' does not carry the usual rights attached to shares (such as the right to vote), it entitles the government of Ghana to receive notice of, and attend and speak at, the general meetings of the shareholders of the company.[186] A mining company to whom the

180 *Mines and Minerals Development Act, 2015* (Zambia), s. 32.(2).
181 *Ibid.* s. 35.(1)(d).
182 Kendra E. Dupuy, "Community Development Requirements in Mining Laws" (2014) 1 *The Extractive Industries and Society* 200 at 201.
183 *Minerals and Mining Act, 2006* (Ghana), s. 43.(1). See also *Mining Act, 2016* (Kenya), s. 48; *Mining Act, 2012* (South Sudan), s. 64.(3)
184 *Minerals and Mining Act, 2006* (Ghana), s. 43.(2).
185 *Ibid.* s. 60.(1).
186 *Ibid.* s. 60.(2)(a).

Minister issues such written notice must grant a special share to the government within two months of the notice, failing which they are liable to a fine not exceeding the local currency (cedi) equivalent of US$10,000.[187]

Similarly, the government of Botswana may, at its option, acquire up to 15% working interest in mining projects other than diamond-mining projects.[188] Where the government of Botswana exercises its option to participate in a mining project, the mining licence holder must issue to the government "a single P1.00 special share at par".[189] This special share entitles the government to "appoint up to two directors, with alternates, and to receive all dividends or other distributions in respect of its working interest percentage".[190] However, the government is

> obliged in the same manner as other shareholders to contribute its working interest percentage of – (i) all audited arms-length expenditure incurred by the company to which the licence was issued that is directly attributable to the acquisition of the licence, including relevant prospecting expenditure; and (ii) all expenditure on the mine incurred subsequent to the issue of the mining licence.[191]

In relation to diamonds, Botswana grants an application for a licence to mine diamonds only if the applicant negotiates in good faith and agrees with the government on all technical, financial, and commercial aspects of the diamond-mining project, including government participation in the project.[192] In fact, an application for a licence to mine diamonds automatically triggers negotiations between the applicant and the government regarding state participation in the project.[193] Thus, the government's working interest in a diamond-mining project may surpass the 15% threshold if the applicant and the government reach an agreement to that effect.[194] Where such negotiations do not lead to an agreement within six months or such extended period as the government may allow, the application for a licence automatically fails.[195] In effect, successful negotiation regarding state participation in a mining project is a statutory condition precedent to the grant of a mining licence for diamonds in Botswana.[196]

With regard to state participation, a unique feature of mineral rights in Zambia is the obligation to insure mining operations and indemnify the government against mining-related loss or injury. Holders of mining rights (that is, exploration licence and mining licence) in Zambia are required to obtain and maintain, at all

---

187 *Ibid.* s. 60.(3).
188 *Mines and Minerals Act*, 1999 (Botswana), s. 40.(1).
189 *Ibid.* s. 40.(1).
190 *Ibid.* s. 40.(1).
191 *Ibid.* s. 40.(1).
192 *Ibid.* s. 51.(1).
193 *Ibid.* s. 51.(1).
194 *Ibid.* ss. 40.(3) & 51.
195 *Ibid.* s. 51.(2).
196 *Ibid.* s. 51.(3).

56 *Types of mineral rights in Africa*

times during the lifetime of the mining right and for the prescribed period thereafter,

> insurance coverage, within the Republic, in such amounts and against such risks as may be prescribed by the Minister, by statutory instrument, and shall furnish to the Minister the certificates evidencing that such coverage is in effect and provide copies of any policies requested.[197]

They must equally ensure that their contractors obtain and maintain such insurance coverage at all times.[198] Perhaps more significantly, holders of mining licences in Zambia are required to "indemnify, defend and hold the Republic [of Zambia] harmless against all actions, claims, demands, injury, losses or damages of any nature".[199] This indemnification includes "claims for loss or damage to property or injury or death to persons, resulting from any act or omission in the conduct of mining operations or mineral processing operations by or on behalf of the holder".[200]

The requirements of insurance and indemnification appear necessitated by the fact that the Zambian government sometimes participates jointly with mining companies in the exploitation of mineral resources. These requirements ensure that the government is not directly or vicariously liable for any injury, loss, or damage arising from mining-related accidents. That said, the indemnity provisions in Zambia do not provide a blanket cover for the government. The indemnity provided by mineral right holders does not apply to any "action, claim, demand, loss, damage or injury [resulting] from any direction given by, or wrongful act committed on behalf of, the Republic" of Zambia.[201]

6.3 Duration, renewal, and amendment of a mining lease

As indicated in Table 2.1, the duration of mining leases in Africa is 20 years or more (for large-scale mining) and ten years for small-scale mining. Most African countries prescribe a 25-year duration for large-scale mining leases, although in a few countries, such as Kenya and Tanzania, a mining lease could last for the life of the mine.[202] However, a mining lease is renewable provided the prescribed conditions are satisfied. In Botswana, a mining licence is valid for a period not exceeding 25 years and it is renewable for a further period of 25 years.[203] The holder may apply for renewal of the licence at least one year prior to its expiry.[204] To be entitled to a renewal, the licence holder must demonstrate that: (a) they are

197 *Mines and Minerals Development Act, 2015* (Zambia), s. 63.(1).
198 *Ibid.* s. 63.(1).
199 *Ibid.* s. 63.(4).
200 *Ibid.* s. 63.(4).
201 *Ibid.* s. 63.(5).
202 *Mining Act, 2016* (Kenya), s. 107; *Mining Act, 2010* (Tanzania), s. 43.
203 *Mines and Minerals Act,* 1999 (Botswana), s. 42.(1) & (6).
204 *Ibid.* s. 42.(2).

not in default with regard to the terms and conditions of the licence; (b) development of the mining area has proceeded with reasonable diligence; (c) the proposed programme of mining operations ensures the most efficient and beneficial use of the mineral resources in the mining area; and (d) in the case of an application for the renewal of a licence to mine diamonds, that they have negotiated and reached an agreement with the government with regard to all technical, financial, and commercial aspects of the diamond-mining project, including government participation in the project.[205]

A mining lease in Ghana is granted for an initial term of 30 years or for a lesser period as mutually agreed by the applicant and the Minister.[206] The holder may apply for renewal or extension of the lease for a further period of up to 30 years.[207] Such extension may be granted in respect of all or part of the area and minerals covered by the lease.[208] An application for renewal of a mining lease must be made not later than three months prior to the expiry of the initial term of the lease or a shorter period, as the Minister may allow.[209] The application must be accompanied by a proposed programme of mineral operations.[210] In practice, an application for renewal of a lease is granted if the Minister is satisfied that the applicant has materially complied with their statutory and contractual obligations regarding the lease.[211] In order to ensure that administrative delays in the lease renewal process do not cause the expiry of a mining lease, the duration of a mining lease in Ghana is deemed extended until the application for renewal is determined:

> Where the holder has made an application for an extension of the term of the lease, and the term of the lease would but for this subsection, expire, the lease shall continue in force in respect of the land the subject of the application until the application is determined.[212]

In South Africa, a mining right is valid for a maximum period of 30 years[213] and may be renewed multiple times provided that each renewal period does not exceed 30 years.[214] In that country, the right to renew a mining right in respect of the minerals and land covered by the mining right is exclusive to holder of the mining right.[215] An application for the renewal of a mining right must be prepared in the prescribed manner, lodged at the office of the Regional Manager in whose region

---

205 Ibid. s. 42.(4).
206 *Minerals and Mining Act, 2006* (Ghana), s. 41.(1).
207 Ibid. s. 44.
208 Ibid. s. 44.(1).
209 Ibid. s. 44.(1).
210 Ibid. s. 44.(2).
211 Ibid. s. 44.(3).
212 Ibid. s. 44.(4).
213 MPRDA, s. 23.(6).
214 Ibid. s. 24.(4).
215 Ibid. s. 25.(1).

58 Types of mineral rights in Africa

the land is situated, and accompanied by the prescribed application fee.[216] The application for renewal of a mining right must indicate the reasons and the period for which the renewal is required, the extent of compliance with the requirements of the approved environmental authorization, and a detailed mining work programme for the renewal period.[217] The Minister must renew a mining right if the holder complies with the statutory requirements, the terms and conditions of the mining right, and the prescribed social and labour plan.[218]

In some instances, the holder of mining lease may need to amend the lease with regard to the minerals and the area covered by the lease. In the course of mining operations, the leaseholder may discover some minerals not covered by the lease or they may wish to voluntarily relinquish a portion of the area covered by the lease because their exploration activities did not reveal any mineral deposit in the area. They may also wish to add contiguous or adjoining areas to the lease, particularly where their exploration activities indicate mineral deposit close to the contiguous or adjoining areas.

Thus, in Botswana the holder of a mining licence may apply to the Minister for amendment of the licence where, in the course of exercising the rights conferred under the licence, the holder discovers any mineral not included in such licence.[219] The holder of the licence must notify the Minister within 30 days after such discovery and must give particulars of the mineral discovered as well as the site and circumstances of the discovery.[220] An application for amendment of a licence must include a proposed programme of mining operations in respect of the mineral.[221] If granted, the amendment has the effect of adding the newly discovered mineral to the mining licence.[222] Similarly, the holder of a mining licence in Botswana may apply to enlarge the area covered by the licence, provided the area to be added is contiguous to the area covered by the mining licence. Such application is granted if the Minister is satisfied that an enlargement of the mining area "will ensure the most efficient and beneficial use of the mineral resources of Botswana".[223] However, the Minister lacks the power to enlarge a mining licence "to include any area which is not contiguous to the mining licence".[224] With regard to diamonds, an amendment of a mining licence or the enlargement of the mining area is granted only if the holder of the licence negotiates and agrees with the government with regard to state participation in the diamond-mining project.[225]

---

216 *Ibid.* s. 24.(1).
217 *Ibid.* s. 24.(2)
218 *Ibid.* s. 24.(3).
219 *Mines and Minerals Act*, 1999 (Botswana), s. 44.(2).
220 *Ibid.* s. 44.(2).
221 *Ibid.* s. 44.(2).
222 *Ibid.* s. 44.(2).
223 *Ibid.* s. 44.(5).
224 *Ibid.* s. 44.(6)
225 *Ibid.* ss. 44.(2) & (5) & 51.

7 Small-scale mining lease

Some African countries distinguish between large-scale mining and small-scale mining operations by requiring small-scale miners to obtain a permit or lease that is different from the lease governing large-scale mining. In Botswana, for example, a minerals permit is required where a person wishes to conduct small-scale mining operations (for any mineral other than diamonds) covering an area not exceeding 0.5 square kilometres.[226] In Nigeria, a small-scale mining lease covers an area not less than 5 acres and not more than 3 square kilometres,[227] while in South Africa a mining permit is granted if the mineral in question can be mined optimally within a period of two years, and where the mining area does not exceed 5 hectares.[228]

7.1 Rights conferred by a small-scale mining lease

A small-scale mining lease (referred to as minerals permit in Botswana and mining permit in South Africa) confers a right on the holder to mine and dispose of the minerals specified in the lease or permit.[229] For example, in Ghana the holder of a small-scale mining lease "may win, mine and produce minerals by an effective and efficient method and shall observe good mining practices".[230] In addition, the holder of a small-scale mining lease has ancillary rights that are identical to those vested in other mineral rights holders, including the right to enter the land and erect temporary structures on the land for mining-related purposes.[231] However, the holder of a small-scale mining lease must undertake their mining activities in an environmentally sustainable manner.[232]

The rights of the holder of a small-scale mining lease are exclusive in some countries while in other countries the holder's rights may co-exist with the rights of a third party to mine the same land. In both Kenya and Sierra Leone, the holder of a small-scale mining lease enjoys the exclusive right to carry out mining operations in the area specified in the lease.[233] In South Africa, a mining permit is not granted in relation to a mineral or land over which there is a subsisting mineral right, thus suggesting that a mining permit grants exclusive rights to the holder to mine the mineral and land specified in the permit.[234] Interestingly, the contrary position holds in Botswana where a minerals permit may be granted over an area

---

226 *Ibid.* s. 52.(1).
227 *Nigerian Minerals and Mining Act, 2007*, s. 90.
228 *MPRDA*, s. 27.(1).
229 *Mines and Minerals Act, 1999* (Botswana), s. 56; *Minerals and Mining Act, 2006* (Ghana), s. 93; *MPRDA*, s. 27.(7).
230 *Minerals and Mining Act, 2006* (Ghana), s. 93.
231 *Mines and Minerals Act, 1999* (Botswana), s. 56; *MPRDA*, s. 27.(7); *Mining Act, 2016* (Kenya), s. 139.
232 *Mines and Minerals Act, 1999* (Botswana), s. 57(b); *Minerals and Mining Act, 2006* (Ghana), s. 93.
233 *Mining Act, 2016* (Kenya), s. 139.(1); *Mines and Minerals Act, 2009* (Sierra Leone), s. 102.(1).
234 *MPRDA*, s. 27.(3b).

60 *Types of mineral rights in Africa*

covered by a subsisting mineral right with the consent of the holder of the mineral right.[235] However, in Botswana the Minister may grant a minerals permit without the consent of the holder of the existing mineral right "if such holder will not be prejudiced by the grant of a permit, and the holder has been given a reasonable opportunity to make representations to the Minister as to why a permit should be refused".[236] The grant of a minerals permit over an area covered by a subsisting mineral right is apparently meant to cater to situations where the holder of a mining licence (for large-scale mining), having conducted prospecting and exploration activities, is uninterested in mining a particular area covered by their licence because the area does not contain mineral deposits large enough to sustain its operations. In such instances, a minerals permit (for small-scale mining) covering that specific area may be granted to a third party with the consent of the holder of mining licence. That said, in Botswana the holder of minerals permit does not have the right to mine diamonds in the area covered by the permit because a minerals permit does not cover diamonds.[237]

7.2 Terms and conditions of a small-scale mining lease

The terms and conditions of a small-scale mining lease in Africa are similar to those governing mining leases for large-scale mining and they include obligations to comply with the approved programme of mining operations; observe environmental laws and regulations; keep accurate records; and employ citizens and procure goods and services from domestic suppliers.[238] In addition, holders of small-scale mining leases in Africa bear reporting obligations identical to those imposed on the holder of a mining lease or mining right (discussed above).[239] For example, holders of small-scale mining leases in Africa must submit periodic reports to the government regarding the volume and value of mineral production during the preceding year; the average number of employees during the preceding year; a description of their plant, vehicles, and equipment; and the extent to which they have complied with the policy objectives of the mining statute.[240] In some African countries, the holder of a small-scale mining lease must commence mineral production within the stipulated timeframe. In Ethiopia, for example, the holder must commence mining operations within one year from the date on which the lease became effective.[241]

---

235 *Mines and Minerals Act, 1999* (Botswana), s. 52.(3)(c).
236 *Ibid.* s. 52.(5).
237 *Ibid.* s. 52.(1) & (7).
238 *Mining Act, 2016* (Kenya), s. 140; *Mines and Minerals Act, 2009* (Sierra Leone), s. 102.(2).
239 *MPRDA*, s. 28; *Mining Act, 2016* (Kenya), s. 140; *Mines and Minerals Act, 2009* (Sierra Leone), s. 102.(2).
240 *Mines and Minerals Act, 1999* (Botswana), s. 57; *MPRDA*, s. 28.(2).
241 *Proclamation No. 678/2010* (Ethiopia), art. 30.

7.3 Duration and renewal of a small-scale mining lease

As indicated in Table 2.1, small-scale mining licences in Africa are valid for periods ranging between two years and ten years and are renewable.[242] More specifically, a small-scale mining lease is valid for a maximum period of two years in South Africa,[243] three years in Sierra Leone,[244] five years in Botswana, Kenya, and Ghana,[245] and ten years in Ethiopia.[246] A small-scale mining lease may be renewed for a further period as prescribed in the mining statute. In Botswana, a minerals permit is renewable for further periods not exceeding five years at a time,[247] while South Africa allows the renewal of a mining permit for three periods, each of which may not exceed one year.[248] In Sierra Leone, the renewal period does not exceed three years at a time, while in Kenya the renewal period is five years or the remaining life of the mine, whichever is the shorter.[249] However, in Ghana, the period of renewal is not prescribed by statute, but is left to the discretion of the Minister.[250] To be entitled to a renewal, the holder of a small-scale mining lease must establish that they have complied with statutory and regulatory requirements, as well as the terms and conditions of the lease.

8 Common features of mineral rights

Although mineral rights in Africa differ depending on the legislative scheme in a country, they exhibit some common features and attributes, which, as discussed in Chapter 4, enhance the security of mineral tenure. Mineral rights in Africa confer the exclusive right to prospect, explore, and recover minerals; are granted for a specified duration; are registrable and renewable, and can be sold, assigned, or mortgaged.[251] In some countries, mineral rights are capable of retention. However, mineral rights may be surrendered and, as discussed above, are subject to statutory terms and conditions, including conditions regarding the obligations of mineral right holders.

In most African countries, mineral rights are layered and linked in the sense that some mineral rights predicate other mineral rights. For example, depending on the

---

242 *Mines and Minerals Development Act, 2015* (Zambia), s. 34; *Proclamation No. 678/2010* (Ethiopia), art. 29; *Mining Law No. 20/2014* (Mozambique), art. 45.(3); *Mining Act, 2012* (South Sudan), s. 58; *MPRDA*, s. 27.(8); *Minerals and Mining Act, 2006* (Ghana), s. 85.
243 *MPRDA*, s. 27.(8).
244 *Mines and Minerals Act, 2009* (Sierra Leone), s. 101.(1).
245 *Mines and Minerals Act, 1999* (Botswana), s. 55; *Mining Act, 2016* (Kenya), s. 138; *Minerals and Mining Act, 2006* (Ghana), s. 85.
246 *Proclamation No. 678/2010* (Ethiopia), art. 29.(1).
247 *Mines and Minerals Act, 1999* (Botswana), s. 55.
248 *MPRDA*, s. 27.(8)(a).
249 *Mines and Minerals Act, 2009* (Sierra Leone), s. 101.(1); *Mining Act, 2016* (Kenya), s. 142.
250 *Minerals and Mining Act, 2006* (Ghana), s.85.(1).
251 P.J. Badenhorst, "Security of Mineral Tenure in South Africa: Carrot or Stick?" (2014) 32(1) *Journal of Energy & Natural Resources Law* 5 at 14.

characterization of mineral rights in a country, a prospecting licence or an exploration licence may be the predicate for a retention licence and a mining lease. In Botswana, only the holder of a prospecting licence, the holder of a retention licence, and persons to whom a waiver is issued by the Minister are entitled to hold a mining licence.[252] In effect, the holding of a prospecting licence or retention licence is a condition precednt to the grant of a mining licence, although this condition precedent may be waived by the Minister. Thus, in practice, the Minister requires applicants for mining licences to prove that they own a prospecting licence or a retention licence over the area covered by their application. In Zambia, only the holder of an exploration licence is eligible to apply for a mining licence with regard to the area covered by the exploration licences.[253] Similarly, in Ghana, unless the Minister determines otherwise, the holding of a reconnaissance licence or a prospecting licence is a requirement for the grant of a mining lease. In that country, a mining lease is granted only to the holder of a reconnaissance licence, the holder of a prospecting licence, and "any other person" qualified to hold a mining lease.[254] The phrase "any other person" signifies discretionary applicants (that is, persons other than the holder of a reconnaissance licence or a prospecting licence) as determined by the Minister.

8.1 Transferability of mineral rights

Perhaps the most profound attribute of mineral rights is the ability of mineral right holders to assign, sell, or otherwise encumber the mineral rights. The ability to transfer mineral rights is the "very life-blood of the mining industry"[255] because not only does it enhance the economic value of mineral rights, but the financing of mineral development projects by banks and other institutions partly depends on it. The transferability of mineral rights enables financial arrangements such as project loans, which are usually granted by banks on the security of the mineral rights held by the borrower. In this context, the transferability of mineral rights constitutes a critical aspect of security of mineral tenure.

8.1.1 Statutory conditions for transfer of mineral rights

Holders of mineral rights in Africa are statutorily and contractually empowered to transfer, sell, assign, or otherwise encumber their mineral rights provided they satisfy the statutory conditions. Most African countries prescribe two conditions for the assignment or transfer of mineral rights: (1) the third-party assignee must satisfy the eligibility requirements for holding mineral rights; and (2) the prior

---

252 *Mines and Minerals Act, 1999* (Botswana), s. 37.(1).
253 *Mines and Minerals Development Act, 2015* (Zambia), s. 30.(1)
254 *Minerals and Mining Act, 2006* (Ghana), ss. 39 & 40.
255 Barry Barton, "Title Registration in Common Law Jurisdictions" in Elizabeth Bastida et al., eds, *International Comparative Mineral Law and Policy: Trends and Prospects* (The Hague: Kluwer Law International, 2005) 375 at 385.

written consent of the government must be obtained.[256] With regard to the first condition, a mineral right cannot be assigned to a person who is not qualified to hold a mineral right. The transferee or assignee of a mineral right must be an individual over 18 years of age or, in the case of a corporate entity, a company incorporated under the domestic law of the host country. In addition, the transferee must possess the financial and technical competency to exploit minerals, and they must be persons who are not disqualified under the mining statute from holding mineral rights.[257]

The second condition regarding the consent of the Minister enables African countries to supervise the acquisition of mineral rights through corporate mergers and acquisition transactions. It preserves the government's ability to exercise oversight power with regard to the transfer of a controlling stake in mining companies through mergers and acquisitions. For example, in South Africa:

> A prospecting right or mining right or an interest in any such right, or a controlling interest in a company or close corporation, may not be ceded, transferred, let, sublet, assigned, alienated or otherwise disposed of without the written consent of the Minister, except in the case of change of controlling interest in listed companies.[258]

Likewise, in Zambia:

> A person shall not transfer, assign, encumber or otherwise deal with a mining right or a mineral processing licence; or an interest in a mining right or a mineral processing licence, without the approval of the Minister and the production of a tax clearance certificate.[259]

In addition, in Zambia the holder of a mining right or a mineral-processing licence cannot

> without the prior written approval of the Minister – (a) register the transfer of any share or shares in the company to any person or that person's nominee if the effect of doing so would give that person control of the company; or (b) enter into an agreement with any person, if the effect of doing so would be to give that person control of the company.[260]

---

256 *Mines and Minerals Act, 1999* (Botswana), ss. 23, 36, 50, & 59; *Minerals and Mining Act, 2006* (Ghana), s. 14; *Nigerian Minerals and Mining Act, 2007*, s. 147; *MPRDA*, s. 11; *Mines and Minerals Development Act, 2015* (Zambia), ss. 66 & 67; *Mining Act, 2010* (Tanzania), s. 9; *Mines and Minerals Act, 2009* (Sierra Leone), ss. 83 & 119.
257 See Chapter 3 for a discussion of the eligibility requirements.
258 *MPRDA*, s. 11.(1).
259 *Mines and Minerals Development Act, 2015* (Zambia), s. 66.(1).
260 *Ibid.* s. 67.(1).

In this regard, a person is assumed to have control of a company if the person and/or their nominee holds 50% or more of the equity shares of the company; or if the cumulative shareholding of the person and their nominee is 50% or more of the shares of the company; or if the person is entitled to appoint, or to prevent the appointment of, half or more than half of the number of directors of the company.[261]

Mining statutes in Africa vest discretion in the Minister regarding the requisite consent for the assignment or transfer of mineral rights; however, the Minister must exercise this discretion objectively based on the facts of each case. The Minister must act reasonably in deciding whether to consent to the assignment, but such consent is not to be unreasonably withheld or delayed.[262] The Minister must consent to an assignment where it complies with statutory and regulatory requirements. In South Africa, for example, the Minister's consent

> must be granted if the cessionary, transferee, lessee, sublessee, assignee or the person to whom the right will be alienated or disposed of – (a) is capable of carrying out and complying with the obligations and the terms and conditions of the right in question; and (b) satisfies the requirements contemplated [in the MPRDA].[263]

However, the Minister may withhold their consent where they believe that the transaction is not an arm's length transaction or where the transaction is designed to evade taxes. In that regard, Kenya requires that transactions "for the transfer of a mineral or mineral product shall be deemed to have occurred at the point of sale and shall be equal to the arm's length value of the mineral or mineral product".[264] The consent of the Minister can also be withheld where the transaction would cede the controlling stake in a mining company to a foreign entity, thus contravening the indigenization provisions in the mining statute.[265] In countries such as Tanzania, the requisite consent "shall not be given unless there is proof that substantial developments have been effected by the holder of a mineral right".[266] This provision is obviously meant to dissuade speculators from acquiring mineral rights. It ensures that companies do not obtain mineral rights solely for the purpose of 'flipping' the rights for profits through assignment or transfer. Thus, the holder of a mineral right in Tanzania must invest some capital in developing their mining operations; otherwise, the government will not consent to such transfer or assignment.

However, there are specific circumstances where the assignment of a mineral right does not require the consent of the government. For example, in Tanzania

---

261 *Ibid.* s. 67.(4).
262 *Ibid.* s. 67.(2).
263 *MPRDA*, s. 11.(2).
264 *Mining Act, 2016* (Kenya), s. 184.
265 *Mines and Minerals Development Act, 2015* (Zambia), s. 67.(1).
266 *Mining Act, 2010* (Tanzania), s. 9.(4) (as amended by *The Written Laws (Miscellaneous Amendments) Act, 2017* (No. 7), s. 8).

the consent of the government is not required for the assignment of a mineral right to an affiliate of the holder of the mineral right where the obligations of the affiliate are guaranteed by the assignor or by a parent company approved by the licensing authority.[267] The consent of the government is also not required for an assignment of a mineral right in two further instances – namely, assignment to a bank or other financial institution by way of a mortgage or charge given as security for any loan or guarantee in respect of mining operations, and assignment to "another person who constitutes the holder of the special mining licence or, as the case may be, the mining licence".[268]

While, as noted above, mining statutes in Africa require ministerial consent for the assignment or transfer of mineral rights, these statutes are silent on the effect of a failure to obtain the Minister's consent prior to the assignment or transfer of a mineral right. The question is, does the failure to obtain ministerial consent render a transfer or assignment void? A failure to obtain the Minister's prior consent could render the assignment or transfer of a mineral right void if the enabling Act so prescribes. However, given the tenor of legislative provisions in Africa, it is arguable that the failure to obtain the prior consent of the Minister does not render the assignment or transfer of a mineral right void. This is because the statutes do not expressly provide that such transfers are void; rather, the statutes merely stipulate that the transfer of a mineral right shall not be undertaken without the Minister's prior consent.[269] Although such transfer or assignment may not be legally void, the transfer of a mineral right without the prior consent of the Minister is ineffective against the government, particularly regarding the vesting of legal rights in the assignee. A statutory condition precedent to the transfer of a mineral right is the prior consent of the Minister, the absence of which renders the transfer legally ineffective as against the government.

In some countries, the transfer or assignment of a mineral right must be registered in the mining register and, upon such registration, the assignee or transferee assumes the rights, liabilities, and obligations of the transferor as prescribed in the granting instrument and the mining statute.[270] In these countries, the agencies responsible for registering mineral rights may decline to register the transfer of a mineral right if the Minister's prior consent is lacking. In such cases, the transfer of the mineral right would be ineffective given that the transfer of a mineral right has no legal effect unless registered by the licensing authority.[271] In South Africa, for example, a mineral right acquires the status of a 'limited real right' only upon registration under the *Mining Titles Registration Act, 1967*.[272] In addition, the transfer or assignment of a mineral right without the prior written consent of the

---

267 *Ibid.* s. 9.(3)(a).
268 *Ibid.* s. 9.(3)(b) & (c). See also *MPRDA*, s. 11.(3).
269 See, for example, *Mines and Minerals Development Act, 2015* (Zambia), s. 66.(1).
270 *Mines and Minerals Act, 2009* (Sierra Leone), ss. 83.(7) & 119.(7); *Mines and Minerals Development Act, 2015* (Zambia), s. 66.(5).
271 *Proclamation No. 678/2010* (Ethiopia), art. 38.(3); *Mines and Minerals Act, 2009* (Sierra Leone), ss. 83.(6) & 119.(6); *MPRDA*, s. 11.(4).
272 *MPRDA*, s. 5.(1).

Minister is grounds for revoking the mineral right.[273] It may also constitute a criminal offence because some mining statutes in Africa criminalize the contravention of the provisions of the statute.[274]

The assignee or transferee of a mineral right must satisfy the terms and conditions of the mineral rights, including the obligations of the transferor. The obligation of the assignee to observe the terms and conditions of the mineral rights is often captured in mining contracts in Africa. In this regard, a mining concession in Sierra Leone stipulates:

> The Company shall be entitled to assign this Agreement, the Lease or any rights, privileges or franchises granted or to be granted herein or hereunder, provided the assignee agrees to be bound by all the terms and conditions contained in each such assigned document.[275]

In countries where the host state participates directly in mineral exploitation through partnership and joint-venture arrangements, the transfer of a mineral right is subject to the terms of the partnership and joint-venture contract between the assignor and the host government. Thus, the assignee of a mineral right is obliged to observe the terms of any partnership and joint-venture agreement between their predecessor in title and the host state. The assignee may be required to negotiate an agreement regarding state participation if their predecessor in title did not enter into such agreement prior to the assignment.[276]

Although most mineral rights in Africa are transferable, some mineral rights are transferable only to citizens of African countries and some mineral rights are inalienable. For example, in countries such as Ghana, Kenya, and Nigeria, a small-scale mining lease is held only by citizens or companies controlled by citizens.[277] Thus, such a lease can only be transferred to a citizen or a citizen-controlled company with the consent of the Minister. Botswana prohibits the transfer of a minerals permit for industrial minerals to non-citizens because, in that country, such a permit can be held only by citizens of Botswana.[278] In Botswana, the transfer of a minerals permit for industrial minerals to a non-citizen is grounds for terminating the permit.[279] As regards mineral rights that are inalienable, a reconnaissance licence is non-transferable in countries such as

---

273 *Mines and Minerals Development Act, 2015* (Zambia), s. 66.(6).
274 See *MPRDA*, s. 98.(a)(viii).
275 *Government of Sierra Leone and Sierra Minerals Holdings 1 Limited: Bauxite Mineral prospecting and Mining Agreement 2012*, art. 9.(a).
276 *Mines and Minerals Act, 1999* (Botswana), s. 51.
277 *Minerals and Mining Act, 2006* (Ghana), ss. 83 & 88; *Mining Act, 2016*, (Kenya), s. 124.(1); *Nigerian Minerals and Mining Act, 2007*, s. 49
278 *Mines and Minerals Act, 1999* (Botswana), s. 53.(1). However, the Minister may allow foreign persons to hold a Minerals Permit for industrial minerals if it is in the public interest. See *ibid.* s. 53.(2).
279 *Mines and Minerals Act, 1999* (Botswana), s. 55.(2).

Ethiopia and South Sudan.[280] South Africa prohibits the transfer, ceding, letting, sub-letting, alienating, disposing, mortgaging, or encumbering of a retention permit.[281] In addition, in South Africa a mining permit for small-scale mining "may not be transferred, ceded, let, sublet, alienated or disposed of, in any way whatsoever, but may be encumbered or mortgaged only for the purpose of funding or financing of the mining project in question with the Minister's consent".[282]

8.2 Surface rights ancillary to mineral rights

In the context of mining, surface rights mean the right to use the surface of land for mining-related purposes. Minerals are by nature situated below the surface of the land; therefore, "access from the surface is often necessary to explore for and certainly necessary to access these minerals".[283] In Africa, the government owns most of the land on which mining operations take place. As owners of land, African governments often grant surface rights in conjunction with the grant of mineral rights; hence, mineral rights in Africa empower the holder to enter and use the surface of the land for mining-related purposes. The surface rights vested in the holder of a mineral right include the right to drill and excavate the surface of the land; erect temporary structures, plant, and buildings on the land; construct roads; cut and use timber for mining purposes; and use water from streams and rivers in accordance with the environmental authorization.

While African governments own most of the land in their jurisdictions, African legal systems also permit private and communal ownership of land. In much of rural Africa, rural communities own land or have some interest in land. These communities farm the land, fish on rivers and streams, and depend on the land for firewood for cooking and other domestic uses. However, because ownership of minerals in all lands (both public lands and private lands) vests in the government, the government can grant a mineral right over private or communal land.[284] The grant of mineral right over private lands could adversely affect the rights of the owner or lawful occupier to use the surface of the land for non-mining purposes. Even in the case of public lands, a third party may have previously acquired the surface rights from the government for non-mining purposes such as farming. In other instances, the right or interest of third parties in land outside of the area covered by a mineral right may be adversely impacted by mining operations. The area covered by a mineral right may only be accessible through adjoining land privately owned by a third party. In such instances, the interest of the owner or occupier of the adjoining land may be injuriously affected by the operations of the mineral right holder who would need to create an access road through such adjoining land.

280 *Proclamation No. 678/2010* (Ethiopia), art. 38.(1); *Mining Act, 2012* (South Sudan), s. 33.
281 *MPRDA*, s. 36. See also *Proclamation No. 678/2010* (Ethiopia), art. 38.(1).
282 *MPRDA*, s. 27.(8)(b).
283 Newman, *supra* note 84 at 95.
284 *Mining Act, 2016* (Kenya), ss. 37 & 38; *Mining Act, 2012* (South Sudan), ss. 22 & 26.

68 Types of mineral rights in Africa

Hence, holders of mineral rights in Africa are obliged to obtain the consent of, and pay compensation to, the owner or lawful occupier of land for any loss, damage, or disturbance they may cause. The holder of a mineral right must obtain the written consent of the owner or lawful occupier of land prior to accessing and using the land for mining purposes.[285] In addition, mineral right holders may be required to obtain the consent of local communities prior to the commencement of mining operations. In Zambia, a holder of a mining right or a mineral-processing licence cannot legally enter "upon land occupied as a village, or other land under customary tenure without the written consent of the chief and the local authority for the district in which the village is situated".[286] Community consent may be obtained through a CDA, an access agreement, or a land use agreement.[287] These contractual agreements usually contain provisions regarding the amount of compensation payable by mineral right holders for use of communal land.

The basis of compensation for surface rights is damage, loss, or disturbance of the surface rights of the owner or lawful occupier of land or the likelihood of such damage, loss, or disturbance.[288] In Nigeria, for example, mineral right holders must pay to the owner and occupier of land "reasonable compensation" for any damage done to the surface of the land, including damage to economic crops, trees, buildings, or works.[289] Similarly, Zambia requires the holder of a mining right or a mineral-processing licence to pay to the owner or lawful occupier of land "fair and reasonable compensation" for any damage or disturbance caused to the surface of the land.[290] The amount of compensation is determined by considering the extent to which the market value of the land is diminished by the operations of the mineral right holder.[291] Hence, compensable items include loss of earnings and income, loss of use of the natural surface of land, cost of property, cost of economic crops and trees, disruption of the socio-economic activities of the owner or lawful occupier of land, and the cost of relocation or resettlement.[292]

---

285 *Mines and Minerals Act*, 1999 (Botswana), s. 60.(1)(b); *Mines and Minerals Development Act, 2015* (Zambia), s.52.(1)(b–c); *Mining Act, 2010* (Tanzania), s. 95.(1)(b).
286 *Mines and Minerals Development Act, 2015* (Zambia), s. 52.(1)(c).
287 See, for example, the *Mines and Minerals Development Act, 2015* (Zambia), s. 53 (which provides that the rights conferred by a mining right or a mineral processing licence are subject to any access agreement between the holder and local host communities).
288 *Minerals and Mining Act, 2006* (Ghana), s. 73.(1); *Mining Act, 2016* (Kenya), s. 153.(1); *Nigerian Minerals and Mining Regulations 2011*, reg. 11.(3–4); *Mines and Minerals Development Act, 2015* (Zambia), s. 57.(1); *Mines and Minerals Act, 1999* (Botswana), s. 63.
289 *Nigerian Minerals and Mining Regulations 2011*, reg. 11.(3).
290 *Mines and Minerals Development Act, 2015* (Zambia), s. 57.(1).
291 *Ibid*. s. 57.(3); *Mines and Minerals Act, 1999* (Botswana), s. 63.(1)(iii).
292 *Minerals and Mining (Compensation and Resettlement) Regulations*, 2012 (L.I. 2175) (Ghana), reg. 3; *Minerals and Mining Act, 2006* (Ghana), s. 74; *Mining Act, 2016* (Kenya), s. 153.(1); *Mines and Minerals Development Act, 2015* (Zambia), s. 57.(1); *Mining Act, 2010* (Tanzania), s. 96.(3).

Compensation must be paid to the owner or lawful occupier prior to the commencement of mining operations.[293]

Mining regimes in Africa require the mineral right holder and the lawful occupier of land to mutually negotiate the amount of compensation.[294] In Ghana, for example, "the holder of the mineral right shall on receipt of the compensation claim enter into negotiations with the claimant for the settlement of the amount of compensation", but the parties "may appoint a committee to negotiate the amount of compensation and any amount that is agreed on shall be in the form of a written agreement and approved by each claimant before the compensation is paid".[295] In South Africa, a mineral right holder and the owner or lawful occupier of land shall endeavour to reach an agreement for the payment of compensation for loss or damage to the surface of the land.[296] Where the parties fail to reach agreement on compensation, the matter may be referred to a quasi-adjudicatory body, an arbitral panel, or a court.[297] In Zambia, for example, the Mining Appeals Tribunal has jurisdiction to "inquire into, and make awards and decisions relating to, any dispute of compensation to be paid under this Act".[298] For its part, South Africa requires that where the parties fail to reach an agreement, compensation must be determined by arbitration in accordance with the *Arbitration Act, 1965* or by a competent court.[299] Unlike the position in Zambia, in Ghana any dispute between the claimant and the holder of the mineral right regarding the amount of compensation is referred to the Minister who shall determine the compensation payable.[300] However, the Minister's decision regarding compensation is subject to judicial review by the High Court of Ghana.[301] Compensation may be achieved through resettlement of the host community.[302] Although there are mechanisms for settling disputes regarding compensation, in countries such as Botswana a dispute over the amount of compensation does not preclude exercise of the mineral rights while the dispute remains unsettled.[303]

---

293 *Mining Act, 2016* (Kenya), s. 153.(7); *Mines and Minerals Act, 2009* (Sierra Leone), s. 37.(2).
294 See *Mining Act, 2016* (Kenya), s. 153.(5); *Minerals and Mining Act, 2006* (Ghana), s. 73.(3).
295 *Minerals and Mining (Compensation and Resettlement) Regulations*, 2012 (L.I. 2175) (Ghana), reg. 2.(2–3).
296 *MPRDA*, s. 54.(3).
297 *Mines and Minerals Act, 1999* (Botswana), s. 63.(2).
298 *Mines and Minerals Development Act 2015* (Zambia), s. 98.(3)(b).
299 *MPRDA*, s. 54.(4)
300 *Minerals and Mining Act, 2006* (Ghana), s. 73.(3). See also *Mining Act, 2016* (Kenya), s. 153.(6) (which requires such disputse to be resolved by the Cabinet Secretary).
301 *Minerals and Mining Act, 2006* (Ghana), s. 75.
302 *Mines and Minerals Act, 2009* (Sierra Leone), s. 38.
303 *Mines and Minerals Act, 1999* (Botswana), s. 63.(1)(v).

8.3 Priority of mining operations over other uses of land

Across Africa, the use of land for mining purposes takes priority over other uses of land. In effect, the mineral right holder's right to use the surface of land takes precedence over the interest of the owner or lawful occupier of the land. Thus, where the owner or lawful occupier of land refuses to consent to the use of the surface by a mineral right holder, or where they demand unreasonable terms for the acquisition of their surface rights, the government can compulsorily acquire the land and surface rights from the owner or lawful occupier in order to enable mining operations on the land.[304] In South Africa, for example, the Minister may expropriate any land or any interest in land if it is necessary for the achievement of the objects of the MPRDA, provided they pay compensation to the owner or lawful occupier of the land.[305] More specifically, where a mineral right holder and the owner or occupier of land fail to reach an agreement regarding the use of the surface, and the Minister determines that any further negotiation will detrimentally affect the object of the MPRDA, the Minister may expropriate the land in order to allow the mineral right holder access to the surface of the land.[306] Similarly, in Nigeria the "use of land for mining operations shall have a priority over other uses of land", and use and occupation of land for mining operations constitute an overriding public interest.[307] Thus, where a mineral right is granted over land that is subject to a statutory or customary right of occupancy held by a third party, such right of occupancy is revoked by the government in order to allow the mineral right holder unfettered access to the surface of the land.[308] The superiority of the mineral right holder's interest is equally reflected in statutory provisions prohibiting the owner or lawful occupier of the land from erecting any structures or buildings on the land without the consent of the mineral right holder.[309]

However, the right of a mineral right holder to access and use the surface of land is circumscribed in certain circumstances. First, as discussed above, mineral right holders are prohibited from accessing certain lands such as cemeteries and lands containing cultural and religious monuments and artifacts.[310] Second, in some countries the owner or lawful occupier of land within the area covered by a mineral right retains the right to graze stock upon or to cultivate the surface of

---

304 *MPRDA*, ss. 54.(5) & 55; *Minerals and Mining Act, 2006* (Ghana), s. 2; *Nigerian Minerals and Mining Act, 2007*, s. 22.(2); *Mining Act, 2012* (South Sudan), ss. 140 & 141; *Mines and Minerals Act, 1999* (Botswana), s. 64; *Mines and Minerals Act, 2009* (Sierra Leone), s. 36.
305 *MPRDA*, s. 55.
306 *MPRDA*, ss. 54.(5) & 55.
307 *Nigerian Minerals and Mining Act, 2007*, s. 22.(1).
308 *Ibid.* s. 22.(2).
309 *Minerals and Mining Act, 2006* (Ghana), s. 72.(4).
310 *Mines and Minerals Act, 1999* (Botswana), s. 60; *Mines and Minerals Development Act, 2015* (Zambia), s.52.(1).

such land, provided that such grazing or cultivation does not interfere with the proper use of the area for mining purposes.[311] A mineral right holder may acquire the exclusive right to use the land within the area covered by their concession by obtaining a lease of the surface rights from the owner or lawful occupier of the land.[312] Third, mineral right holders are enjoined to use the surface of land in a reasonable manner in order not to affect injuriously the interest of the owner or lawful occupier of the land.[313]

9 Conclusion

An interesting observation arising from the discussion in this chapter is that there is a striking similarity regarding the types of mineral rights in African countries, particularly in relation to the classification of mineral rights, the terms and conditions of mineral rights, and the rights and obligations attached to mineral rights. As discussed above, holders of mineral rights in Africa have common rights and obligations, including the exclusive right to prospect, explore, and recover minerals, as well as the right to the renewal of mineral rights. In addition, mineral right holders have certain ancillary rights, including the right to enter land, drill and excavate land, and erect temporary buildings and structures for mining purposes. The specificity of the statutory and contractual rights vested in the holders of mineral rights in Africa circumscribes ministerial and administrative discretion. For example, as discussed above, the Minister is obliged to renew a mineral right if the applicant complies with the relevant statutory and regulatory requirements. Even then, mining statutes across Africa prescribe specific timeframes for granting renewal of mineral rights, thus further circumscribing administrative discretion and ensuring the expeditious renewal of mineral rights. The curtailment of ministerial discretion is not only a disincentive for corruption, but may also improve security of tenure for mineral right holders, thus boosting investor confidence in the African mining industry.

---

311 *Mines and Minerals Act*, 1999 (Botswana), s. 61.(1); *Minerals and Mining Act, 2006* (Ghana), s. 72.(3); *Mining Act, 2016* (Kenya), s. 152; *Mines and Minerals Development Act, 2015* (Zambia), s. 54.
312 *Mines and Minerals Act*, 1999 (Botswana), s. 62; *Mines and Minerals Development Act, 2015* (Zambia), s. 55; *Mines and Minerals Act, 2009* (Sierra Leone), s. 34.
313 *Mining Act, 2010* (Tanzania), s. 96; *Mines and Minerals Development Act, 2015* (Zambia), s. 53.

3 Acquisition of mineral rights

1 Introduction

As discussed in Chapter 1, ownership of mineral resources in Africa is vested in the government of African countries. Thus, across the continent mining operations cannot be undertaken unless a mineral right is granted by the government. In fact, the undertaking of mining operations without the authorization of the government is expressly prohibited. In South Africa, for example, "no person may prospect for or remove, mine, conduct technical cooperation operations, reconnaissance operations, explore for and produce any mineral" without the authorization of the government.[1] Similarly, in Botswana, Ghana, Kenya, Nigeria, Tanzania, and Zambia,[2] no person may prospect for or mine minerals except as provided for in the mining statute. The undertaking of mining operations without the consent of the government is an offence for which the guilty party is liable to a fine, or imprisonment, or both, as well as confiscation of any minerals obtained in the course of such unauthorized mining operations.[3] However, Botswana has created an exception for indigenous tribes to mine minerals without the prior authorization of the government. In Botswana, tribal peoples are allowed to take,

> subject to such conditions and restrictions as may be prescribed, minerals from any land from which it has been the custom of members of that tribe to take minerals and to the extent that this is permissible under the customary law of that tribe.[4]

---

1 *MPRDA*, s. 5A.
2 *Mines and Minerals Act, 1999* (Botswana), s. 5.(2); *Minerals and Mining Act, 2006* (Ghana), s. 9.(1); *Mining Act, 2016* (Kenya), s. 10; *Nigerian Minerals and Mining Act, 2007*, s. 2.(1); *Mining Act, 2010* (Tanzania), s. 6.(1); *Mines and Minerals Development Act 2015* (Zambia), s. 12.(1).
3 *Mining Act, 2010* (Tanzania), s. 6.(3–4); *Mines and Minerals Development Act, 2015* (Zambia), s. 12.(3).
4 *Mines and Minerals Act, 1999* (Botswana), s. 5.(3).

2 Land available for mining operations

As a necessary legal incident of government ownership and custodianship of all mineral resources in African countries, mining operations can occur on all lands, including public land, private land, and communal land. In effect, African governments can grant mineral rights in relation to public land, private land, and communal land, provided the land is not the subject of a subsisting mineral right. In Ghana, for example, any land in the country may the subject of an application for a mineral right, provided the land is not the subject of an existing mineral right or expressly exempted from mining operations.[5] In fact, in exercising ownership rights over mineral resources, African governments may acquire compulsorily any land or authorize the occupation of any land to secure the development or utilization of mineral resources.[6] Although private lands are available for mining, as discussed in Chapter 2, mineral right holders are required to obtain the prior consent of, and pay compensation to, the private owner or lawful occupier of land prior to commencing mining operations.[7]

However, the government may restrict or prohibit mining operations on certain lands based on public interest.[8] For example, in South Africa,

> the Minister may after inviting representations from relevant stakeholders, from time to time by notice in the Gazette, having regard to the national interest, the strategic nature of the mineral in question and the need to promote the sustainable development of the nation's mineral resources – (a) prohibit or restrict the granting of any reconnaissance permission, prospecting right, mining right or mining permit in respect of land identified by the Minister for such period and on such terms and conditions as the Minister may determine.[9]

In practice, lands exempted from mining include culturally and environmentally sensitive lands such as nature reserves and forest reserves, lands containing ancient monuments, lands consisting of cemeteries, lands that are sacred to ethnic or tribal communities, and lands containing public utilities and national parks.[10] In addition, lands containing residential dwellings and water sources, lands used for agricultural purposes, and lands set side or used for public purposes may be exempt from mining.[11] In countries such as Botswana, indigenous communities have a

---

5 *Minerals and Mining Act, 2006* (Ghana), s. 3.
6 *Ibid.* s. 2; *MPRDA*, ss. 54.(5) & 55; *Nigerian Minerals and Mining Act, 2007*, s. 22. (2); *Mining Act, 2012* (South Sudan), ss. 140 & 141; *Mines and Minerals Act, 1999* (Botswana), s. 64; *Mines and Minerals Act, 2009* (Sierra Leone), s. 36.
7 *Mines and Minerals Act, 1999* (Botswana), s. 60; *Mines and Minerals Development Act, 2015* (Zambia), s. 52.(1).
8 *Minerals and Mining Act, 2006* (Ghana), s. 4; *Mining Act, 2012* (South Sudan), s. 24.
9 *MPRDA*, s. 49.
10 *Mines and Minerals Act, 1999* (Botswana), s. 60.(1); *Mines and Minerals Development Act, 2015* (Zambia), s. 52.(1)(a).
11 See *Mines and Minerals Act, 1999* (Botswana), s. 60.(1).

74 *Acquisition of mineral rights*

right to mine customary lands, meaning that such lands may not be available for mining by non-indigenous persons.[12] However, the government may permit mining operations in these lands if it is in the public interest. Thus, the government may permit a mineral right holder to enter, prospect, and produce minerals from these lands based on specified conditions designed to preserve the cultural value of the lands.[13]

3 Requirements for grant of mineral rights

The process and methods of mineral rights acquisition across the globe differ depending on the legal regime governing the mining of minerals in individual countries. There are two dominant methods for acquiring mineral rights: the 'free entry' method and the 'sovereign discretion' method. The 'free entry' method, sometimes referred to as the 'free miner' method, allows the staking and recording of mineral claims on land belonging to the government without any prior application to the government, provided that the land is open for staking and the person staking the claims holds a prospector's licence or a 'free miner' certificate. In essence, the 'free entry' method permits the staking of lands belonging to the government without prior authorization "and obliges the government to grant exploration and development rights if the miner applies for them".[14] Under the 'free entry' method, the government has no discretion regarding the grant of mineral rights. Thus, where a miner has complied with the legal requirements for the grant of a mineral right, the government has a legal duty to grant the right.[15] Unlike the 'free entry' method, the 'sovereign discretion' method requires anyone desiring to acquire mineral rights to submit an application to the government, and the government, as a sovereign, retains the discretion to grant or refuse such application. The acquisition method prevalent in Africa is the 'sovereign discretion' method.

3.1 Eligibility requirements

Mineral rights may be granted to legal persons and entities, including individuals, companies, partnerships, and cooperative associations, provided they meet the requirements for the grant of mineral rights and they are not disqualified from holding mineral rights. However, Ghana appears to preclude individual persons from holding mineral rights by requiring mineral rights holders to be partnerships and corporations.[16] The Ghanaian provision is meant to dissuade the indiscriminate grant of mineral rights to politicians and other persons connected with

---

12 *Ibid.* s. 5.(3).
13 See *ibid.* s. 60.(1).
14 Barry J. Barton, *Canadian Law of Mining* (Calgary: Canadian Institute of Resources Law, 1993) at 151.
15 *Ibid.* at 151.
16 *Minerals and Mining Act, 2006* (Ghana), s. 10.

government officials, thus helping to curb influence peddling and corruption within government agencies.

The qualification threshold for holding mineral rights in Africa is expressed in the form of restrictions on acquisition of mineral rights, as encapsulated in Botswana's *Mines and Minerals Act, 1999*, which provides that:

No mineral concession shall be granted to or held by –

(a) an individual who –

   (i) is under the age of 18 years;
   (ii) not being a citizen of Botswana, has not been ordinarily resident in Botswana for a period of four years or such other period as may be prescribed;
   (iii) is or becomes an undischarged bankrupt, having been adjudged or otherwise declared bankrupt, whether under the laws of Botswana or elsewhere; or
   (iv) has been convicted, within the previous 10 years, of any offence of which dishonesty is an element, or of any offence under this Act, any related or similar Act, or any similar written law in force in Botswana, and has been sentenced to imprisonment without the option of a fine or to a fine exceeding P1000 or the equivalent thereof;

(b) a company –

   (i) which has not establish a *domicilium citandi et executandi* in Botswana;
   (ii) unless, in the case of a mining licence, such company is incorporated under the Companies Act, and intends to carry on the sole business of mining under the mining lease;
   (iii) which is in liquidation or under judicial management except where such liquidation or judicial management is a part of a scheme for the reconstruction or amalgamation of such company; or
   (iv) which has among its directors or shareholders any person who would be disqualified in terms of paragraphs (a)(iii) and (iv).[17]

As is evident from the provision above, some persons are statutorily prohibited from holding mining rights, including persons who are minors and persons who have been convicted of an offence. Any person below the age of 18 years cannot hold a mineral right in Botswana, Nigeria, Kenya, Sierra Leone, Tanzania, and Zambia.[18] Similarly, a person or company who is bankrupt, in liquidation, or

---

17 *Mines and Minerals Act, 1999* (Botswana), s. 6.
18 Ibid. s. 6.(a)(i); *Nigerian Minerals and Mining Regulations 2011*, reg. 23; *Mining Act, 2016* (Kenya), s. 11.(1)(b); *Mines and Minerals Act, 2009* (Sierra Leone), s. 26.(a)(i); *Mining Act, 2010* (Tanzania), s. 8.(1)(a)(i) (as amended); *Mines and Minerals Development Act, 2015* (Zambia), s. 14.(3).

76 Acquisition of mineral rights

under judicial management is prohibited from holding mineral rights in many African countries.[19] However, in countries such as Botswana, companies in liquidation or under judicial management may be granted mineral concessions where such liquidation or judicial management is undertaken as part of a restructuring or amalgamation scheme.[20] Likewise, across Africa a company cannot hold mineral rights if any of its directors or a shareholder holding a majority of the controlling shares is an undischarged bankrupt or is convicted of an offence involving dishonesty or an offence under the mining statute.[21] The prohibition is stricter in Zambia where the bankruptcy or conviction of a minority shareholder that holds more than 10% of the issued equity of a company disqualifies the company from holding mineral rights.[22]

Some countries, including Nigeria, Botswana, and Zambia, impose a time limit on the prohibition of companies from holding mineral rights based on the criminal conviction of their shareholders. The time limit in Nigeria and Zambia is five years, while Botswana imposes a time limit of ten years. In this regard, the *Nigerian Minerals and Mining Regulations* provide that a mineral right shall not be granted to a company "if it is shown that within a period of five years before the date of the application a shareholder holding a controlling share of the applicant has been convicted of an offence under this Act".[23] Similarly, in Botswana a mineral concession cannot be granted to, or held by, a company which has among its directors or shareholders any person who

> has been convicted, within the previous 10 years, of any offence of which dishonesty is an element, or of any offence under this Act, any related or similar Act, or any similar written law in force outside Botswana, and has been sentenced to imprisonment without the option of a fine or to a fine exceeding P1000 or the equivalent.[24]

The prohibition under the Nigerian statute is problematic because, unlike the situation in Botswana and Zambia, it only caters to the conviction of shareholders for offences under the mining statute. It does not cover the conviction of shareholders under any other statute in Nigeria or conviction for offences in foreign countries. Conceivably, then, a company whose controlling shareholder was previously convicted under a statute other than the *Nigerian Minerals and Mining Act, 2007*, could apply for and be granted a mineral right in Nigeria. It is equally

---

19 *Mines and Minerals Act*, 1999 (Botswana), s. 6; *Nigerian Minerals and Mining Regulations 2011*, reg. 23; *Mining Act, 2010* (Tanzania), s. 8.(1) (as amended); *Mining Act, 2016* (Kenya), s. 11; *Mines and Minerals Development Act, 2015* (Zambia), s. 14.
20 *Mines and Minerals Act, 1999* (Botswana), s. 6.(b)(iii).
21 *Ibid.* s. 6; *Nigerian Minerals and Mining Regulations 2011*, reg. 23; *Mining Act, 2010* (Tanzania), s. 8.(1) (as amended).
22 *Mines and Minerals Development Act, 2015* (Zambia), s. 14.(2).
23 *Nigerian Minerals and Mining Regulations 2011*, reg. 53. See also *Mines and Minerals Development Act, 2015* (Zambia), s. 14.(2)(d)(ii).
24 *Mines and Minerals Act, 1999* (Botswana), s. 6.(a)(iv) & (b)(iv).

possible for a company whose controlling shareholder was previously convicted of an offence in a foreign country to apply for and obtain a mineral right in Nigeria.

Some African countries impose restrictions on the size or magnitude of mineral rights held by individuals, while other countries exclude small-scale miners from exploiting certain minerals. In Zambia, for example, mineral rights over an area exceeding two cadastre units are granted only to companies,[25] implying that individuals cannot hold mineral rights over an area exceeding two cadastre units. Furthermore, individuals who have been convicted (within a specified period) of an offence involving fraud, dishonesty, or other statutory offences cannot hold mineral rights in some countries.[26] In addition, a person may be ineligible to apply for a mineral right if, being the holder of a subsisting mineral right, they are in default of the terms and conditions of the mineral right.[27] In Botswana, a minerals permit cannot be granted in relation to diamonds, and therefore small-scale miners in Botswana cannot exploit diamonds.[28] In addition, some African countries reserve small-scale mining leases for citizens, thus barring foreign entities from engaging in small-scale mining.[29]

3.2 Indigenization and domestic incorporation requirements

Some African countries require that applicants for mineral rights be citizens of the country or a company or business entity incorporated or registered in the country.[30] In Botswana, for example, an applicant for a mining licence must be "a company incorporated under the *Companies Act*, which intends to carry on the sole business of mining under the mining licence applied for".[31] Thus, foreign companies wishing to apply for mineral rights in Africa must incorporate a subsidiary company under domestic corporate statutes. The indigenization requirements appear stricter in Tanzania where holders of a primary mining licence must be citizens of Tanzania, or a partnership composed exclusively of citizens of Tanzania, or a company whose membership is composed exclusively of citizens of Tanzania.[32] In fact, in Tanzania a corporate entity is eligible to hold a primary mining licence only if its directors are citizens of Tanzania and if control over the company, both direct and indirect, is exercised from within Tanzania by persons who are also citizens of Tanzania.[33] In addition, prior to commencing mining

---

25 *Mines and Minerals Development Act, 2015* (Zambia), s. 13.(4).
26 *Mines and Minerals Act, 1999* (Botswana), s. 6.(a)(iv); *Nigerian Minerals and Mining Regulations 2011*, reg. 23; *Mines and Minerals Act, 2009* (Sierra Leone), s. 26.(a)(iv); *Mines and Minerals Development Act, 2015* (Zambia), s. 14.(3)(c).
27 *Mining Act, 2010* (Tanzania), s. 31(b).
28 *Mines and Minerals Act, 1999* (Botswana), s. 52.(1) & (7).
29 See, for example, *Minerals and Mining Act, 2006* (Ghana), s. 83.
30 See *Mines and Minerals Act, 1999* (Botswana), s. 6; *Minerals and Mining Act, 2006* (Ghana), s. 10; *Nigerian Minerals and Mining Act, 2007*, ss. 47–51; and the *Mines and Minerals Development Act, 2015* (Zambia), s. 14.(2).
31 *Mines and Minerals Act, 1999* (Botswana), s. 37.(4).
32 *Mining Act, 2010* (Tanzania), s. 8.(2).
33 *Ibid.* s. 8.(2).

operations foreign companies may be required to register with government agencies responsible for foreign investment, such as the Nigerian Investment Promotion Commission.[34]

The strictness of the indigenization requirements is cushioned by the discretion vested in the Minister to permit foreign legal persons and entities to hold mineral rights in specific instances. For example, foreign legal persons and entities may be granted a mining licence for gemstones in Tanzania where the Minister, in consultation with the Mining Advisory Board, "determines that the development of gemstone resources in an area of land subject to a mineral right, is most likely to require specialised skills, technology or high level of investment".[35] Countries such as the DRC permit foreign legal persons to hold mineral rights if they elect domicile in the DRC either by acting through an authorized mining agent based in the DRC or incorporating a local subsidiary.[36] In addition, some countries exempt foreign entities from the requirement of domestic incorporation.

Additional restrictions on the holding of mineral rights may be based on the ownership structure of a company, the size of the area covered by the mineral right, and the type of mineral right in question. For example, in Tanzania a prospecting licence cannot be granted to any person or entity that owns more than 20 other valid prospecting licences, unless the cumulative prospecting areas of the other prospecting licences do not exceed 2,000 square kilometres.[37] Restrictions based on the size of the area covered by mineral rights are aimed at preventing the dominance of the domestic mining industry by a few powerful companies, thus promoting competition within the mining industry. Similarly, in Zambia a mining right covering areas ranging between two cadastre units and 120 cadastre units is granted only to companies that qualify as a "citizen-influenced company", "citizen-empowered company", and "citizen-owned company".[38] A 'citizen-empowered company' means a company with 25% to 50% equity held by Zambian citizens,[39] while a 'citizen-influenced company' is defined as "a company where five to twenty-five percent of its equity is owned by citizens and in which citizens have significant control of the management of the company".[40] Likewise, a 'citizen-owned company' is "a company where at least fifty point one percent of its equity is owned by [Zambian] citizens and in which citizens have significant control of the management of the company".[41]

34 See *Nigerian Investment Promotion Commission Act*, CAP. N117, LFN 2004, s. 20; *Ghana Investment Promotion Centre Act, 2013 (Act 865)*, s. 24.
35 *Mining Act, 2010* (Tanzania), s. 8.(5) (as renumbered by the *Written Laws (Miscellaneous Amendments) Act, 2017* (No. 7), s. 7.(d)).
36 *Law No. 007/2002 of July 11, 2002 Relating to the Mining Code* (DRC), art. 23 (as amended by *Law No. 18/001 of 9 March 2018*).
37 *Mining Act, 2010* (Tanzania), s. 8.(7) (as renumbered by the *Written Laws (Miscellaneous Amendments) Act, 2017* (No. 7), s. 7.(d)).
38 *Mines and Minerals Development Act, 2015* (Zambia), s. 13.(3).
39 *Ibid.* s. 2; *Citizens Economic Empowerment Act, 2006*, s. 3.
40 *Citizens Economic Empowerment Act, 2006*, s. 3.
41 *Ibid.* s. 3.

Foreign legal persons and entities may be prohibited from holding specific types of mineral rights such as a small-scale mining lease and artisanal rights. In Zambia, for example, foreign companies can only engage in large-scale mining and are expressly prohibited from engaging in artisanal mining and small-scale mining.[42] In that country, a mining licence for artisanal mining can be held only by a citizen of Zambia or a cooperative wholly composed of Zambian citizens, while a mining licence for small-scale mining can be held only "by a citizen-owned, citizen-influenced or citizen-empowered company".[43]

Although companies are eligible to hold mineral rights in Africa, only specific types of companies can hold mineral rights in some African countries. For example, in Botswana and the DRC mineral rights are granted exclusively to companies and other business entities established specifically to undertake mining operations. Thus, in order for companies to hold mineral rights in these countries, they must demonstrate that their sole business is mining.[44]

3.3 Financial and technical competency

Applicants for mineral rights in Africa are required to satisfy certain financial and technical competency requirements designed partly to prevent speculators from gaining access to minerals. A major factor considered by rights-granting agencies is whether the applicant possesses the financial resources and technical competence to undertake mining operations.[45] Thus, in African countries, applicants for mineral rights are required to demonstrate that they have the financial capital and technical expertise and experience to undertake mining operations. For example, Botswana requires applicants for a prospecting licence to prove that they have, or have "secured access to, adequate financial resources, technical competence and experience to carry on effective prospecting operations".[46] In fact, the Minister may conduct an investigation to ascertain whether the applicant possesses the requisite financial resources and technical competence and experience.[47] Similarly, Nigeria requires applicants for mineral rights to "provide proof of sufficient working capital for the exploration or mining of the area applied for and of technical competence to carry on the proposed exploration or mining operation",[48] while in Ghana an applicant must furnish "particulars of the financial and technical resources available to the applicant for the proposed mineral operations".[49]

42 *Mines and Minerals Development Act, 2015* (Zambia), s. 29.(2) & (3).
43 *Ibid.* s. 29.(2) & (3).
44 See *Mines and Minerals Act, 1999* (Botswana), s. 6.(b)(ii); *Law No. 007/2002 of July 11, 2002 Relating to the Mining Code* (DRC), art. 23 (as amended by *Law No. 18/001 of 9 March 2018*).
45 *Mines and Minerals Development Act, 2015* (Zambia), s. 31.(1)(f); *Mines and Minerals Act, 1999* (Botswana), s. 14.(1)(a).
46 *Mines and Minerals Act, 1999* (Botswana), s. 14.(1)(a).
47 *Ibid.* s. 14.(2).
48 *Nigerian Minerals and Mining Act, 2007*, s. 54.(1)(a).
49 *Minerals and Mines Act 2006* (Ghana), s. 11.(a). See also *Mining Act, 2010* (Tanzania), s. 49.(2)(g).

80 *Acquisition of mineral rights*

In effect, mineral rights are granted only if the applicant has the financial resources and the technical ability to conduct the proposed mining operations optimally.[50]

Relatedly, some countries impose minimum investment requirements on foreign investors. In Ghana, for example, foreign investors are required to invest a minimum of US$200,000 where they engage in a joint venture or partnership with a citizen of Ghana.[51] However, where an enterprise is wholly owned by a foreign investor, the foreign investor must invest at least US$500,000 in cash or capital goods relevant to the investment or a combination of cash and equity capital in the enterprise.[52]

Where an applicant for a mineral right is a subsidiary of a parent company, the parent company may be required to sign a contractual guarantee regarding the financial and technical competency of the subsidiary company to undertake mining operations. For example, a mining contract in Liberia provides that, on the effective date of the contract, the concessionaire shall provide a guarantee executed by its parent company "guaranteeing the obligations of the concessionaire under Article V, Section 2 (Capital Expenditures), Article XV as amended (Environmental Protections and Management) and Article XVII as amended (provision of funds to the concessionaire)".[53] Likewise, in Botswana the parent company must unconditionally and irrevocably guarantee that

> it will make available, or cause to be made available, to the company or its permitted assignees, such financial, technical, managerial and other resources as are required to ensure that the company and/or any such assignee is able to carry out the obligations of the company or any permitted assignee as set forth in the licence and under the Act.[54]

The parent company's guarantee covers all obligations of the subsidiary company such that if the subsidiary company or its assignees

> fail to perform its obligations under the licence or commits any breach of such obligations under the licence or the Act, then the guarantor shall fulfil or cause to be fulfilled the said obligations in place of the company or any permitted assignees, and will indemnify the Government against all losses, damages, costs, expenses or otherwise which may result directly from such failure to perform or breach on the part of the company or any permitted assignees.[55]

---

50 *MPRDA*, ss. 17.(1) & 23.(1).
51 *Ghana Investment Promotion Centre Act, 2013 (Act 865)*, s. 28.(1).
52 *Ibid.* s. 28.(1).
53 *An Act Ratifying the Amendments to the Mineral Development Agreement (MDA) Dated August 17, 2005 between the Government of the Republic of Liberia (the Government) and Mittal Steel Holding A.G. and Mittal Steel (Liberia) Holdings Limited (the Concessionaire)*, art. 32.
54 *Mines and Minerals Act, 1999* (Botswana), 1st Schedule, Annexure 1, Parent Company Guarantee.
55 *Ibid.*

This guarantee remains in effect throughout the duration of the mineral right, including the renewal period.[56]

Regarding technical competency, an applicant for a mineral right may be required to demonstrate that they have in their employment persons possessing adequate professional qualification and experience in mining.[57] In Nigeria, for example, the employment of professionally qualified persons is a condition for the continuation of both the mining lease and the mining operations undertaken pursuant to the lease.[58] Thus, a lessee cannot undertake mining operations in the absence of professionally qualified persons.[59]

These financial and technical competency requirements ensure that only companies with the requisite capacity to conduct mining operations are granted mineral rights. Hence, the failure of an applicant to satisfy these financial and technical competency requirements is grounds for rejecting the application.[60] These competency requirements are also necessary for the preservation and maintenance of mineral rights. For example, an application for the renewal of a mineral right may be rejected where the applicant is unable to prove that they retain the requisite financial resources and technical competence to conduct mining operations. In practice, the financial and technical competency requirements discourage speculators from applying for mineral rights.

4 Methods of acquisition of mineral rights

Mining statutes in Africa set out the parameters for the acquisition of mineral rights, including the manner in which the government disposes of minerals. Mineral rights in Africa can be acquired through a prescribed application process, public auction, or by way of corporate mergers and other takeover transactions. However, as noted below, the direct application method is the primary method for the acquisition of minerals rights, while the public auction method is rarely used in Africa.

4.1 Discretionary grant

4.1.1 Application for a mineral right

Mineral rights in Africa are acquired primarily through an application to the Minister or mineral rights-granting agency. The acquisition of mineral rights in Africa begins with the submission of an application to a designated government agency, such as Botswana's Ministry of Minerals, Energy and Water Resources; the Ghana Minerals Commission; Nigeria's Mining Cadastre Office; the Regional Manager of

---

56 Ibid.
57 *Nigerian Minerals and Mining Act, 2007*, s. 73.
58 Ibid. s. 73.(2).
59 Ibid. s. 73.(3).
60 Ibid. s. 54.(2).

South Africa's Department of Minerals and Energy; and the Mining Licensing Committee of Zambia. Although applications are submitted to state agencies for administrative and processing purposes, in many African countries the ultimate authority for granting mineral rights vests in the Minister responsible for mineral resource development. In practice, the Minister may delegate this power to subordinate officers within the ministry responsible for mineral development. However, some African countries circumscribe the rights-granting authority of the Minister by subjecting it to parliamentary oversight. In Ghana, for example, mineral rights are granted by the Minister on the recommendation of the Ghana Minerals Commission,[61] but such grants are subject to ratification by the Parliament of Ghana.[62] However, the Parliament may, by two-thirds majority vote, exempt from the requirement of parliamentary ratification "a particular class of transaction, contract or undertaking".[63] In some countries, such as Zambia and Zimbabwe, some mineral rights are granted by semi-autonomous bodies such as a licensing committee or board.[64]

An application for mineral rights is submitted in a prescribed form and accompanied by requisite documents and the application fee. The documents accompanying an application vary depending on the type of mineral right in question. These documents are generally intended to demonstrate that the applicant has satisfied the statutory requirements for the grant of mineral rights. Some African countries require corporate applicants to submit documents attesting to their financial and technical competencies, as well as documents showing that they and their controlling shareholders and directors have not been previously convicted of an offence.[65] More specifically, applicants for a prospecting licence and an exploration licence are required to submit documents establishing that they have, or have secured access to, the capital, equipment, and expertise to undertake prospecting and exploration activities.[66]

Applicants are also required to submit a proposed programme of prospecting or exploration operations indicating the prospecting or exploration activities the applicant intends to undertake, as well as an estimate of the applicant's proposed expenditure.[67] For example, in Ghana an applicant for a mineral right must specify: (a) the particulars of the financial and technical resources available to the applicant for the proposed mineral operations; (b) an estimate of the amount of money the applicant proposes to spend on the mineral operations; (c) the

61 *Minerals and Mining Act, 2006* (Ghana), ss. 31.(1), 34.(1) & 39.(2).
62 *Ibid.* s. 5.(4).
63 *Ibid.* s. 5.(5).
64 See, for example, *Mines and Minerals Development Act, 2015* (Zambia), s. 6 (establishing the Mining Licensing Committee); *Mines and Minerals Act* (Zimbabwe), s. 6 (establishing the Mining Affairs Board).
65 See *Nigerian Minerals and Mining Regulations 2011*, regs. 32.(3), 35.(3c) & 57.(3h).
66 See *MPRDA*, s. 17.(1); *Mines and Minerals Development Act, 2015* (Zambia), s. 22. (1); *Mines and Minerals Act, 1999* (Botswana), s. 14.(1); *Minerals and Mines Act, 2006* (Ghana), s. 11.
67 See *Mines and Minerals Act, 1999* (Botswana), s. 14; *MPRDA*, s. 17.(1); and *Mines and Minerals Development Act, 2015* (Zambia), s. 22.(1).

particulars of the programme for the proposed mineral operations; and (d) the particulars of the applicant's proposal for the employment and training of Ghanaians in the mining industry.[68] The programme of prospecting operations is not sacrosanct and can be amended with the approval of the Minister.[69]

Similarly, an application for a mining lease or mining licence must be accompanied by a proposed programme of mining operations and, in some cases, a bankable feasibility study.[70] The programme of mining operations specifies the mining scheme, as well as details regarding mineral deposit, including probable and proven mineral reserves. The feasibility study must demonstrate the viability of the proposed mining operations. In Nigeria, for example, an applicant for a mining lease must submit a pre-feasibility study containing a general description of the proposed mining scheme; details regarding the scale of operation; possible location of all major operation facilities, pits, shafts, dumps, and dams; the anticipated commencement date of commercial production; the planned production profile and capacity; and the characteristics and nature of the final products.[71] In Botswana, the feasibility study submitted by an applicant for a mining licence must indicate:

(a) Details of mineral deposit (including all known, proven, indicated, inferred ore reserves and mining conditions)
(b) Technical report on mining and treatment possibilities and the applicant's intention in relation thereto
(c) Proposed programme of mining operations including
  (i) estimated date by which applicant intends to work for profit
  (ii) estimated recovery rate (s)
  (iii) nature of product
  (iv) envisaged marketing arrangements for sale of mineral product(s)
  (v) environmental impact assessment study
  (vi) environmental management programme
(d) Forecast of capital investment, cashflow and details of anticipated financing plan
(e) Outline of proposed employment level and training program
(f) Outline of proposed sources of goods and services
(g) Details of expected infrastructure requirements
(h) Attach audited statement of relevant exploration and arms length acquisition expenditure incurred prior to this application on the area applied for.[72]

---

68 *Minerals and Mines Act, 2006* (Ghana), s. 11.
69 *Mines and Minerals Act, 1999* (Botswana), s. 22.
70 See *Mines and Minerals Act, 1999* (Botswana), s. 39.(1); *Minerals and Mines Act, 2006* (Ghana), s. 11; *Mines and Minerals Development Act, 2015* (Zambia), s. 31.(1).
71 *Nigerian Minerals and Mining Regulations 2011*, reg. 57.(3)(h)(iii).
72 *Mines and Minerals Act, 1999* (Botswana), 1st Schedule, Form V (Mining Licence Application Form: Issue/Renewal).

4.1.2 Order of processing of applications

In many African countries, the processing of applications for mineral rights is undertaken on a first-come, first-served basis. Thus, where there are multiple applications for a mineral right over the same land, the applications are disposed in the order in which they were received by the licensing or rights-granting agency.[73] In Angola, for example, "applications for access to mining rights shall be registered and addressed in accordance with their order of reception, within legally established deadlines".[74] In South Africa, applications for mineral rights are determined in the order in which they were received, thus according priority to applications on a first-come, first-served basis.[75] Ethiopia operates a slightly different regime for determining priority of applications. In that country, applications are determined on a first-come, first-served basis only where there are multiple applications "for licences of the same status covering the same mineral and area".[76] Thus, in Ethiopia an application for a large-scale mining licence takes precedence over an application for a small-scale mining licence even though the latter application was submitted prior to the former.[77]

In order to determine the order of receipt of applications, some countries require the licensing agency to keep a register containing a record of all applications for mineral rights, including the chronological order of receipt of the applications.[78] In Nigeria, for example, the MCO maintains a chronological record of all applications for mineral titles in a Priority Book, which is "specifically used to ascertain the priority and registration of applications for exclusive rights on vacant areas".[79] Thus, where there are multiple applications regarding the same area, the MCO disposes of the applications in the order in which they were registered in the Priority Book, thus according priority to the earliest application.[80] Similarly, in Tanzania, where two or more persons apply for mineral rights over the same area, the person whose application was first registered has priority over the other applicants and is entitled to be granted the mineral right provided they satisfy the statutory criteria for the grant of the mineral right.[81] Nigeria requires its MCO to issue a receipt to all applicants for mineral rights evidencing the date and time of their application.[82] In Sierra Leone, applications for mineral rights are numbered

---

73 *MPRDA*, s. 9.(1); *Mining Code* (Angola), art. 91; *Proclamation No. 678/2010* (Ethiopia), art. 13; *Mining Act, 2016* (Kenya), s. 56; *Mines and Minerals Development Act, 2015* (Zambia), s. 15; *Mining Act, 2010* (Tanzania), s.14.(1); *Nigerian Minerals and Mining Act, 2007*, s. 8.
74 *Mining Code* (Angola), art. 91.
75 *MPRDA*, s. 9.(1)
76 *Proclamation No. 678/2010* (Ethiopia), art. 13.(1b).
77 *Ibid.* art. 13.(1a).
78 See *Nigerian Minerals and Mining Act, 2007*, s. 4.(5c); *Mines and Minerals Act, 2009* (Sierra Leone), s. 41; *Mining Act, 2010* (Tanzania), s. 106.(1); *Mines and Minerals Development Act, 2015* (Zambia), s. 79; *Mining Act, 2016* (Kenya), ss. 191 & 192.
79 *Nigerian Minerals and Mining Act, 2007*, s. 4.(5c).
80 *Ibid.* s. 8.
81 *Mining Act, 2010* (Tanzania), s. 14.
82 *Nigerian Minerals and Mining Act, 2007*, s. 8.(3).

serially in the Register of Mineral Rights Applications, which contains the date, hour, and minute the application was registered.[83]

4.1.3 Statutory criteria for grant of mineral rights

In most African countries, the Minister or rights-granting agency is vested with broad discretionary powers to determine whether to grant mineral rights. The Minister's discretion is exercised based on the statutory criteria for the grant of mineral rights, but in some countries the Minister may conduct investigations to determine whether an applicant has complied with the statutory requirements.[84] Generally speaking, an application for a mineral right is granted where the applicant has satisfied the statutory requirements and the land is not the subject of a subsisting mineral right. In fact, statutory provisions prohibit acceptance of an application for mineral rights covering an area over which there is a subsisting mineral right held by another entity.[85]

The criteria for granting mineral rights are similar across Africa and they relate generally to the financial and technical competency of the applicant to conduct mining operations, environmental protection, and compliance with the terms of extant mineral rights held by the applicant. For example, in many African countries prospecting and exploration licences are granted if the Minister is satisfied that

> (a) the applicant has, or has secured access to, adequate financial resources, technical competence and experience to carry on effective prospecting operations; (b) the proposed programme of prospecting operations is adequate and makes proper provision for environmental protection; (c) the proposed prospecting area is not the same as, nor does it overlap an existing prospecting area, retention area, mining area or minerals permit area in respect of the same mineral or associated mineral; and (d) the applicant is not in default [of statutory requirements].[86]

A prospecting licence is granted in Ghana if the Minister determines that the applicant has materially complied with the statutory criteria, including the discharge of any obligations arising from a prior or subsisting mineral right held by the applicant.[87] Similarly, South Africa's MPRDA enjoins the Minister to grant a prospecting licence if:

---

83 *Mines and Minerals Act, 2009* (Sierra Leone), s. 41.
84 See *Mines and Minerals Act, 1999* (Botswana), s.14.(2); and *Mines and Minerals Development Act, 2015* (Zambia), s. 22.(2) &. 31.(2).
85 *MPRDA*, ss. 13(2)(b), 16(2)(b), 22(2)(b), & 27(3)(b).
86 *Mines and Minerals Act, 1999* (Botswana), s. 14. See also *Mining Act, 2012* (South Sudan), s. 43; *Mines and Minerals Development Act, 2015* (Zambia), s. 22.(1).
87 *Minerals and Mining Act, 2006* (Ghana), s. 34.(4).

(a) the applicant has access to financial resources and has the technical ability to conduct the proposed prospecting operation optimally in accordance with the prospecting work programme;
(b) the estimated expenditure is compatible with the proposed prospecting operation and duration of the prospecting work programme;
(c) the prospecting will not result in unacceptable pollution, ecological degradation or damage to the environment and an environmental authorization is issued;
(d) the applicant has the ability to comply with the relevant provisions of the *Mine Health and Safety Act*, 1996 (Act 29 of 1996);
(e) the applicant is not in contravention of any relevant provision of this Act [and]
(f) in respect of prescribed minerals the applicant has given effect to the objects referred to in section 2(d).[88]

The criteria for granting a mining lease are more elaborate than the criteria for other mineral rights, but even here there is considerable similarity across Africa. In Zambia, the discretionary power of the Mining Licensing Committee (MLC) to grant a mining licence (that is, a mining lease) is exercised based on specific criteria, including:

(a) whether there are sufficient deposits or resources of minerals to justify their commercial exploitation;
(b) that the area of land over which the licence is sought is not in excess of the area required to carry out the applicant's proposed programme for mining operations;
(c) that the proposed programme of mining operations is adequate and compliant with the decision letter in respect of the environmental project brief or environmental impact assessment approved by the Zambian Environmental Management Agency;
(d) where consent is required for the area under any written law, that the applicant has submitted evidence of that consent;
(e) the standards of good mining practice and the applicant's proposed programme for development, construction and mining operations in order to ensure the efficient and beneficial use of the mineral resources for the area over which the licence is sought;
(f) in respect of large-scale mining –
  (i) whether the applicant has the financial resources and technical competence and the financing plan is compatible with the programme of mining operations;
  (ii) the applicant's undertaking for the employment and training of citizens and promotion of local business development;

---

88 *MPRDA*, s. 17.(1).

(iii) whether the applicant's feasibility study report is bankable; and
(iv) the applicant's capital investment forecast; and
(g) that the applicant is not in breach of any condition of the exploration licence or any provision of this Act.[89]

Likewise, in Botswana a mining licence is granted if the Minister is satisfied that the proposed programme of mining operations will ensure the most efficient and beneficial exploitation of minerals; the proposed mining area is not the subject of an existing mineral right; the applicant has or has secured access to adequate financial resources, technical competence, and experience to carry on effective mining operations; the parent company has issued a guarantee in the prescribed form; the applicant is not in default of their statutory and contractual obligations; and, in the case of an application to mine diamonds, that the applicant has reached an agreement with the government regarding state participation in the mining project.[90]

However, some African countries curtail ministerial discretion regarding the grant of mineral rights. In South Africa, for example, mineral rights are granted as of right where the applicant has satisfied the statutory and regulatory requirements. In that country, the Minister is statutorily obliged to grant a mining right if the applicant complies with statutory and regulatory requirements.[91] More specifically, the Minister must grant a mining right if the applicant complies with statutory and regulatory requirements and

(a) the mineral can be mined optimally in accordance with the mining work programme; (b) the applicant has access to financial resources and has the technical ability to conduct the proposed mining operation optimally; (c) the financing plan is compatible with the intended mining operation and the duration thereof; (d) the mining will not result in unacceptable pollution, ecological degradation or damage to the environment and an environmental authorization is issued; (e) the applicant has provided for the prescribed social and labour plan; (f) the applicant has the ability to comply with the relevant provisions of the *Mine Health and Safety Act, 1996* (Act No. 29 of 1996); (g) the applicant is not in contravention of any provision of this Act; and (h) the granting of such right will further the objects referred to in section 2 (d) and (f) and in accordance with the charter contemplated in section 100 and the prescribed social and labour plan.[92]

---

89 *Mines and Minerals Development Act, 2015* (Zambia), s. 31.(1).
90 *Mines and Minerals Act, 1999* (Botswana), s. 39.(1).
91 *MPRDA*, ss. 14.(1), 17.(1), & 23.(1). See *Minister of Mineral Resources v Mawetse (SA) Mining Corporation (Pty) Ltd* [2015] ZASCA 82 at para. 27 where the Supreme Court of Appeal held that "a prospecting right must be granted once there has been compliance with s. 17(1)" of the *MPRDA*.
92 *MPRDA*, s. 23.(1).

88 Acquisition of mineral rights

Aside from these statutory criteria, applicants for mineral rights in Africa may be required to satisfy the policy objectives underlying the mining regime, including equitable access to mineral resources, employment of citizens, and the procurement of domestic goods and services. In South Africa, for example, applicants for mineral rights must satisfy the Minister that they have given effect to the objects referred to in section 2(d) of the MPRDA.[93] These objects are the promotion of mineral development in ways that "substantially and meaningfully expand opportunities for historically disadvantaged persons, including women and communities, to enter into and actively participate in the mineral and petroleum industries and to benefit from the exploitation of the nation's mineral and petroleum resources".[94] Section 2(d) of the MPRDA is aimed at redressing the historical inequities in South Africa's mining industry.[95] Compliance with section 2(d) of the MPRDA is thus a prerequisite for the grant of a mineral right in South Africa. As the Supreme Court of Appeal has noted, compliance with section 2(d) is not optional because the grant of a mineral right is expressly made subject to compliance with the subsection.[96] Thus, it is lawful for the Minister to refuse to grant a mineral right if the applicant fails to comply with section 2(d) of the MPRDA.[97]

African governments also require applicants for mineral rights to comply with policy objectives regarding employment of citizens and the procurement of local goods and services as a condition precedent to the grant of a mineral right. In Botswana, for example, applicants for a mining licence must satisfy the government that they would employ and train local personnel, as well as source goods and services from domestic suppliers. They must submit details regarding proposed employment level and training programme, proposed sources of goods and services, as well as details of expected infrastructure required for their mining operations.[98] In South Africa, the Minister may require an applicant for a prospecting right to diversify its workforce so as to achieve the objective of section 2(d) of the MPRDA (as indicated above).[99] In addition, where the application for a prospecting right "relates to land occupied by a community, the Minister may impose such conditions as are necessary to promote the rights and interests of the community, including conditions requiring the participation of the community" in the mining project.[100] In that regard, the Minister may require an applicant for a prospecting right to enter into a community development agreement (CDA) with the community as a condition for granting the right.[101]

93 Ibid. s. 17.(1)(f).
94 Ibid. s. 2(d).
95 *Minister of Mineral Resources v Mawetse (SA) Mining Corporation (Pty) Ltd* (20069/14) [2015] ZASCA 82 at para. 17.
96 Ibid. at para. 17.
97 Ibid. at para. 17.
98 *Mines and Minerals Act, 1999* (Botswana), 1st Schedule, Form V (Mining Licence Application Form: Issue/Renewal).
99 *MPRDA*, s. 17.(4).
100 Ibid. s. 17.(4A).
101 See Chapter 8 for a discussion of CDAs.

4.1.4 Rejection of an application

The grant of mineral rights is prohibited under certain enumerated circumstances, including where the applicant is disqualified from holding a mineral right; the applicant has previously acted in breach of any conditions of a mineral right or any provision of the mining statute; the land covered by the application is the subject of a subsisting mineral right; and where the land covered by the application is the subject of a prior application by a third party.[102] An application for a mineral right is rejected where the applicant does not satisfy the criteria for eligibility to hold mineral rights. For example, in Zambia the MLC can reject an application that does not satisfy statutory requirements, but it must give the applicant a written notice indicating the reasons for rejecting the application.[103] Similarly, in South Africa the Regional Manager may reject an application that does not satisfy the legal requirements, but the Regional Manager is obliged to notify the applicant of the reasons for the rejection within 14 days of the receipt of the application.[104] In addition, an application for a prospecting right is refused where the Minister is satisfied that the applicant does not have the financial and technical ability to conduct the proposed prospecting operation optimally in accordance with the prospecting work programme, or where "the granting of such right will result in the concentration of the mineral resources in question under the control of the applicant and their associated companies with the possible limitation of equitable access to mineral resources".[105]

In Nigeria, an application for a mineral right is rejected if the application is defective or incomplete; or the applicant fails to provide any requisite information; or any members or directors or a controlling shareholder of the applicant has been convicted of a felony or an offence under the Mining Act or the Regulations.[106] Similarly, an application for a mining lease is denied if the applicant does not employ "a person who possesses adequate professional qualification and experience in mining and the Minister is not satisfied that the company shall, during the currency of the lease, have such qualified person in its employment".[107] However, an applicant whose application for a mining lease is rejected may, within 45 days of the rejection, file a suit at the Federal High Court for redress.[108] Where the Federal High Court determines that the mining lease ought to be granted, the Minister is obliged to grant the lease within seven days of the court's decision.[109]

Some countries afford an opportunity to applicants for mineral rights to correct defects in their application. For example, Nigeria's MCO could request an

---

102 *Mines and Minerals Development Act, 2015* (Zambia), s. 22.(3); *Mines and Minerals Act, 2009* (Sierra Leone), s. 73.
103 *Mines and Minerals Development Act, 2015* (Zambia), s. 31.(3).
104 MPRDA, s. 13.(3).
105 *Ibid.* s. 17.(2).
106 *Nigerian Minerals and Mining Regulations 2011*, regs 32.(5), (9–10) & 57.(5) & (12).
107 *Ibid.* reg. 57.(14).
108 *Ibid.* reg. 57.(20).
109 *Ibid.* reg. 57.(21).

applicant to correct any defects or omissions in their application in order to make it compliant.[110] Similarly, in Ghana the Minister may issue a written notice to an applicant regarding defects in their application and they must give the applicant an opportunity to rectify those defects and make appropriate amendments to the application or the proposed programme of mineral operations.[111] However, the Minister may reject an application if the applicant fails to address the Minister's concerns within a reasonable time as prescribed by the Minister.[112]

4.1.5 Timeframe for determining an application

Modern mining statutes in Africa provide specific timeframes within which applications for mineral rights are to be processed and determined. These timelines are designed to ensure expeditious determination of applications. In Ghana, for example, the Minerals Commission must submit to the Minister its recommendation regarding an application for a mineral right within 90 days of the receipt of the application.[113] The Minister must make a decision within 60 days of the receipt of the Minerals Commission's recommendation.[114] Tanzania requires the licensing authority to grant or reject an application for a prospecting licence within four weeks of the date of registration of the application,[115] while in Zambia the licensing authority is obliged to determine an application for an exploration licence within 60 days.[116] In fact, where the application complies with statutory requirements, Zambia's MLC is obliged to grant an exploration licence within 60 days of receipt of the application,[117] and, in the case of a mining licence, within 90 days of the receipt of the application.[118] Similarly, in Nigeria applications for a reconnaissance permit and exploration licence are determined within 30 days of the receipt of the applications, while the Minister is obliged to grant a mining lease within 45 days of the receipt of an application.[119]

South Africa requires any administrative process or decision regarding mineral rights to be undertaken "within a reasonable time and in accordance with the principles of lawfulness, reasonableness and procedural fairness".[120] Thus, the Regional Manager must accept an application for a mining right within 14 days of its submission if the application is made in the prescribed manner; a non-refundable application fee is paid; no other person holds a mineral right for the same land; and no prior application for a mineral right has previously been accepted for

---

110 *Ibid.* regs 32.(7) & 57.(10).
111 *Minerals and Mining Act, 2006* (Ghana), s. 42.(1).
112 *Ibid.* s. 42.(1).
113 *Ibid.* s. 12.
114 *Ibid.* s. 13.(1).
115 *Mining Act, 2010* (Tanzania), s. 33.(1)
116 *Mines and Minerals Development Act 2015* (Zambia), s. 23.(1).
117 *Ibid.* s. 23.(1).
118 *Ibid.* s. 32.(1).
119 *Nigerian Minerals and Mining Act, 2007*, ss. 55, 64.(4), & 65.(1).
120 *MPRDA*, s. 6.(1).

the same mineral and land.[121] Upon receipt and acceptance of an application, the Regional Manager must,

> within 14 days from the date of acceptance, notify the applicant in writing – (a) to submit the relevant environmental reports, as required in terms of Chapter 5 of the National Environmental Management Act, 1998, within 180 days from the date of the notice.[122]

In addition, the Regional Manager's written notice must request the applicant "to consult in the prescribed manner with the landowner, lawful occupier and any interested and affected party and include the result of the consultation in the relevant environmental reports".[123] Subsequently, within 14 days of receipt of the environmental reports and results of the consultation, the Regional Manager forwards the application to the Minister for their consideration.[124]

4.2 Public auction of mineral rights

Although competitive auctions and bids are often utilized in relation to petroleum resources, African countries seldom grant hard-rock mineral rights through the auction method.[125] In the recent past, however, a number of African countries have enacted statutory provisions enabling the acquisition of mineral rights through a competitive auction process. Angola, the DRC, Ethiopia, Kenya, Mozambique, Nigeria, Tanzania, Sierra Leone, and South Sudan allow competitive auction of hard-rock minerals.[126] Competitive auctions may be mandatory or discretionary, depending on the statutory scheme in a country. In Angola, public tender for mineral rights is mandatory where "in light of studies realized or approved by the entity responsible for geology, the area is considered to be of enhanced geological potential" or the area involves a mineral considered strategic in accordance with the Mining Code.[127] In that country, the regulatory authority is obliged to publish at least once a year the list of areas and mineral resources that may be awarded by public tender.[128]

---

121 *Ibid.* s. 22.(2).
122 *Ibid.* s. 22.(4)(a).
123 *Ibid.* s. 22.(4)(b).
124 *Ibid.* s. 22.(5).
125 African Union, *African Mining Vision* (February 2009) at 16, www.africaminingvision.org/amv_resources/AMV/Africa_Mining_Vision_English.pdf
126 See *Mining Code* (Angola), art. 98; *Mining Code, Law No. 007/2002 of July 11, 2002* (DRC), art. 33 (as amended by *Law No. 18/001 of 9 March 2018*); *Proclamation No. 678/2010* (Ethiopia), art. 13.(4); *Mining Act, 2016* (Kenya), ss. 14 & 41; *Mining Law No. 20/2014* (Mozambique), art. 10; *Nigerian Minerals and Mining Act, 2007*, s. 9; *Mines and Minerals Act, 2009* (Sierra Leone), s. 25; *Mining Act, 2012* (South Sudan), s. 25; *Mining Act, 2010* (Tanzania), s. 71.
127 *Mining Code* (Angola), art. 98.
128 *Ibid.* art. 98.

92 *Acquisition of mineral rights*

For the most part, however, competitive auction of mineral rights in Africa is at the discretion of the Minister responsible for mining. In the DRC, Nigeria, Tanzania, Kenya, Mozambique, and South Sudan, the Minister has power to determine the areas in respect of which mineral rights may be granted on the basis of a competitive auction.[129] While the Minister retains a discretion to undertake public auction of mineral rights, in some African countries the Minister's discretion is exercisable only in exceptional circumstances where the area has already been prospected by the government or the mineral right covering the area has been returned or reverted to the government. For example, the DRC empowers the Minister to call for tenders for mineral rights where the area "has been studied, documented or possibly worked on by the state or its entities and which is considered as an asset with considerable known value".[130] Likewise, competitive auction of mineral rights in Nigeria is reserved for areas in which minerals classified as "security" minerals have been found and areas which the advisory committee recommends to the Minister as areas appropriate for competitive bidding.[131] In Sierra Leone, public auction of mineral rights is restricted to areas "in which minerals have been discovered by a government survey", suggesting that the government would already have conducted exploration activities in the area.[132] In countries such as Kenya, the Minister's discretionary power to call for tender is restricted to large-scale mining.[133]

Given the nascent history of competitive auction in the African mining industry, clear rules have yet to emerge regarding the conduct of auctions in Africa. However, some mining statutes in Africa provide general guidelines regarding the auction of mineral rights. For example, Nigeria requires the bidding exercise to be both open and transparent.[134] The bidding process must result in the selection of a winning bid which promotes the expeditious and beneficial development of mineral resources, taking into account the programme of exploration and mining operations submitted by the bidder, including their undertakings and commitments as regards expenditure; the financial and technical resources of the bidder; and the previous experience of the bidder in the conduct of reconnaissance and mining operations.[135] Tanzania applies similar guidelines by requiring the Minister to select a winning bid that will most likely promote the expeditious and beneficial development of the mineral resources in question.[136] In Tanzania, the Minister must also consider the programme of prospecting operations submitted by the applicant; their commitments regarding expenditure; the financial and technical

---

129 See *Nigerian Minerals and Mining Act, 2007*, s. 9; *Mining Act, 2010* (Tanzania), s. 71; *Mining Act, 2016* (Kenya), ss. 14 & 41; *Mining Law No. 20/2014* (Mozambique), art. 10; *Mining Act, 2012* (South Sudan), s. 25.
130 *Mining Code, Law No. 007/2002 of July 11, 2002* (DRC), art. 33 (as amended by *Law No. 18/001 of 9 March 2018*).
131 *Nigerian Minerals and Mining Regulations 2011*, reg. 4.(2).
132 *Mines and Minerals Act, 2009* (Sierra Leone), s. 25.
133 *Mining Act, 2016* (Kenya), ss. 14 & 41.
134 *Nigerian Minerals and Mining Act, 2007*, s. 9.
135 *Ibid.* s. 9.(2).
136 *Mining Act 2010* (Tanzania), 71.(3).

resources of the applicant; and the previous experience of the applicant in the conduct of prospecting and mining operations.[137]

Competitive auction promotes transparency in the acquisition of mineral rights by ensuring that mineral rights are granted through an open bidding process, as opposed to the secrecy of the extant discretionary method in Africa. It could also enhance the economic benefits accruing to host African governments by ensuring that a fair market value is paid for mineral rights. This is particularly so where the bidding process is undertaken after geological surveys of the area, in which case the bidding companies would be aware of the potential for commercial deposits in the area. However, industry proponents have argued that the public auction of mineral rights discourages investors because it not only increases the cost of obtaining mineral rights, especially up-front costs such as signing bonus, but imposes additional costs on investors.[138] According to Ken Haddow, "[t]endering can involve requesting competitive bids with substantial up-front payment to obtain rights, significant ongoing payments, substantial annual work commitments and achievement hurdles, and payments (such as the level of royalty) during mining".[139] Hence, proponents argue that if public auction is to be undertaken at all, it must be in exceptional circumstances "such as where an advanced mineral resource or reserve (as defined by international mineral resource standards) has reverted to state hands and a developer is being sought for it".[140] Notwithstanding the scepticism of industry proponents, auctions and competitive bids are potentially advantageous to African countries because they are transparent mechanisms for allocating natural resources to those best able to produce the resources.[141] A transparent auction process could also discourage corruption, which plagues Africa's mining industry.[142]

4.3 Mergers and takeover transactions

Mineral rights can be acquired through corporate arrangements such as mergers and acquisitions. In fact, modern mining statutes in Africa envisage and make provisions for mergers and acquisitions in the mining industry. These provisions enable the transfer, assignment, and sale of mineral rights. As discussed in Chapter 2, mining statutes in Africa allow holders of mineral rights to transfer or assign their title with the prior consent of the Minister.[143] Although mining statutes in Africa provide for

---

137 *Ibid.* s. 71.(3).
138 Ken Haddow, "Should Mineral Rights for Hard-Rock Minerals Be Awarded by Tender?" (2014) 32(3) *Journal of Energy & Natural Resources Law* 335 at 341.
139 *Ibid.* at 336.
140 *Ibid.* at 346.
141 Peter Cramton, "How Best to Auction Natural Resources", in Philip Daniel, Michael Keen, & Charles McPherson, eds, *The Taxation of Petroleum and Minerals: Principles, Problems and Practice* (London: Routledge 2010) 289 at 291.
142 *Ibid.* at 300.
143 See *Mines and Minerals Act, 1999* (Botswana), s. 50; *Minerals and Mining Act, 2006* (Ghana), s. 14; *Nigerian Minerals and Mining Act, 2007*, s. 147; *Mines and Minerals Development Act, 2015* (Zambia), s. 66; *Mining Act, 2010* (Tanzania), s. 9.

the acquisition of mineral rights through mergers, the merger and acquisition process is governed primarily by corporate law and securities law in individual African countries. Mergers and acquisitions in the African mining industry may also be impacted by competition laws in developed Western economies, given that companies based in these countries are the drivers of mergers and acquisition transactions in the African mining industry.[144]

Mergers and acquisitions may be undertaken for a multitude of reasons, including operational and financial reasons, diversification, expansion of market share, and tax considerations.[145] Major mining companies acquire junior mining companies particularly in instances where the area covered by the mineral title held by the junior company holds a proven reserve of minerals. There is a paucity of data regarding the volume of corporate mergers and acquisitions in Africa, but there is evidence that mergers and acquisitions are common in the mining industry. In this regard, about 236 mergers and acquisition transactions occurred in Africa between September 2011 and March 2012, most of which were in the energy, mining, and utilities sectors.[146] In South Africa, about 130 mergers and acquisition transactions were undertaken in the mining industry between 2003 and 2008.[147] During the last two decades, notable mergers and acquisition transactions in Africa's mining industry include the merger between AngloGold (South Africa) and Ashanti Goldfields (Ghana), culminating in the creation of AngloGold Ashanti in 2003; the merger between Harmony Gold Mining Company Limited (South Africa) and African Rainbow Minerals Gold Limited (South Africa) in 2003; the acquisition of De Beers Consolidated Mines (South Africa) by DB Investments (Luxembourg) in 2001; and the 2011 acquisition of Metorex, a South African mining company, by China's Jinchuan Group International Resources Co. Ltd for US$1.32 billion.[148]

Mergers and acquisitions in the African mining industry are spurred by the economic liberalization policies adopted by African countries desirous of attracting FDI, which in the last few years has equally prompted the divestment and privatization of state-owned mining enterprises across the continent. The economic liberalization policies in Africa have encouraged the proliferation of junior mining

---

144 Francis N. Botchway, "Mergers and Acquisitions in Resource Industry: Implications for Africa" in Francis N. Botchway, ed., *Natural Resource Investment and Africa's Development* (Cheltenham: Edward Elgar Publishing, 2011) 159 at 160.
145 W.K. Osae, C.J. Fauconnier, & R.C.W. Webber-Yougman, "A Value Assessment of Mergers and Acquisitions in the South African Mining Industry – the Harmony ARMgold Example" (2011) 111 *Journal of the Southern African Institute of Mining and Metallurgy* 857 at 858.
146 African Development Bank, "Mergers and Acquisitions in Africa" (20 December 2012), https://blogs.afdb.org/fr/blogs/afdb-championing-inclusive-growth-across-africa/post/mergers-and-acquisitions-in-africa-10163
147 Osae *et al.*, *supra* note 145 at 860.
148 Ed Stoddard, "China's Jinchuan Trumps Vale's Metorex Bid", *Reuters*, July 5, 2011, www.reuters.com/article/us-metorex-jinchuan/chinas-jinchuan-trumps-vales-metorex-bid-idUKTRE76436Z20110705

companies in the continent.[149] Quite often, junior mining companies lack the financial capacity to undertake mineral development and production. Hence, they engage solely in mineral exploration with the hope of entering into strategic alliances with major and financially resourced companies to develop their mineral discoveries. These alliances manifest in the form of partnerships, joint ventures, mergers, and, in some cases, outright acquisition of the junior company by the major company.

Mergers and acquisitions in the African mining industry are likely to increase in the near future given China's growing reliance on Africa for the supply of minerals and metals for its booming economy. China's appetite for minerals and metals appears insatiable due, in part, to the industrial and technological revolution that the country has been undergoing for the past three decades. China consumes more than 40% of the world's copper supply and it is the leading importer of iron ore.[150] The number of Chinese companies investing in the African mining industry has increased exponentially in the last three decades. Some of these investments occurred in the form of the outright acquisition of mining assets by Chinese companies.[151] China's growing participation in Africa's mining industry is fuelled partly by the China-Africa Development Fund and the China Development Bank, both of which are actively providing capital to Chinese companies investing in Africa.[152]

5 Registration of mineral rights

Mining statutes in Africa require that mineral rights be registered with a designated government agency. Some African countries require mineral right holders to take positive steps to register their mineral rights, while other countries impose the duty of registering mineral rights on rights-granting agencies by requiring these agencies to maintain a register of mineral rights. In South Africa, mineral right holders must, within 60 days of acquiring the mineral right, register the mineral right with the Mineral and Petroleum Titles Registration Office.[153] In Ghana, the Minerals Commission is mandated to "maintain a register of mineral rights in which shall be promptly recorded applications, grants, variations and dealings in, assignments, transfers, suspensions and cancellation of the rights".[154] In Nigeria, the MCO is obliged to maintain a separate register for each mineral right provided for under the mining statute.[155] Similarly, mining statutes in Botswana, Kenya, and Tanzania require a register of all mineral rights, including a record of all

149 Botchway, *supra* note 144 at 178.
150 PriceWaterhouse, "Stop. Think… Act – Mine 2017", at 17, www.pwc.com/gx/en/mining/assets/mine-2017-pwc.pdf
151 *Ibid*. at 17.
152 UNECA & African Union, *Minerals and Africa's Development: The International Study Group Report on Africa's Mineral Regimes* (Addis Ababa: UNECA, 2011) at 32.
153 See *Mining Titles Registration Act, 1967*, s. 5. (as amended by the *Mining Titles Registration Amendment Act, 2003*). The 60-day period is prescribed under the *Amended Procedure for Lodgement of Rights at the Mineral and Petroleum Titles Registration Office*, www.dmr.gov.za/mineral-regulation/mining-titles
154 *Minerals and Mining Act, 2006* (Ghana), s. 103.(1).
155 *Nigerian Minerals and Mining Act, 2007*, s. 7.

96 Acquisition of mineral rights

applications, date of receipt of applications, grants, assignments, transfers, suspension, and cancellation of mineral rights.[156] The register of mineral rights in individual African countries is accessible to the public on payment of a prescribed fee. The register of mineral rights contains detailed information, including the names of the holder of the mineral right, type of mineral right, duration of the mineral right, the area covered by the mineral right, and the date of registration of the mineral right.[157]

The registration of mineral rights is significant in several respects. First, it signals the commencement of the duration or term of the mineral right, as is the case in Sierra Leone where "the granting of a mineral right takes effect upon the registration date of the licence granting the right".[158] Second, the registration of mineral rights is an important part of the security of mineral tenure because it not only states the present ownership of the mineral right, but also protects against any adverse claims that third parties might make against the registered holder of the mineral right.[159] In essence, the registration of a mineral right ensures security of tenure for the mineral right holder as against third parties. In South Africa, for example, the registration of a mining right transforms the mining right into a limited real right binding on third parties.[160] Thus, where there are two or more competing claims over the same land, the person whose mineral right was registered first has priority unless, prior to the registration, they had notice of the existence of a competing (albeit unregistered) mineral right over the land.[161] Third, in some jurisdictions the assignment or transfer of a mineral right is ineffective unless it is registered. In Sierra Leone, for example, the register of mineral rights must indicate certain memorials, including any renewals, transfers, surrenders, pledges, or encumbrances, and until any such memorial has been so entered in the register, the renewal, transfer, surrender, pledge, or encumbrance "shall have no effect".[162] Finally, the registration of a mineral right is conclusive proof of the validity of the mineral right; that the mineral right vests in the holder; and that the conditions precedent to the grant of the mineral right have been satisfied by the holder.[163] Thus, the registration of a mineral right protects the holder against

---

156 *Mines and Minerals Act, 1999* (Botswana), s. 85; *Mining Act, 2016* (Kenya), ss. 191–2; *Mining Act, 2010* (Tanzania), s. 106.
157 See *Mines and Minerals Act, 1999* (Botswana), s. 85; *Mines and Minerals Act, 2009* (Sierra Leone), s. 42.(3); *Mines and Minerals Development Act, 2015* (Zambia), s. 79.(1c).
158 *Mines and Minerals Act, 2009* (Sierra Leone), s. 42.(6).
159 Barry Barton, "Title Registration in Common Law Jurisdictions" in Elizabeth Bastida et al., eds, *International Comparative Mineral Law and Policy: Trends and Prospects* (The Hague: Kluwer Law International, 2005) 375 at 375 ["Title Registration"].
160 P.J. Badenhorst & Hanri Mostert, *Mineral and Petroleum Law of South Africa: Commentary and Statutes* (Lansdowne: Juta Law, 2004) at 16–17.
161 Barton, "Title Registration", *supra* note 159 at 382.
162 *Mines and Minerals Act, 2009* (Sierra Leone), s. 43.(2). See also *Proclamation No. 678/2010* (Ethiopia), art. 38.(3).
163 See *Mines and Minerals Act, 2009* (Sierra Leone), s. 44; *Mining Act, 2016* (Kenya), s. 195.

any future adverse claims by third parties regarding title to the minerals in the land. As a South African court recently observed, the "purpose and effect of registration is not only that the right becomes binding on third parties, but it also serves as notice to the general public" regarding the existence and validity of the mineral right.[164]

6 An assessment of the mineral rights acquisition process

Modern mining statutes in Africa contain clear rules for the grant of mineral rights, thus ensuring that qualified applicants who comply with the rules are granted access to mineral rights. As discussed above, these rules relate to the eligibility to hold mineral rights, financial and technical competency, and indigenization and shareholding requirements. If effectively implemented, these rules could ensure consistency and certainty in the administrative process governing the grant of mineral rights in Africa. Consistency in administrative decision-making would in turn promote security of tenure and instil investor confidence in the African mining industry. However, as discussed below, the mineral rights acquisition process in Africa is fraught with problems, some of which have the potential to undermine Africa's desire to harness mineral resources for economic development.

6.1 Paucity of domestic and indigenous participation

The vast majority of companies involved in mineral exploration and production in Africa are foreign companies. While, as noted above, African countries require domestic incorporation in order to hold a mineral right, domestic subsidiaries are owned and controlled by parent companies based in foreign countries. The fact that the African mining industry is dominated by foreign companies hardly comes as a surprise. Mineral exploitation is both capital-intensive and technology-driven. Africans lack the financial resources and the technological expertise required for the successful exploitation of mineral resources. Moreover, Africa lacks venture capital sources for large-scale mining operations, and even in the case of small-scale mining, banks and other financial institutions are reluctant to lend capital to African entrepreneurs, apparently due to the risky nature of the mining business. Thus, at this moment, the mineral infrastructure financing constraints in Africa represent a major challenge for domestic private participation in mining projects.[165] The few Africans who are rich enough to afford investing in mining operations are fearful of losing their investment due to the risky nature of the mining business.

---

164 *Minister of Mineral Resources v Mawetse (SA) Mining Corporation (Pty) Ltd* (20069/14) [2015] ZASCA 82 at para. 19.
165 UNECA, *Africa Review Report on Mining (Executive Summary)* at 14, www.uneca.org/sites/default/files/PublicationFiles/aficanreviewreport-on-miningsummary2008.pdf

98 *Acquisition of mineral rights*

The paucity of indigenous participation in African mining is attributable to a number of factors, including the dearth of policy initiatives for promoting local participation in mining operations. Until very recently, indigenous participation in mining operations in Africa was not seriously embedded in domestic law, and even where it was so embedded, African governments did not enforce the provisions. However, as mentioned in Chapter 1, a common objective informing modern mining statutes in Africa is the promotion of domestic participation in mineral extraction. Africa's desire to indigenize the mining industry accounts for citizen-empowerment initiatives such as South Africa's Broad-Based Socio-Economic Empowerment Charter for the Mining and Minerals Industry; the Transformation of Economic and Social Empowerment Framework of Namibia; and Zimbabwe's Indigenization and Economic Empowerment Regulations. In addition, the shareholding requirements discussed above are part of a broader effort by Africa to indigenize the mining industry. In that regard, Kenya's *Mining Act, 2016* requires mining companies whose capital expenditure exceeds the prescribed amount to "list at least twenty percent of its equity on a local stock exchange within three years after commencement of production".[166] Such listing would allow citizens of Kenya to purchase an ownership stake in the companies, thus institutionalizing indigenous participation in mineral exploitation.

The paucity of indigenous participation in mineral exploitation in Africa has certain deleterious effects on the mining industry in particular and African economies more generally. Foreign MNCs that hold the bulk of mineral rights in Africa are primarily engaged in the production and exportation of raw minerals and metals from Africa.[167] Upon their exportation, these raw minerals and metals are then refined and processed in foreign countries, thus denying Africa any value addition in the form of job creation in the mineral-processing sub-sector. The reality, then, is that Africa has yet to take full advantage of her mineral resource endowment particularly with regard to the domestic processing and refining of mineral resources. If mineral resources are to be catalysts for economic development in Africa as envisaged under the AMV, African countries must create critical linkages between the mining industry and other sectors through domestic beneficiation of minerals – that is, the local processing and refining of raw mineral ores.[168] As argued in Chapter 7, such linkages can be achieved through the imposition of minimum levels of beneficiation in mining contracts between African governments and MNCs.[169]

Indigenous participation in resource exploitation can be enhanced by empowering African citizens to pool their resources, in the form of cooperatives and partnerships, with a view to collective ownership of mineral rights, particularly with regard to small-scale mining. In addition, the introduction of flow-through share arrangements in domestic tax legislation in Africa could encourage local

---

166 *Mining Act, 2016* (Kenya), s. 49.(2).
167 *African Mining Vision, supra* note 125 at 2.
168 *Ibid*. at 13–14.
169 *Ibid*. at 14.

participation in mineral exploitation. Flow-through share arrangements allow mining companies to incur exploration and production expenses using investors' funds. Consequently, the companies would renounce those expenses to the investors so that the investors can deduct the expenses from their personal taxable income, thus reducing the investors' tax liability. A flow-through share arrangement is implemented by way of a 'flow-through share agreement' between the investor and the resource company, and it works as follows. First, the investor signs a flow-through share agreement with the company. Under the agreement, the investor purchases 'flow-through shares' from the company and the purchase price is paid into a trust account. Second, the company incurs an exploration or production expense using the funds in the trust account (that is, the purchase price paid by the investor). Third, the company renounces the exploration and production expense to the investor. Fourth, the investor claims tax deductions representing the expense amount renounced to them by the company. Under this arrangement, the investor not only purchases an equity stake in the company but also reduces or eliminates their tax liability to the government on the basis of the renounced expense. This arrangement makes investment in mineral development more attractive to investors.

The flow-through share arrangement cannot exist in a vacuum; rather, it is predicated on a provision in the tax legislation enabling mining companies to deduct exploration and production expenses from their taxable income. In Canada, for example, expenses incurred in exploration and production of minerals within Canada are recognized under the *Income Tax Act* as capital expenditures which are deductible for income tax purposes.[170] Under that Act, "[w]here a person gave consideration under an agreement to a corporation for the issue of a flow-through share of the corporation" and the corporation incurred Canadian exploration and production expenses, the corporation may renounce to the person certain amounts specified in the Act.[171] For this purpose, Canadian exploration and production expenses are any expense incurred in drilling or exploration activities in Canada, as well as expenses incurred in bringing a mine into production in Canada.[172]

In the context of Africa, Zambia's mining regime is unique for its recognition of the concept of 'expense renunciation'. In Zambia, mining companies may irrevocably elect to renounce their deductible prospecting expenditures to their shareholders, in which case

> the deductions shall be allowed, not to the company but its shareholders instead, in proportion to the calls on shares paid by them during the relevant accounting period or in such other proportions as the Commissioner-General having regard to any special circumstances, may determine.[173]

---

170 *Income Tax Act*, Revised Statutes of Canada, 1985, c. 1, ss. 66.(1).
171 *Ibid.* ss. 66.(12.6) & 66.(12.62).
172 *Ibid.* s. 66.(15).
173 *Income Tax Act*, Chapter 323, Statutes of Zambia, 5th Schedule, para. 21.(2).

African countries may want to emulate Zambia's expense renunciation programme, but, to be successful, the scope of 'flow-through share' provisions in Africa must be clear and precise in order to prevent abuse of the provisions by companies and investors alike. In Canada, for example, a company can only renounce an expense to the investor if the expense qualifies as a Canadian exploration or production expense. Furthermore, a company can renounce such expenses "only to the extent that, but for the renunciation, it would be entitled to a deduction in respect of the expenses in computing its income".[174] In effect, a company cannot renounce an expense that the company is not entitled to deduct from its income taxes. In addition, the Canadian *Income Tax Act* prohibits companies from 'warehousing' their expenses for the purpose of renunciation.[175] The Act prohibits companies from renouncing expenses incurred prior to the signing of a flow-through share agreement with the investor.

Local participation in mining operations in Africa may also be incentivized through the introduction of preferential rights with regard to minerals in land occupied by local communities. A preferential right enables communities to hold mineral rights as a collective, rather than as individuals. A preferential mineral right gives the community preference if it applies for a mineral right "before, or at the same time as, another party for the same right to prospect on the same land".[176] In South Africa, for example, local communities have a preferential right with regard to minerals in their land provided they meet the eligibility threshold. In that country, a mineral right may be granted and registered in the name of a community if the Minister is satisfied that the mineral right will contribute towards the socio-economic development of the community, the community submits a development plan indicating the manner in which the right will be exercised, and the envisaged benefits of the mineral right will accrue to the community as a whole.[177] For the purpose of preferential mineral rights, South Africa allows communities to incorporate juristic persons known as Community Property Associations (CPA). CPAs have statutory power to "acquire, hold and manage property in common" on behalf of members of the community and "in a manner which is non-discriminatory, equitable and democratic".[178] Because communities consist of individual members, a preferential mineral right allows individual South African citizens to own an indirect stake in mining operations, thus enhancing indigenous participation in the mining industry.

6.2 Rarity of public auctions

As noted earlier, a few African countries permit the auctioning of mineral rights. In practice, however, the auctioning of mineral rights is rare in Africa. Unlike the

---

174 *Income Tax Act*, Revised Statutes of Canada, 1985, c. 1, s. 66.(12.71).
175 *Ibid.* s. 66.(19).
176 P.J. Badenhorst & N.J.J. Olivier, "Host Communities and Competing Applications for Prospecting Rights in Terms of the Mineral and Petroleum Resources Development Act 28 of 2002" (2011) 44 *De Jure* 126 at 141.
177 *MPRDA*, s. 104.
178 *Communal Property Associations Act, 1996*, Preamble.

situation in Africa, public auction of mineral rights is common in Latin American countries such as Ecuador where mineral rights are acquired primarily through public auctions.[179] The Ecuadorian *Mining Law* obliges the Ministry of Mining to "convene a public bidding process for the grant of all metal mining concessions", including concessions relating to "expired concession areas or those which have been returned, or have reverted, to the State".[180]

A primary reason for the lack of auctions and competitive bids in Africa's mining industry is the lack of geological data regarding mineral deposits in the continent. Many areas have yet to be geologically surveyed, and therefore there is a lack of geological data regarding the existence of commercial quantities of minerals in Africa. While African countries have established geological survey departments and agencies, the reality is that these agencies lack financial and technological competence to undertake complex geological survey which is the springboard for mining operations. Thus, Africa must devise strategies and policies that incentivize the geological mapping and surveying of the continent.

Africa need not rely solely on private companies to conduct geological surveys and other prospecting and exploration activities. Rather, African governments should embark on publicly funded geological survey and exploration of their territories with a view to auctioning off areas that have the potential to be commercially viable. African governments can undertake geological survey in partnership with private entities. The African Union has observed that geological survey is best undertaken through public–private partnerships, such that "if a viable resource is delineated, the private exploration company is guaranteed step-in rights, when the resource is eventually auctioned".[181] The African Union further recommends that the "size of the 'earned' step-in rights" should "be determined by the cost and extent of the exploration programme, as well as the prospectivity of the terrain".[182]

Public–private partnership regarding exploration can be undertaken under the current mining regimes in many African countries, which vest discretion in the Minister responsible for mineral development to grant exploration licences. Thus, African governments could ask for expression of interest from exploration companies and thereafter select companies with whom the government could sign contracts for the exploration of their territories. The contractual arrangement could be such that the exploration companies bear the cost of exploration or, in appropriate cases, the cost of exploration could be borne jointly by the companies and the government. The exploration company would then be rewarded with a 'step-in right' in the event that viable resources are discovered.

---

179 See *Mining Law* (Ecuador), art. 29. See also *General Mining Regulations Decree No. 119* (Ecuador), arts 27–43.
180 *Mining Law* (Ecuador), art. 29.
181 *African Mining Vision, supra* note 125 at 16.
182 *Ibid.* at 16.

6.3 Consultation with host communities

Community consultation is a multifaceted process that involves establishing the parameters for dialogue; the full disclosure of information regarding proposed projects, including the potential impacts of the projects; affording host communities the opportunity to make representations regarding a project; and taking into account the concerns of host communities in the decision-making process. Consultation also encompasses disclosure of applications for mineral rights, including any prior records of the applicants in relation to mining operations in the country. Perhaps more significantly, the consultation process must be both 'meaningful' and 'legitimate'. A meaningful consultation process is one in which the government takes into account the concerns of host communities in deciding whether to approve a project. In appropriate cases, the government may need to accommodate the concerns of host communities by revising the project to mitigate those concerns. With regard to legitimacy, consultation with the host community can be considered legitimate only if it is undertaken through the authorized representatives of the community, rather than self-serving members of the community selected at random by the government and mining companies.

The need to consult host communities has long been recognized by the international community through instruments such as the United Nations Declaration on the Rights of Indigenous Peoples, which requires states to consult in good faith with indigenous peoples "before adopting and implementing legislative or administrative measures that may affect them".[183] Such prior consultation is particularly imperative "in connection with the development, utilization or exploitation of mineral, water or other resources".[184] Regrettably, at the continental level not much has been done by the African Union to incorporate community consultation into mineral development. However, a regional economic bloc, the Economic Community of West African States (ECOWAS), has issued a policy directive requiring companies to "obtain the free, prior and informed consent of local communities before exploration begins and prior to each subsequent phase of mining and post-mining operations".[185]

Many African countries do not consult with host communities either prior to the grant of mineral rights or after the grant of mineral rights.[186] However, some mining statutes in Africa require consultation of host communities, but the degree and timing of consultation differ considerably across the continent. The consultation regimes in Africa can be divided into three groups – namely, countries requiring consultation prior to the grant of mineral rights; countries requiring that

---

183 *United Nations Declaration on the Rights of Indigenous Peoples*, art. 19, www.un.org/development/desa/indigenouspeoples/wp-content/uploads/sites/19/2018/11/UNDRIP_E_web.pdf
184 *Ibid*. art. 32.(2).
185 ECOWAS, *Directive C/DIR. 3/05/09 on the Harmonization of Guiding Principles and Policies in the Mining Sector*, art. 16.(3), https://documentation.ecowas.int/wpfb-file/ecowas-directive-and-policies-in-the-minning-sector2-pdf [*ECOWAS Directive on Mining*].
186 UNECA & African Union, *supra* note 152 at 198.

communities be notified of applications for mineral rights; and countries requiring consultation after mineral rights are granted but before commencement of mining activities.

6.3.1 Countries requiring consultation prior to grant of mineral rights

Within the first group of countries, the Kenyan regime on community consultation stands out because it provides specific requirements regarding the consultation process. Kenya's *Mining Act, 2016* provides that prior to recommending the grant of a mineral right to the Cabinet Secretary, the Mineral Rights Board shall require applicants for mineral rights to seek the approval of any person who may be affected by the grant of the mineral rights, including the owner of private land or the community occupying the land.[187] The Act also requires applicants for mineral rights to obtain the consent of communities and other private owners of land prior to the grant of mineral rights.[188] With regard to community land, the Act provides that prospecting and mining rights

> shall not be granted under this Act or any other written law over community land without the consent of – (a) the authority obligated by the law relating to administration and management of community land to administer community land; or (b) the National Land Commission in relation to community land that is unregistered.[189]

In Kenya, the "authority obligated by the law" to manage registered community land is the Community Land Management Committee elected by the Community Assembly, which itself consists of all adult members of the community.[190] Thus, the consent of the community is obtained through direct negotiation with the elected representatives of the community. The consent of the host community is deemed to be given if the registered owners of the community land enter into

> (a) a legally binding arrangement with the applicant for the prospecting and mining rights or with the Government, which allows the conduct of prospecting or mining operations; or (b) an agreement with the applicant for the prospecting and mining rights concerning the payment of adequate compensation.[191]

However, except where otherwise provided by a law relating to community land, the consent of the community, once granted, is valid for as long as the prospecting and mining rights subsist.[192]

187 *Mining Act, 2016* (Kenya), s. 36.(2h).
188 *Ibid.* ss. 37 & 38.
189 *Ibid.* s. 38.(1).
190 *The Community Land Act*, No. 27 of 2016, s. 15.
191 *Mining Act, 2016* (Kenya), s. 38.(2).
192 *Ibid.* s. 38.(3).

Unlike the position in Kenya, mining statutes in other African countries contain general and vague provisions regarding consultation of host communities prior to the grant of mineral rights. For example, South Africa requires Regional Managers to disclose all applications for mineral rights, and invite interested and affected persons to submit their comments regarding such applications.[193] Where an interested and affected person objects to the granting of a mineral right, the Regional Manager must refer the objection to the Regional Mining Development and Environment Committee who shall advise the Minister on the merit of the objection.[194]

Aside from providing host communities a formal process for objecting to the grant of mineral rights, South Africa imposes on applicants for mineral rights a duty to consult with interested and affected persons.[195] In South Africa, the Regional Manager must, within 14 days of accepting an application for a prospecting right, mining right, or mining permit, issue a written notice to the applicant instructing them "to consult in the prescribed manner with the landowner, lawful occupier and any interested and affected party and include the result of the consultation in the relevant environmental reports".[196] In Zambia, the MLC may, in considering an application for a mineral right, consult any person who may be affected by the grant of the mineral right, particularly persons living in the area to which the application relates.[197] In the DRC, an applicant for an exploration licence is required to attach to their application a "report on the consultation with the authorities of the local administrative entities and with the representatives of the surrounding communities".[198]

While it is commendable for these countries to provide for community consultation prior to the grant of mineral rights, the problem is that the statutory provisions regarding community consultation in most of these countries are vague. For example, the DRC provision does not impose an obligation on the government to consult with local communities; rather, the provision merely requires an applicant to show proof that they have consulted with the community. Similarly, the language of the provision regarding consultation in Zambia is permissive rather than mandatory. In that country, the MLC "may" consult with project-affected communities, suggesting that the Committee has discretionary powers with regard to community consultation.[199] While the Kenya regime requires prior consultation, the burden of consultation is placed wholly on applicants for mineral rights rather than the government. Given the power disparity between mining companies and local communities, mining companies would easily be able to manipulate the consultation process to their advantage by prevailing on local

---

193 *MPRDA*, s. 10.(1).
194 *Ibid.* s. 10.(2).
195 *Ibid.* ss. 16.(4)(b), 22.(4)(b), & 27.(5)(a).
196 *Ibid.* ss. 16.(4), 22.(4), & 27.(5).
197 *Mines and Minerals Development Act, 2015* (Zambia), s. 6.(6).
198 *Mining Code, Law No. 007/2002 of July 11, 2002* (DRC), art. 69 (as amended by *Law No. 18/001 of 9 March 2018*).
199 See *Mines and Minerals Development Act, 2015* (Zambia), s. 6.(6).

communities to agree to terms that might not serve the long-term interest of the communities. Moreover, communities in Kenya cannot withdraw their consent or renegotiate the terms of their consent, even if the circumstances change significantly. This is because, as noted above, the consent of communities subsists throughout the duration of the prospecting and mining rights of the project proponent.[200]

Likewise, the South African provision neither expressly obliges the Regional Manager to take into account the objections of local communities regarding applications for mineral rights, nor provides any legal recourse for communities whose objections are overlooked by the Regional Manager. Hence, it has been argued that the South African provision "does not contain a minimum standard for sufficient consultation" with host communities.[201] However, the Constitutional Court has held that while consultation under the MPRDA does not require the parties to reach an agreement, it is not a mere formality but one that involves engaging in good faith to ascertain whether an accommodation can be reached in respect of the impact on the landowner's right to use their land.[202] The Constitutional Court held further that:

> The consultation process required by section 16(4)(b) of the Act thus requires that the applicant must: (a) inform the landowner in writing that his application for prospecting rights on the owner's land has been accepted for consideration by the Regional Manager concerned; (b) inform the landowner in sufficient detail of what the prospecting operation will entail on the land, in order for the landowner to assess what impact the prospecting will have on the landowner's use of the land; (c) consult with the landowner with a view to reach an agreement to the satisfaction of both parties in regard to the impact of the proposed prospecting operation; and (d) submit the result of the consultation process to the Regional Manager within 30 days of receiving notification to consult.[203]

Although the consultation regime under the MPRDA is vague, recent jurisprudence indicates that, in the specific context of South Africa's *Interim Protection of Informal Land Rights Act* 31 of 1996 (IPILRA), a mineral right cannot be granted in relation to land over which a community has an informal right unless the prior consent of the community is obtained. For example, the Constitutional Court of South Africa has elaborated on community 'consent' within the context of the IPILRA thus:

> [I]n instances where land is held on a communal basis, affected parties must be given sufficient notice of and be afforded a reasonable opportunity to

---

200 *Mining Act, 2016* (Kenya), s. 38.(3).
201 Badenhorst & Olivier, *supra* note 176 at 141.
202 *Bengwenyama Minerals (Pty) Ltd and Others v Genorah Resources (Pty) Ltd and Others* 2011 (4) SA 113 (CC) at para. 65.
203 *Ibid.* at para. 67.

participate, either in person or through representatives, at any meeting where a decision to dispose of their rights to land is to be taken. And this decision can competently be taken only with the support of the majority of the affected persons having interest in or rights to the land concerned, and who are present at such a meeting.[204]

In the more recent case of *Baleni and Others v Minister of Mineral Resources and Others*, [205] the Minister granted to Transworld Energy and Mineral Resources, an Australian mining company, a mining right on land that the Umgungudlovu community had lived on and occupied for centuries. The Minister granted the mining right without obtaining the consent of the community. The Umgungudlovu community alleged that the Minister's decision to grant a mining right constituted a deprivation of their informal rights in land and that, in the context of section 2 of the IPILRA, such deprivation is unlawful where the prior consent of the community was not obtained. The community sought a declaration that, as holder of rights in land under the IPILRA, the full and informed consent of the community must be obtained prior to the grant of a mining right on their land. The North Gauteng High Court held that:

> In keeping with the purpose of IPILRA to protect the informal rights of customary communities that were previously not protected by the law, the applicants in this matter therefore [have] the right to decide what happens with their land. As such they may not be deprived [of] their land without their consent. Where the land is held on a communal basis – as in this matter – the community must be placed in a position to consider the proposed depravation and be allowed to take a communal decision in terms of their custom and community on whether they consent or not to a proposal to dispose of their rights to their land.[206]

The court then made a consequential declaration that where a community holds an informal right to land, the Minister cannot lawfully grant a mining right in relation to that land unless they comply with the IPILRA which provides that "no person may be deprived of any informal right to land without his or her consent".[207] The court declared further that the Minister "is obliged to obtain the full and informed consent of the Applicants and the Umgungundlovu Community, as holder of rights in land, prior to granting any mining right" over the land.[208]

---

204 *Maledu and Others v Itereleng Bakgatla Mineral Resources (Pty) Limited and Another* [2018] ZACC 41 at para. 97.
205 *Baleni and Others v Minister of Mineral Resources and Others* [2018] ZAGPPHC 829; [2019] 1 All SA 358 (GP) [*Belani v Minister*].
206 *Ibid.* at para. 83.
207 *Interim Protection of Informal Land Rights Act 31 of 1996*, s. 2.(1).
208 *Baleni v Minister, supra* note 205 at para. 84.

6.3.2 Countries requiring that communities be notified of applications for mineral rights

The second group consists of countries requiring simply that host communities be notified of applications for mineral rights. In Ghana, prior to the grant of a mineral right, the Minister is obliged to give a written notice of the application to the local chief of the area where the land is situated or the allordial owner of the land, as well as the relevant District Assembly.[209] In addition, the Minister's written notice must be published in a manner customarily acceptable to the area concerned, and gazetted and exhibited at the offices of the District Assembly where the land is situated.[210] Statutory provisions requiring community notification of applications for mineral rights neither guarantee nor mandate actual consultation with host communities. Thus, in practice, "potentially affected communities do not take part" in the process leading to the grant of mineral rights in Ghana.[211] Although the Minister is obliged to issue a written notice of applications for mineral rights, the views and concerns of host communities appear to carry little, if any, weight. Worse still, Ghana's mining statute does not expressly vest a right in host communities to make submissions regarding mining projects on their land. This position is contrary to the *African Charter on Human and Peoples' Rights*, which requires Africa governments to provide "meaningful opportunities for individuals to be heard and to participate in development decisions affecting their communities".[212]

6.3.3 Countries requiring consultation prior to the commencement of mining operations

The third group of African countries require mining companies to consult with host communities after mineral rights are granted but before commencement of mining operations. In Angola, for example:

> The regulatory body, in coordination with local State entities and the holder of the mining rights, must create consultation mechanisms that allow local communities affected by mining projects to actively participate in decisions relating to the protection of their rights, within constitutional limits.[213]

In fact, consultation is mandatory in Angola in situations where "the implementation of mining projects may result in the destruction or damage of material, cultural or historical assets belonging to the local community as a whole".[214]

---

209 *Minerals and Mining Act, 2006* (Ghana), s. 13.(2).
210 *Ibid.* s. 13.(3)(b).
211 UNECA & African Union, *supra* note 152 at 198.
212 *The Social and Economic Rights Action Centre & Another v Nigeria* (155/96, October 27, 2001) at para. 53, www.achpr.org/sessions/descions?id=134
213 *Mining Code* (Angola), art. 16.(2).
214 *Ibid.* art. 16.(4).

108 *Acquisition of mineral rights*

Similarly, Mozambique requires consultation with communities "before the granting of an authorization for the beginning of mining exploration".[215] The reference to "the beginning of mining exploration" suggests that community consultation is required in Mozambique after the mineral right is granted but prior to the approval of the exploration operations. Within this third group, some African countries, including Nigeria, Guinea, Sierra Leone, and South Sudan, require mineral right holders to consult with host communities within the context of CDAs.[216] In Nigeria, for example, mining leaseholders are required to negotiate and sign CDAs which "specify appropriate consultative and monitoring frameworks between the mineral title holder and the host community, and the means by which the community may participate in the planning, implementation, management and monitoring of activities carried out under the agreement".[217] The problem, however, is that the consultation envisaged in these countries occurs after the mineral rights have been granted, rather than prior to the grant of mineral rights. In effect, host communities in these countries are denied participation in the mineral rights acquisition process.

6.3.4 *The need for community consultation in Africa*

Although some African countries require community consultation prior to the grant of mineral rights, overall Africa appears to disregard the interest of host communities in the mineral rights acquisition process. Many African countries do not accord communities likely to be impacted by mining projects the opportunity to participate in the statutory process leading to the acquisition of mineral rights.[218] Given the adverse social and environmental impacts often associated with the mining industry in Africa, community consultation ought to be an integral component of the mineral rights acquisition process. Consultation not only allows citizens to voice their concerns prior to the execution of mining projects, but enables decision-makers to take the concerns of citizens into account in making decisions regarding mineral resource development. This position is acknowledged by international instruments such as the ILO's *Indigenous and Tribal Peoples Convention, 1989 (No. 169)* which provides:

> In cases in which the State retains the ownership of mineral or sub-surface resources or rights to other resources pertaining to lands, governments shall establish or maintain procedures through which they shall consult these peoples, with a view to ascertaining whether and to what degree their interests

---

215 *Mining Law No. 20/2014* (Mozambique), art. 32.(2).
216 *Nigerian Minerals and Mining Act, 2007*, s. 116; *Amended 2011 Mining Code* (Guinea), arts 37.(II) & 130; *Mines and Minerals Act, 2009* (Sierra Leone), ss. 138–41; *Mining Act, 2012* (South Sudan), s. 68.
217 *Nigerian Minerals and Mining Act, 2007*, s. 117.
218 IANRA, *African Mining and Mineral Policy Guide: A Resource for Non-Governmental Organisations, Activists, Communities, Governments and Academics* (2016) at chapter 10, www.ianra.org/images/images/PDFs/Policy-Guide.pdf

would be prejudiced, before undertaking or permitting any programmes for the exploration or exploitation of such resources pertaining to their lands.[219]

In the African context, community consultation can be predicated on two factors: the government's custodial ownership of mineral resources for the benefit of citizens, and community ownership of land or interest in land. Regarding the first factor, there are statutory and constitutional grounds for the duty to consult local communities prior to resource development in Africa. As discussed in Chapter 1, ownership of mineral resources in Africa is statutorily and constitutionally vested in the government in trust for the citizens. For example, in South Africa, "mineral and petroleum resources are the common heritage of all the people of South Africa and the State is the custodian thereof for the benefit of all South Africans".[220] Similarly, in Ghana, Kenya, Nigeria, and Tanzania, mineral resources in, under, or upon any land are vested in the government in trust for the citizens of these countries.[221] These statutory and constitutional provisions impose a burden on African governments to exploit mineral resources for the overall benefit of their citizens. In this sense, African governments occupy a position akin to that of a trustee that owes a fiduciary duty to the beneficiaries of the trust to act with utmost good faith for the benefit of the beneficiaries. To paraphrase Elmarie van der Schyff, these statutory provisions impose on African governments a 'fiduciary responsibility' regarding the exploitation of mineral resources "with the sole aim of protecting intergenerational interests".[222] Thus, as custodians of mineral resources, African governments bear the "responsibility of public trusteeship" to protect and preserve these resources "and to manage resource use in a sustainable and equitable manner for the benefit of current and future users and stakeholders".[223] Given that mineral resources are held by governments ("trustees") in trust for their citizens ("beneficiaries"), it is logical to require African governments to consult with citizens prior to the exploitation of mineral resources.

In addition to domestic regimes, the *African Charter on Human and Peoples' Rights* (African Charter) provides a continent-wide basis for the duty to consult local communities. The African Commission on Human and Peoples' Rights has held that Article 16 (right to enjoy the best attainable state of physical and mental health) and Article 24 (right to a satisfactory environment) of the African Charter enjoin African governments to provide "meaningful opportunities" for citizens to

---

219 *Indigenous and Tribal Peoples Convention, 1989 (No.169)*, art. 15.(2); (adopted in Geneva at the 76th ILC Session, 27 June 1989; entry into force: 5 September 1991).
220 *MPRDA*, s. 3.(1). See also *Minerals and Mining Act, 2006* (Ghana), s. 1; *The Constitution of the Federal Republic of Nigeria 1999*, s. 44.(3).
221 *Minerals and Mining Act, 2006* (Ghana), s. 1; *Mining Act, 2016* (Kenya), s. 6; *Nigerian Minerals and Mining Act, 2007*, s. 1.(1); *The Natural Wealth and Resources (Permanent Sovereignty) Act, 2017* (Tanzania), s. 4.
222 Elmarie van der Schyff, "South Africa: Public Trust Theory as the Basis for Resource Corruption Litigation" (Open Society Foundations, August 2016), www.justiceinitiative.org/uploads/61421070-3832-402a-b258-befa48b2a6cf/legal-remedies-6-schyff-20160802.pdf
223 *Ibid*.

voice their concerns regarding development projects in so far as these projects affect or are likely to affect their communities.[224] The Commission has equally held that, in relation to "any development or investment projects that would have a major impact within" a community, "the State has a duty not only to consult with the community, but also to obtain their free, prior, and informed consent, according to their customs and traditions".[225]

In addition, a number of soft law instruments in Africa recognize the right of local communities to consultation prior to the extraction of natural resources. In 2012, the African Commission on Human and Peoples' Rights adopted a resolution exalting African countries to adopt policies that ensure community participation, including the free, prior, and informed consent of communities regarding exploitation of natural resources.[226] Likewise, the ECOWAS has issued a policy directive urging mining companies to "conduct their mining activities in a manner that respects the right to development in which peoples are entitled to participate in, contribute to, and enjoy economic, social, cultural, and political development in a sustainable manner".[227] It also exalts mining companies to respect the rights of local communities, particularly "the rights of local people and similar communities to own, occupy, develop, control, protect, and use their lands, other natural resources, and cultural and intellectual property".[228] The ECOWAS Directive states further that "[c]ompanies shall obtain free, prior, and informed consent of local communities before exploration begins and prior to each subsequent phase of mining and post-mining operations", while also maintaining "consultations and negotiations on important decisions affecting local communities throughout the mining cycle".[229]

Consultation can equally be predicated on historical antecedents regarding ownership of land and mineral resources in Africa. Prior to the arrival of European colonialists, land in sub-Saharan Africa and the minerals situated in the land were owned and exploited by local African communities. This prior ownership of land forms at least a historical basis for the imposition on African governments of a statutory duty to consult local communities prior to the grant of mineral rights. This is particularly so because the governments of modern African states are the successors to European colonial governments in Africa. This is the case in Canada where the government's assumption of sovereignty over lands and natural resources previously held by Aboriginal communities is one of the predicates for the government's constitutional duty to consult with these communities.[230]

---

224 *The Social and Economic Rights Action Centre & Another v Nigeria*, supra note 212 at para. 53.
225 *Centre for Minority Rights Development (Kenya) and Minority Rights Group (on behalf of the Endorois Welfare Council) v Kenya* (276/03, November 25, 2009) at para. 291, www.achpr.org/sessions/descions?id=193
226 African Commission on Human and Peoples' Rights, *Resolution on a Human Rights-Based Approach to Natural Resources Governance*, ACHPR/Res. 224 (LI) 2012.
227 *ECOWAS Directive on Mining*, supra note 185, art. 16(1).
228 *Ibid*. art. 16.(2).
229 *Ibid*. art. 16.(3) & (4).
230 *Haida Nation v British Columbia (Ministry of Forestry)*, 2004 S.C.C. 73 at para. 53.

6.3.5 Nature of consultation

Consultation can be meaningful and effective only if host communities are able to freely participate in the consultation process; have access to requisite information regarding the proposed project, particularly the harmful effects of the project; and make informed choices regarding the project. Community participation in the consultation process must not be coerced by the government. In addition, persons engaged in consultation on behalf of host communities must be chosen or elected by members of the community, rather than by the government and mining companies. The consultation process would be illegitimate from the perspective of the host communities if the representatives of these communities are chosen by mining companies or the government. There is, however, the problem that, because of the rampant poverty and illiteracy pervading much of rural Africa where most mining projects are situated, members of the host community may be vulnerable to the powerful influence of governments and mining companies. Rural communities in Africa lack the knowledge and skills to understand complex mining projects, "which impedes their proper and substantive participation in complex and asymmetric negotiations with companies and governments over natural resource projects".[231] Non-governmental organizations (NGOs) can provide independent advice to host communities and they sometimes participate in the consultation process on behalf of the communities.[232] Even then, host communities must be knowledgeable enough to be able to choose which NGO to work with. Some NGOs are totally opposed to mining projects irrespective of the economic benefits that such projects could bring to the host community. A host community that relies on such NGOs may find that its interests are inadequately protected in the course of the consultation with the government and mining companies.

Although this book advocates for community consultation prior to the grant of mineral rights, this author does not subscribe to the idea that host communities should have veto power over mining projects, a view propounded by some proponents of the concept of free, prior, and informed consent (FPIC). In the author's opinion, the right to consultation should not be so broad as to confer a veto power on local communities over mining projects. Any such veto power on the part of host communities would disrupt the African mining industry, and few mining projects would see the light of day if communities were empowered to veto such projects. Moreover, "the recognition of the right to a veto that derives from FPIC may not be consistent with the legal regime of the African

---

231 Pacifique Manirakiza, "Asserting the Principle of Free, Prior and Informed Consent (FPIC) in Sub-Saharan Africa in the Extractive Industry", in Isabel Feichtner, Markus Krajewski, & Ricarda Roesch, eds, *Human Rights in the Extractive Industries: Transparency, Participation, Resistance* (Cham, Switzerland: Springer, 2019) 219 at 241.
232 See Evaristus Oshionebo, "Transnational Corporations, Civil Society Organizations and Social Accountability in Nigeria's Oil and Gas Industry" (2007) 15(1) *African Journal of International & Comparative Law* 107 at 113–8.

Charter".[233] Put simply, the right to consultation advocated here encompasses a duty on the part of African governments to disclose information regarding all applications for mineral rights; engage in good faith dialogue with communities regarding the potential impacts of mining projects; afford an opportunity to communities to make representations regarding their concerns; take into account the concerns of communities; compensate host communities for the adverse effects of mining operations; and, in appropriate cases, accommodate the interest of communities by revising and amending project plans to cater to the interest of the communities.

7 Conclusion

This chapter analyzes the statutory processes governing the grant of mineral rights in Africa, including the eligibility threshold for holding mineral rights. A prominent feature of modern mining statutes in Africa is the clarity of the rules governing the grant of mineral rights. The primary objective of these rules is to attract foreign investors – hence the streamlining of the Minister's discretionary power to grant and renew mineral rights. In many African countries, an application for a mineral right is determined within a prescribed timeframe. While the Minister retains a discretion to grant mineral rights, the Minister must exercise their discretionary power within the statutory guidelines. Thus, where an applicant has satisfied the statutory requirements, the Minister usually grants a mineral right to the applicant. In fact, in South Africa, ministerial discretion is circumscribed altogether such that the Minister is obliged to grant a mineral right where the applicant complies with statutory requirements.

However, the analysis in this chapter exposes the deficiencies in the mineral right acquisition process, including the lack of indigenous participation in mineral exploitation, the rarity of public auction of mineral rights, and the utter neglect of the opinions and concerns of host communities regarding the grant of mineral rights. Until recently, African countries did little to encourage their citizens to invest in mining operations. It is, of course, the case that mining operations are highly capital-intensive. It is also true that most Africans do not have the financial resources to invest in mining operations. These obstacles notwithstanding, African countries can promote indigenous participation in mineral extraction through legislative and tax incentive schemes, such as the 'flow-through share' arrangement. African citizens could be encouraged to invest in mining projects if they are allowed to deduct mining-related investments from their taxes. Moreover, African countries could maximize their resource rents by auctioning mineral rights. While some mining statutes contain provisions enabling competitive auction of mineral rights, the reality is that competitive auctions are rare in the African mining industry. The rarity of competitive auctions is due primarily to the lack of geological data in Africa, but, as argued in this chapter, this obstacle can be ameliorated through governments' participation in geological survey of their territories. African

---

233 Manirakiza, *supra* note 231 at 234.

countries ought to enter into strategic partnerships with private companies with regard to the generation of geological data.

Finally, in order to conduct mining operations in a sustainable manner, local host communities should be consulted prior to the grant of mineral rights in Africa. The consultation process should afford communities an opportunity to express their opinions, and these opinions must be taken into account in the decision-making process. If nothing else, such prior consultation would ensure that host communities buy into the project, thus fostering a conducive atmosphere for mining operations.

4 Security of mineral tenure

1 Introduction

This chapter analyzes the security of mineral tenure in Africa – that is, the stability, continuity, and protection of mineral rights throughout the life circle of a mining project. As a precursor to the discussion of security of mineral tenure, this chapter discusses two related concepts with significant implications for the security of mineral tenure – namely the source of mineral rights and the legal nature of mineral rights. The source of mineral rights is significant because it bears directly on the scope of protection afforded to the holders of mineral rights. Hence, this chapter addresses the question of whether mineral rights in Africa are statutory or contract-based. Where mineral rights are statutory, the holder enjoys the rights, privileges, and protections prescribed in the mining statute because the holder's rights derive from the statute. However, where mineral rights are contractual, the holder's rights derive primarily from the investment contract with the host government, although the holder also enjoys and bears the general rights and obligations prescribed in relevant statutes. Moreover, the duration of a mineral right hinges on the source of the mineral right. The second concept, the legal nature of mineral rights, is equally significant because the degree to which mineral rights are protected in a given country depends partly on the legal characterization of mineral rights. For example, mineral right holders enjoy constitutional protection in countries that characterize mineral rights as 'property'. In such countries, the legality of the host government's actions in expropriating mineral rights may be questioned based on the constitutional provisions protecting property rights.[1]

2 Sources of mineral rights

As discussed in Chapter 1, mining regimes in Africa are state-oriented in the sense that ownership of minerals vests in the governments of African countries. Thus, in order to explore and extract minerals, mining companies are required to obtain the consent and authorization of the government. The authorization of the

---

1 See, for example, *Newcrest Mining (WA) Ltd & Another v The Commonwealth of Australia* (1997) 190 CLR 513.

government manifests in the form of a state-investor contract. In practice, African governments grant mineral rights through contract mechanisms reflective of the particular mineral right (prospecting licence, exploration licence, and mining licence or lease) granted by the government. However, as discussed below, the terms and conditions of mineral rights in Africa are often predicated on statutory provisions. Hence, depending on the legal regime in a given country, mineral rights in Africa are statute-based, contract-based, or both.

2.1 Statute-based mineral rights

Mineral rights are statutory where the mineral rights flow from a statute or where the terms of the mineral rights are dictated by statute. This is the case in South Africa, where the Supreme Court of Appeal held in *Minister of Mineral Resources v Mawetse (SA) Mining Corporation (Pty) Ltd* [2] that a prospecting right is statutory in nature because the terms and conditions of the right are prescribed by the MPRDA. According to the court, section 17(6) of the MPRDA "makes a prospecting right subject to the MPRDA, any other applicable law and the terms and conditions stipulated in the right".[3] The MPRDA prescribes that a prospecting right is granted subject to the condition that the applicant must, prior to the notarial execution of the right, submit to the Minister evidence indicating compliance with the policy objective regarding 'substantial and meaningful' expansion of opportunities for historically disadvantaged persons, as encapsulated in the Broad-Based Socio-Economic Empowerment Charter for the Mining and Minerals Industry (Mining Charter).[4] In effect, under the MPRDA compliance with the Mining Charter is a precondition for the grant of a prospecting right. The court held that although the Minister had approved the grant of a prospecting right to the fifth appellant, the approval was conditional on the appellant's compliance with the Mining Charter. Because the appellant did not comply with the Mining Charter, the Minister was lawfully entitled to refuse to execute the prospecting right.[5] The court elaborated thus:

> The objects set out in s. 2(d) [of the MPRDA] are of cardinal importance when the purpose of the MPRDA and the Mining Charter is borne in mind and, as our courts have stressed, are essential to redress the historical inequalities in the mining industry. Compliance with the request is not merely optional and the grant of the prospecting right was expressly made subject to such compliance. Absent compliance, the [Minister] was lawfully entitled to refuse to execute the right.[6]

---

2 [2015] ZASCA 82.
3 *Ibid.* at para. 17.
4 *MPRDA*, s. 2.(d).
5 [2015] ZASCA 82 at para. 17.
6 *Ibid.* at para. 17.

116 Security of mineral tenure

Mineral rights in South Africa are not contractual because the terms of the mineral rights are not based on the mutual consensus of the parties.[7] An applicant for a mineral right is not required to reach a consensus with the Minister regarding the terms and conditions of the right. Rather, the terms and conditions of mineral rights are prescribed by the MPRDA and any other applicable law. The grant of a mineral right under the MPRDA is a unilateral administrative act by the Minister who must act on the basis of the terms prescribed by the MPRDA. As the court observed in *Mawetse*:

> [T]he granting of a prospecting right, as is the case with all other rights under the MPRDA, is not contractual in nature, but a unilateral administrative act by the Minister or her delegate in terms of their statutory powers under the MPRDA. It occurs outside the ambit of and regardless of the existence of a contract between the Minister and a successful applicant.[8]

The court continued:

> The granting of a prospecting right does not admit of a legal relationship where consensus between the parties is a legal requirement. I disagree with the dictum in *Meepo* that an applicant for a prospecting right has to reach consensus with the Minister or has to consent to the terms and conditions of the right. The MPRDA sets out in no uncertain terms the respective parties' rights and obligations. It is plainly an authoritative, unilateral administrative act whereby rights are granted with or without conditions and in terms whereof rights accrue to and obligations are imposed upon the successful applicant.[9]

The decision in *Mawetse* is significant because it overruled an earlier case, *Meepo v Kotze*.[10] In *Meepo*, the High Court of South Africa (Northern Cape Division) held that the legal nature of a prospecting right under the MPRDA is contractual because

> the Minister, as the representative of the State as the custodian of the mineral resources of the Republic of South Africa, consensually agrees to grant to an applicant a limited real right to prospect for a mineral or minerals on specified land for a specified period and subject to such conditions as may be determined or agreed upon.[11]

---

7 *Ibid.* at para. 24.
8 *Ibid.* at para. 24.
9 *Ibid.* at para. 26.
10 2008 (1) SA 104 NC.
11 *Ibid.* at para. 46.3.

The *Meepo* court relied on *Ondombo Beleggings (EDMS) Beperk v Minister of Mineral and Energy Affairs*,[12] even though the *Precious Stones Act 73 of 1964* (now repealed) on which *Ondombo Beleggings* was decided differs significantly from the statutory schemes under the MPRDA. As the Supreme Court of Appeal observed in *Mawetse*, because of the significant differences between the statutory schemes under the MPRDA and the *Precious Stones Act 73 of 1964*, *Ondombo Beleggings* is distinguishable from *Meepo*, and hence it was inappropriate for the *Meepo* court to rely on *Ondombo Beleggings*.[13] That being the case, it appears settled law that, in South Africa, the legal nature of mineral rights is statutory.

2.2 Contract-based mineral rights

Mineral rights are contractual where the state, as owners of mineral resources, grants interest in minerals to investors by means of a contract, the terms of which are mutually agreed between the contracting parties. In other words, mineral rights are contractual where they derive from a contract between the investor and the host state. The host state conveys "interests in such resources by contract, thereby giving rise to contractual rights with obligations binding on it as a matter of contract".[14] Such contracts usually set out the respective rights and obligations of the contracting parties, as well as the terms and conditions under which the mineral rights are granted and maintained. Mining statutes in Africa empower the government to negotiate and enter into mining contracts with persons interested in exploiting mineral resources, but some of these statutes do not dictate the terms and condition of such contracts. In Zimbabwe, for example, the *Mines and Minerals Act* empowers the Minister of Mines to enter into an agreement regarding the issue of a special mining lease, including the terms and conditions of the lease, the renewal of the lease, the liabilities and obligations of the lessee, and any other matter connected with the lease.[15] Likewise, in Guinea the government is obliged to negotiate mining agreements setting out the terms and conditions for operating mineral concessions.[16]

Thus, the common practice in Africa is to grant mineral rights on the basis of terms and conditions consensually negotiated and agreed by the government and the mineral right holder. The terms of such mining contracts usually make clear that the source of the mineral rights granted under the contract is the contract itself. More specifically, the granting clause in a mining contract expresses the exclusivity of the rights conferred on the mineral right holder under the contract. For example, the mineral development agreement between Liberia and Mittal Steel states that "[b]y this Agreement and subject to its terms and conditions, the

---

12 [1991] ZASCA 108; 1991 (4) SA 718 (AD) 724.
13 [2015] ZASCA 82 at para. 27.
14 Martin K. Ayisi, "The Legal Character of Mineral Rights under the New Mining Law of Kenya" (2017) 35(1) *Journal of Energy & Natural Resources Law* 25 at 34 ["Legal Character of Mineral Rights"].
15 *Mines and Minerals Act, Chapter 21:05* (Zimbabwe), s. 167.
16 *Amended 2011 Mining Code* (Guinea), art. 37.III.

118 *Security of mineral tenure*

GOVERNMENT hereby grants to the CONCESSIONAIRE the exclusive right and license to conduct Exploration, Development, Production and marketing of Iron Ore and associated products ... in the Concession Area".[17] Because the contract evidencing the mineral right "involves mutual rights and obligations between the parties", the interest of the holder of a mineral right in these countries flows from the contract "and not from the mining law empowering the state to enter the agreement".[18]

However, mineral rights in Africa are not entirely contract-based. Though granted in contractual forms, mineral rights are anchored on mining statutes which authorize the government to grant mineral rights in the form of a contract. In this sense, both sources of mineral rights are complementary in many African countries. While the terms of contract-based mineral rights are primarily consensual, the standard terms and conditions of mineral rights in most African countries are dictated by mining statutes. These standard terms are not subject to the mutual agreement of the contracting parties, but are statutorily mandated as part of the mineral concession contract. For example, Kenya grants mineral rights subject to specific statutory terms and conditions, including conditions regarding the protection of the environment, community development, safety of mining operations, and the protection of the lawful interests of the holders of any other mineral rights.[19] Likewise, the obligations of mineral right holders are prescribed by multiple statutes, including statutes governing mining, taxation, and environmental protection. These statutory obligations are usually incorporated into mining contracts as terms and conditions of the contracts. Thus, today, mining contracts in Africa contain obligations and covenants relating to the payment of royalty, sustainable environmental practices, the training and employment of citizens, and the use of local goods and services.[20] For example, Mozambique's *Mining Law No. 20/2014* authorizes the government to enter into mining contracts based on the terms prescribed by the statute, including state participation in mining ventures, minimum local content, social responsibility activities to be undertaken by the company, and dispute resolution mechanisms.[21] Relatedly, mining contracts in Africa sometimes incorporate statutory terms and conditions by providing that the terms and conditions of the contracts are in addition to the terms prescribed in the mining statute.

---

17 *An Act Ratifying the Amendment to the Mineral Development Agreement (MDA) Dated August 17, 2005 between the Government of the Republic of Liberia (The Government) and Mittal Steel Holding A.G. and Mittal Steel (Liberia) Holdings Limited (The Concessionaire)*, art. 3, www.leiti.org.lr/uploads/2/1/5/6/21569928/52423536-an-act-ratifying-the-amendment-to-the-mineral-development-agreement-mda-dated-august-17-2005-between-the-government-of-liberia_and_mittal_steel.pdf
18 Ayisi, "Legal Character of Mineral Rights", *supra* note 14 at 34.
19 *Mining Act, 2016* (Kenya), s. 42.(1).
20 See, for example, *Mines and Minerals Development Act, 2015* (Zambia), s. 32.(2) and *Mining Act, 2016* (Kenya), s. 106.
21 *Mining Law No. 20/2014*, dated 18 August (Mozambique), art. 8.

In addition, as discussed in Chapter 5, mining statutes in Africa prescribe the fiscal regimes governing mineral rights, including the royalty rates payable by mineral right holders. The statutory fiscal regimes governing mining operations in Africa are sacrosanct in the sense that the provisions "shall not be subject to negotiations".[22] In South Sudan, a mining contract "may not vary any fiscal matter specified in applicable law and may not alter for any period the tax rate, royalty rate, fee or duty or their method of calculation".[23] Because some of the terms of the contract evidencing mineral rights are consensually agreed while others are mandated by statute, it could be said that mineral rights in some African countries are partly contractual and party statutory.

A corollary observation is that, in countries where mining statutes prescribe some mandatory terms and conditions for mineral rights, mining contracts entered into by the government must comply with the mandated terms, otherwise the validity of the contracts is questionable. The validity of a contract evidencing a mineral right may depend on the inclusion of the terms and conditions prescribed under the mining statute. A mining contract could be invalid if the terms of the contract derogate significantly from the mandatory terms of the enabling statute. For example, where a statute prescribes the royalty rate for minerals, a provision in a mining lease that contravenes the royalty rate prescribed in the statute is invalid because the statute is not subject to the will of the contracting parties. In this regard, the Mining Code of Guinea provides that the "provisions of the Mining Agreement supplement the provisions of this Code but shall not derogate from" the Code,[24] while in South Sudan the provisions in the mining statute prevail over a mining contract in the event of a conflict.[25] Likewise, in Kenya "a term or condition contained in a mineral agreement which is inconsistent with any provision of this Act or the Constitution shall, to the extent of the inconsistency, be void and of no legal effect".[26]

Moreover, even where mineral rights are contractual, the rights of the holders of the mineral rights are not determined exclusively by contract. Across Africa, mineral right holders enjoy certain rights and bear duties outside of the mining contracts signed with host states. For example, holders of mineral rights bear the statutory obligations specified in mining and environmental statutes, and they enjoy constitutional protection against unlawful expropriation of mineral rights. Mining statutes in Africa impose certain general obligations on mineral right holders exclusive of the obligations provided for in the contracts granting the mineral rights. In Kenya, for example, a mineral agreement does not "absolve any party to such agreement from a requirement prescribed by law".[27] In addition, mining statutes prescribe the terms and conditions for maintaining mineral rights. Likewise, the security of mineral tenure is anchored primarily on statutory

22 *Income Tax Act, Chapter 52:01* (Botswana), 12th Schedule, para. 11.
23 *Mining Act, 2012* (South Sudan), s. 108.(2).
24 *Amended 2011 Mining Code* (Guinea), art. 18.
25 *Mining Act, 2012* (South Sudan), s. 108.(1).
26 *Mining Act, 2016* (Kenya), s. 121.(1).
27 *Ibid.* s. 117.(3).

provisions, rather than the contract evidencing mineral rights. The effect, then, is that while mineral rights are granted through contracts, the rights and obligations of mineral right holders are prescribed in multiple instruments, including contracts and mining statutes.

The complementarity of the sources of minerals rights is amplified in countries that require parliamentary ratification of mining contracts. In Ghana, for example, all transactions, contracts, or undertakings involving the grant of mineral rights "shall be subject to ratification by Parliament".[28] Likewise, a stability agreement between the government of Ghana and the holder of a mining lease is subject to parliamentary ratification.[29] In Kenya, agreements relating to large-scale mining operations on terrestrial and marine areas must be submitted to the Parliament (that is, the National Assembly and the Senate) for ratification prior to the execution of the agreement.[30] A similar situation prevails in Liberia, where mining concessions require parliamentary ratification to be effective. Thus, in Ghana, Kenya, and Liberia, large-scale mining contracts are not valid unless ratified by the Parliament. Parliamentary ratification may manifest in a statute encompassing the contractual terms agreed between the state and the investor. In Liberia, for example, the Parliament has ratified several mining contracts through statutes such as the *Act Ratifying the Amendment to the Mineral Development Agreement between the Government of Liberia and Mittal Steel Holding*.[31] In essence, parliamentary ratification of a contract makes the contract an act of parliament, thus conferring legitimacy and validity on the contract. In this sense, the rights of the mineral rights holder flow in part from the ratifying statute enacted by the Parliament.

3 Time of vesting of mineral rights

The time of vesting of a mineral right is significant with regard to the duration of the mineral right. The question is, does a mineral right vest at the time of approval of an application by the Minister or does the right vest only when the contract evidencing the right is signed or executed by the parties? The answer to this question rests on the source(s) of mineral rights in a given country. The instrument through which the host state grants a mineral right usually indicates the date the mineral right legally vests in the holder, and the time of vesting of a mineral right is the date the mineral right commences. Thus, the duration of a mineral right commences on the date prescribed in the granting instrument.

---

28 *The Constitution of the Republic of Ghana, 1992*, art. 268(1); *Minerals and Mining Act, 2006* (Ghana), s. 5.(4). However, the Parliament may exempt a mineral right transaction from the requirement of parliamentary ratification through a resolution supported by not less than two-thirds of all members of parliament. See *Minerals and Mining Act, 2006* (Ghana), s. 5.(5).
29 *Minerals and Mining Act, 2006* (Ghana), s. 48.(2).
30 *Mining Act, 2016* (Kenya), s. 120.(2); *The Constitution of Kenya, 2010*, s. 71.
31 *Supra* note 17.

Generally speaking, where a mineral right is contract-based, the vesting of the mineral right follows the ordinary principles of the law of contract. Thus, the mineral right vests and accrues on the prescribed 'effective date', which may be the date the contract is executed or a future date mutually agreed by the contracting parties. However, where a contract granting a mineral right is conditional or subject to the fulfilment of a condition precedent, the mineral right vests when the condition precedent is fulfilled. In some instances, the principles of contract law regarding the vesting of mineral rights may be expressly displaced by statutory provisions. A statute may provide that a mineral right shall vest at a time other than the time the mining contract is executed. Such is the case in Sierra Leone where the grant of a mineral right takes effect on the date of registration of the mineral right.[32] In that country, effect is given to a mineral right upon registration "without the need for formal acceptance by or on behalf of the person named therein as the licence holder".[33] In other words, the consent of a mineral right holder is not required in order for the mineral right to be legally effective.

However, where a mineral right is statutory, the vesting of the right may commence prior to the execution of the contract evidencing the mineral right. In South Africa, for example, a prospecting right granted under the MPRDA comes into effect on the "effective date", the "effective date" being the date on which the permit is issued or the date on which the relevant right is executed.[34] In effect, a prospecting right in South Africa may be effective prior to the execution of the contract evidencing the right. The case of *Minister of Mineral Resources & Others v Mawetse (SA) Mining Corporation (Pty) Ltd*[35] illustrates this position. In *Mawetse*, the Supreme Court of Appeal of South Africa held that a prospecting right is granted on the date the Minister of Mineral Resources (or their delegate) approves the recommendation of the Regional Manager that the right be granted.[36] The court elaborated thus:

> The period for which Dilokong's prospecting right endured must in my view be calculated from the date on which it was informed of the granting of the right, namely 18 July 2007. On that date Dilokong became the holder of a valid prospecting right, subject to compliance with the request to prove BEE compliance. It matters not, for purposes of computing the period of the duration of the right, that the right still had to be executed and that the right had not yet become effective. Thus construed, Dilokong's prospecting right had expired due to the effluxion of time on 17 July 2011, ie 4 years after the date on which Dilokong had been notified of the granting of the right.[37]

---

32 *Mines and Minerals Act, 2009* (Sierra Leone), s. 42.(6).
33 *Ibid.* s. 42.(5).
34 *MPRDA*, s. 17.(5) read with s. 1.
35 [2015] ZASCA 82
36 *Ibid.* at para. 19.
37 *Ibid.* at para. 21.

122 *Security of mineral tenure*

The justification for the effectiveness of a prospecting right even before the right is contractually executed is the need to prevent the indefinite sterilization or indeterminate reservation of rights. As the court noted in *Mawetse*, indefinite sterilization of mineral rights "offends one of the MPRDA's key objectives, namely that mineral rights must be exploited within stipulated timeframes for the benefit of the public".[38]

4 The legal nature of mineral rights

4.1 Overview of common law and statutory regimes

At common law, the landowner owns the minerals in and underneath the land. However, a landowner may grant to another person the right to enter the land for the purpose of winning, working, and recovering minerals in or under the land. Such grants are referred to as *profit à prendre* because they confer a right on the grantee to enter the land of the grantor to win, work, and recover minerals from the land. Thus, at common law, a mining lease is an interest in land – that is, a *profit à prendre* – because it vests a right on the lessee to enter land belonging to the lessor for the specific purpose of making a profit from the land.[39] In the words of the House of Lords, a lease "is a liberty given to a particular individual for a specific length of time to go into and under the land and to get certain things there if he can find them and to take them away just as if he had bought so much of the soil".[40] The legal characterization of a mineral right as an interest in land is significant because, as an interest in land, the mineral right runs with the land.[41] Thus, a mining leaseholder is able to enforce their rights under the lease "against any subsequent transferee of the mineral fee, except in the case of a bona fide purchaser for value without notice".[42]

The common law position is buttressed by statutory provisions and case law in Canada and Australia, two prominent mining countries where courts have consistently interpreted mineral leases as *profit à prendre*. For example, the *Mineral Tenure Act* of British Columbia, Canada, provides that a mining "lease is an interest in land and conveys to the lessee the minerals or placer minerals, as the case may be, within and under the leasehold, together with the same rights that the lessee held as the recorded holder of the claim or group of claims",[43] while in

38 *Ibid.* at para. 20.
39 See Robert Rennie, *Minerals and the Law in Scotland* (Hertfordshire: EMIS Professional Publishing, 2001) at 105 (describing a lease as a "contract for the creation of an interest in land"). See also J.R.S. Forbes & A.G. Lang, *Australian Mining and Petroleum Laws*, 2nd Edition (Sydney: Butterworths, 1987) at 184 (noting that "a lease confers on its holder an estate or interest in land").
40 *Gowan v Christie* [1873] L.R. 2 Sc. & Div. 273 (H.L.).
41 Fasken Martineau, *Canadian Mining Law* (from the 2nd Edition of *American Law of Mining*) (New Providence, NJ: LexisNexis, 2012) at 212–4.
42 *Ibid.* at 212–4.
43 *Mineral Tenure Act*, Revised Statutes of British Columbia 1996, Chapter 292, s. 48(2).

Alberta, Canada, "a lease conveys the exclusive right to win, work and recover metallic and industrial minerals that are the property of the Crown (a) within the location".[44] In *Berkheiser v Berkheiser*,[45] the Supreme Court of Canada held that a petroleum and natural gas lease which granted to the lessee "the exclusive right and privilege to explore, drill for, win, take, remove, store and dispose of, the leased substances" amounted to a *profit à prendre*.[46] In *British Columbia v Tener*,[47] the Supreme Court of Canada was asked to determine the legal nature of a mining lease for the purpose of ascertaining the amount of compensation payable for expropriation of the lease. The court held that the mining lease at issue constituted "one integral interest in land in the nature of a *profit à prendre* comprising both the mineral claims and the surface rights necessary for their enjoyment".[48] A similar position was adopted in *Amoco v Potash Corporation*,[49] where the Saskatchewan Court of Appeal held that a lease relating to petroleum, natural gas, and related hydrocarbons could be protected by a registered caveat because "the minerals and substances at issue are capable of being the subject of a *profit à prendre*".[50] In fact, in Canada recorded or registered mineral claims constitute an interest in land even prior to the holder of the recorded claims obtaining a mining lease.[51] The Australian case of *Emerald Quarry Industries Pty Ltd v Commissioner of Highways*[52] held that a mineral lease authorizing the quarrying and removal of stone from private land is an interest in land and a *profit à prendre*. A similar conclusion was reached in *Mills v Stokman*[53] where the court held that a contract authorizing the purchaser to enter private lands "for the purpose of cutting or sawing up or carrying away" slate conferred on the purchaser an interest in the nature of a *profit à prendre*.[54]

Where a mineral right is granted based on statutory provisions, the legal nature of the mineral right must be determined by reference to the statute under which the right was granted.[55] To paraphrase the Supreme Court of Canada, the legal nature of a mineral right must be ascertained through a contextual interpretation of the language of the statute, taking into account the overall purpose of the

44 *Metallic and Industrial Minerals Tenure Regulation*, Alberta Regulation 145/2005, s. 39(1).
45 [1957] S.C.R. 387
46 *Ibid.* at 399.
47 [1985] 1 S.C.R. 533.
48 *Ibid.* at para. 11.
49 (1991), 86 D.L.R. (4th) 700; 1991 CanLII 7951 (Sask. C.A.).
50 *Ibid.* at para. 27 [cited to CanLII 7951].
51 See *Uranerz Exploration and Mining Ltd. v Blackhawk Diamond Drilling Inc.* [1989], 63 D.L.R. (4th) 350; 1989 CanLII 5113 (SKQB) at para. 17 (where the court observed that "it is well settled that the interest of a free miner in his claim is an interest in land").
52 (1976) 14 S.A.S.R. 486 (Supreme Court of South Australia).
53 *Mills & Another v Stokman & Another* (1966) 116 C.L.R. 61.
54 *Ibid.* at 71–2 & 77.
55 Barry Barton, "Property Rights Created under Statute in Common Law Legal Systems", in Aileen McHarg *et al.*, eds, *Property and the Law in Energy and Natural Resources* (Oxford: Oxford University Press, 2010) 80 at 82.

statute.[56] This is a significant observation because, generally speaking, the common law does not preclude the legislature from abrogating common law principles or creating exceptions to common law principles. Thus, in exercising its sovereign powers, the legislature may expressly characterize a mineral right as "property", even though the mineral right is a *profit à prendre* at common law. In effect, the legal characterization of a mining lease as a *profit à prendre* at common law does not prevent the lease "from fitting into other legal pigeon-holes under different circumstances".[57]

Such is the experience in Canada where, for the specific purposes of bankruptcy and insolvency, mineral leases are legislatively classified as 'property', even though at common law these leases constitute a *profit à prendre*. In this regard, Canada's *Bankruptcy and Insolvency Act* defines "property" as

> any type of property, whether situated in Canada or elsewhere, and includes money, goods, things in action, land and every description of property, whether real or personal, legal or equitable, as well as obligations, easements and every description of estate, interest and profit, present or future, vested or contingent, in, arising out of or incident to property.[58]

In *Saulnier v Royal Bank*,[59] a question arose as to whether a fishing licence qualified as 'property' under the *Bankruptcy and Insolvency Act*. Answering this question in the affirmative, the Supreme Court of Canada held that the broad definition of property under the Act "unambiguously signalled" the intention of the Parliament "to sweep up a variety of assets of the bankrupt not normally considered 'property' at common law".[60] Similarly, in *Kasten Energy Inc. v Shamrock Oil and Gas Ltd*,[61] the court held that an "oil and gas lease is a proprietary interest within the purposive contemplation of Alberta's *Personal Property Security Act*",[62] which defines 'personal property' as "goods, chattel, investment property, a document of title, an instrument, money or an intangible".[63]

The above discussion of the situation in Canada and Australia illustrates that mining leases are not monolithic in nature. Rather, the legal nature of mining leases across the common law world is context-specific and depends on the particular circumstances of the lease, including the statutory provisions governing the lease, the terms of the lease, and the intention of the parties to the lease. Thus, a mining lease which would qualify as *profit à prendre* in one context may assume

---

56 *Saulnier v Royal Bank of Canada* 2008 S.C.C. 58; [2008] 3 S.C.R. 166 at paras 16–17.
57 John B. Ballem, *The Oil and Gas Lease in Canada*, 4th Edition (Toronto: University of Toronto Press, 2008) at 18.
58 *Bankruptcy and Insolvency Act*, Revised Statutes of Canada, 1985, c. B-3, s. 2.
59 [2008] 3 S.C.R. 166.
60 *Ibid.* at para. 44.
61 2013 ABQB 63.
62 *Ibid.* at para. 32.
63 *Personal Property Security Act*, Revised Statute of Alberta 2000, c. P-7, s. 1(gg).

the character of a personal property in another context. This is particularly so in circumstances where the legislature provides expressly through a statutory enactment that a mining lease is personal property.

4.2 Statutory regimes on the nature of mineral rights in Africa

Most of the countries covered in this book do not expressly classify the legal nature of mineral rights. To make matters worse, there is a dearth of judicial decisions on the legal nature of mineral rights in Africa. The dearth of judicial decisions is perhaps explained by the fact that mineral exploitation in Africa is conducted mainly on the basis of investment contracts between MNCs and African governments that externalize dispute resolution to international arbitral panels. Mining contracts in Africa contain arbitration provisions to the effect that disputes arising from the contracts shall be referred to international arbitral bodies, such as the International Centre for the Settlement of Investment Disputes and the London Court of International Arbitration. These arbitration clauses deny domestic African courts the opportunity to adjudicate disputes arising from mining operations – hence the dearth of African jurisprudence on the legal nature of mineral rights.

The question is, are mineral rights in Africa proprietary or are they simply an interest in land? Although this question is not amenable to easy answers, it could be said with a reasonable degree of certainty that the legal nature of mineral rights in any given country depends on the mining system, including the statutory and constitutional regimes in the country. Where the mining system is privately oriented (that is, a freehold system), "mineral rights are real rights in property, constituting not merely licences but indeed rights analogous to servitudes or easements and are freely transferable".[64] However, the mining system in Africa is not privately oriented; rather, African countries operate state-oriented mining systems in which title to minerals, being owned exclusively by the state, can be acquired only through licences and leases granted by the state. Thus, the question whether mineral rights in Africa confer a proprietary interest on the holder is one that must be answered through a purposeful interpretation of the statutes under which mineral rights are granted, rather than by reliance on common law principles.[65]

In that regard, the mineral rights granted under mining statutes in Africa exhibit certain characteristics from which the legal nature of the mineral rights can be ascertained. Mineral rights in Africa are granted for a fixed term; they are capable of registration; they are alienable by sale, assignment, or disposal; they are capable of cession, transfer, and mortgage; they are renewable; and they are capable of

---

64 Michael O. Dale, "Security of Tenure as a Key Issue Facing the International Mining Company: A South African Perspective" (1996) 14 *Journal of Energy & Natural Resources Law* 298 at 302 ["Security of Tenure"].
65 *Saulnier v Royal Bank of Canada* (2008) 298 D.L.R. (4th) 193 at para. 16 (S.C.C.).

retention during unfavourable market conditions.[66] Generally speaking, the statutory and constitutional regimes regarding the nature of mineral rights in Africa fall into three distinct categories, namely (1) regimes expressly characterizing mineral rights as 'property'; (2) regimes that define 'property' broadly as including an interest in land; and (3) regimes lacking express classification of mineral rights.

4.2.1 Statutory regimes characterizing mineral rights as property

A few African countries expressly classify the legal nature of mineral rights. In South Africa, the MPRDA provides that a prospecting right and a mining right are "a limited real right in respect of the mineral or petroleum and the land to which such rights relates" if the rights are registered under the *Mining Titles Registration Act, 1967*.[67] In effect, prospecting rights and mining rights in South Africa are recognized as 'property' when they are registered at the Mineral and Petroleum Titles Registration Office pursuant to the *Mining Titles Registration Act, 1967*. As argued by Badenhorst and Mostert, "a real right is only created upon registration" of the mineral rights, at which point they become binding on third parties.[68] Judicial decisions in South African confirm the proprietary nature of mining rights. In *Baleni and Others v Minister of Mineral Resources and Others*, the court held that "in granting a mineral right, the State awards limited real rights in respect of the land to which such mining relates".[69] Similarly, in *Maledu and Others v Itereleng Bakgatla Mineral Resources (Pty) Limited and Another*, the Constitutional Court of South Africa held that

> the MPRDA confers on the holder of a mining right a limited real right in respect of the mineral or petroleum and the land to which such right relates. Moreover, and significantly, it grants to the holder a right of access to the land, even against the wishes of the landowner. The mining right holder is free to enter the land and do everything necessary in the exercise of her right, including constructing or laying down any surface or underground infrastructure, which may be required for the purpose of the mining rights holder's rights.[70]

This position was earlier reiterated in *Agri South Africa v Minister for Minerals and Energy* where the Constitutional Court held that mineral rights were "undoubtedly property with economic value" and that they could "be kept as a

---

66 P.J. Badenhorst, "Security of Mineral Tenure in South Africa: Carrot or Stick?" (2014) 32(1) *Journal of Energy & Natural Resources Law* 5 at 14.
67 *MPRDA*, s. 5.(1).
68 P.J Badenhorst & Hanri Mostert, *Mineral and Petroleum Law of South Africa: Commentary and Statutes* (Lansdowne, South Africa: Juta Law, 2004) at 16–17.
69 *Baleni and Others v Minister of Mineral Resources and Others* [2018] ZAGPPHC 829; [2019] 1 All SA 358 (GP) at para. 46.
70 *Maledu and Others v Itereleng Bakgatla Mineral Resources (Pty) Limited and Another* [2018] ZACC 41 at para. 100.

Security of mineral tenure 127

valuable investment or asset, be bequeathed or mortgaged and constituted property as envisaged by section 25 of the Constitution".[71] However, some mineral rights such as reconnaissance permissions, retention permits, and mining permits are not expressly classified as 'limited real rights' under the MPRDA and hence these rights are said to be personal rights.[72]

In the DRC, mineral rights such as exploration licences, exploitation licences, and small-scale exploitation licences are classified under the Mining Code as "real property" which grant exclusive, conveyable, and transferable rights to the holder.[73] Likewise, Guinea's Mining Code describes an Exploration Permit as conferring "on its holder a movable property right which is undivided" and unassignable.[74] Curiously, Guinea's Mining Code does not refer to other mineral rights – that is, 'Operating Permits' and 'Mining Concessions' – as property rights. Rather, the Code provides that an "Operating Permit creates, in favour of its holder, a movable and divisible right that may be sub-leased",[75] while it describes a Mining Concession as "an immovable, divisible, right which can be sub-leased" and mortgaged in order to secure loans to finance mining operations.[76] It is surprising that Guinea's Mining Code does not expressly classify a Mining Concession as property. This is because a Mining Concession is superior to (or a more fundamental right than) an Exploration Permit which is classified as "property" under the Code. The superiority of a Mining Concession under the Code manifests in the fact that the grant of a Mining Concession automatically cancels any prior mineral rights issued in relation to the area, including the Exploration Permit and the Mining Operation Permit.[77]

4.2.2 Statutory regimes defining 'property' in broad terms to include mineral rights

Some African countries do not statutorily characterize mineral rights as 'property', but the constitutions of these countries define 'property' broadly to include an interest in land. For example, the Bill of Rights enshrined in the constitutions of Kenya and Zambia provide, in identical terms, that "property includes a vested or contingent right to, or interest in, or arising from – (a) land, permanent fixtures on, or improvement to, land; (b) goods or personal property; (c) intellectual property; or (d) money, choses in action or negotiable instruments".[78] Because, as discussed previously, mineral rights qualify as an 'interest in land' at common law,

---

71 *Agri South Africa v Minister for Minerals and Energy* [2013] ZACC 9 at paras 44 & 50.
72 Badenhorst & Mostert, *supra* note 68 at 13–13 & 13–15.
73 *Law No. 007/2002 of July 11, 2002 Relating to the Mining Code* (DRC), arts. 51, 65, & 100 (as amended by *Law No. 18/001 of 9 March 2018*).
74 *Amended 2011 Mining Code* (Guinea), art. 19.
75 *Ibid*. art. 28.
76 *Ibid*. art. 35.
77 *Ibid*. art. 36.
78 *The Constitution of Kenya, 2010*, art. 260; *Constitution of Zambia (Amendment) No. 2 of 2016*, art. 266.

128 *Security of mineral tenure*

the constitutional definition of property as including an interest in land means that mineral rights fall within the meaning of property under the Kenyan and Zambian Constitutions.[79]

Although 'property' is defined broadly under these constitutions as including an interest in land, for the purpose of security of mineral tenure, this broad definition must be analyzed in the context of the constitutions and, in particular, the objectives of the constitutions. Property is defined broadly in these constitutions in order to protect owners of property against unlawful deprivation or expropriation of their property rights. Thus, mineral rights, which are ordinarily an interest in land at common law, are constitutionally elevated to the status of 'property' for the specific purpose of protecting mineral right holders against unlawful seizure, deprivation, or expropriation by the government. In effect, mineral rights in these countries are not 'property' for all purposes, but they are property in the context of constitutional protection against unlawful expropriation. This argument is better captured within the context of alienation of mineral rights. As discussed in Chapter 2, mineral rights in Africa can be transferred, sold, assigned, or otherwise encumbered, albeit with the prior approval of the Minister responsible for mineral development. When a mineral right holder sells or assigns their right to a third party, in reality what is sold or assigned is the holder's interest in the minerals – that is, an interest in land. The holder cannot be said to have sold a 'property' because, in these countries, the holder's interest is not statutorily classified as 'property'.

4.2.3 Statutory regimes lacking characterization of mineral rights

The third category consists of countries where mineral rights are not expressly classified as property and the constitution, though protective of property rights, does not provide a definition for 'property'. The non-classification of mineral rights as 'property' in common law countries such as Ghana, Nigeria, and Tanzania means that the common law position prevails. Thus, in these countries, mineral rights constitute an interest in land and, hence, a *profit à prendre*. However, some of these countries describe mineral resources as 'property' of the government. They recognize minerals as 'property' only when ownership of the minerals vests in the government, but they do not extend the proprietary description to the holders of mineral rights.

More specifically, statutory and constitutional provisions in some African countries indicate clearly that yet-to-be-recovered minerals (that is, minerals *in situ* or in place) are proprietary in nature, but ownership of the property in the minerals vests exclusively in the government in trust for their citizens. For example, the *Nigerian Minerals and Mining Act, 2007*, provides:

> The entire *property* in and control of all mineral resources in, under or upon any land in Nigeria, its contiguous continental shelf and all rivers, streams and water courses throughout Nigeria, any area covered by its territorial waters or

---

79 See Ayisi, "Legal Character of Mineral Rights", *supra* note 14 at 35.

constituency and the Exclusive Economic Zone is and shall be vested in the Government of the Federation for and on behalf of the people of Nigeria.[80]

The *Constitution of the Federal Republic of Nigeria* affirms this position by providing that

> the entire *property* in and control of all minerals, mineral oils and natural gas in, under or upon any land in Nigeria or in, under or upon the territorial waters and Exclusive Economic Zone of Nigeria shall vest in the Government of the Federation.[81]

In Ghana, every mineral in, under, or upon land, rivers, streams, and watercourses is "the *property* of the Republic" and is vested in the government in trust for the citizens of Ghana.[82] Similarly, in Tanzania the "entire *property* in and control of all minerals in, and under or upon any land, rivers, streams [and] water courses" is vested in the government.[83] In essence, in these countries minerals *in situ* are legislatively characterized as proprietary. The description of minerals *in situ* as the 'property' of the government is premised on the fact that natural resources are the common heritage of African citizens and, as such, the state reserves for itself the proprietary interest in mineral resources to the exclusion of all other persons.

However, the proprietary interest held by the government in unrecovered minerals in or under land in these countries does not pass to the lessee at the time of execution of the lease. Rather, the proprietary interest remains vested in the government, but is passed to the lessee upon the recovery of minerals.[84] In this regard, the *Nigerian Minerals and Mining Act, 2007*, provides that "[t]he property in mineral resources shall pass from the government to the person by whom the mineral resources are lawfully won, upon their recovery in accordance with this Act".[85] Because 'property' in the minerals vests in the mineral right holder only upon the recovery of minerals in these countries, it is arguable that, prior to the recovery of minerals, the mineral right holder holds an interest in land – that is, an interest in the minerals *in situ*.[86] The idea that mineral rights are an interest in

---

80 *Nigerian Minerals and Mining Act, 2007*, s. 1.(1) [emphasis added].
81 *The Constitution of the Federal Republic of Nigeria 1999*, s. 44.(3) [emphasis added].
82 *Minerals and Mining Act, 2006* (Ghana), s. 1 [emphasis added].
83 *Mining Act, 2010* (Tanzania), s. 5A.(1) (introduced by the *Written Laws (Miscellaneous Amendments) Act, 2017* (No. 7), s. 6) [emphasis added].
84 Latifa Ibraimo, "Comparative Mineral Law in African Portuguese-Speaking Countries", in Elizabeth Bastida, Thomas Wälde, & Janeth Warden-Fernández, eds, *International and Comparative Mineral Law and Policy: Trends and Prospects* (The Hague: Kluwer Law International, 2005) 887 at 897 (asserting that "contrary to what may happen in other jurisdictions where land is privately owned, the issuing of the mineral title [in Africa] does not imply any transfer of ownership on the resources").
85 *Nigerian Minerals and Mining Act, 2007*, s. 1.(3).
86 For a contrary argument, see Martin K. Ayisi, "Ghana's New Mining Law: Enhancing the Security of Mineral Tenure" (2009) 27 *Journal of Energy & Natural Resources Law* 66 at 88–9 (arguing that "mineral rights are regarded as property" in Ghana).

land is buttressed by domestic land law in Africa, which requires any written instrument affecting land to be registered as an interest in land. Mineral rights often manifest in the form of contracts covering a delineated piece of land and, as such, they require registration as an interest in land. In sum, in common law African countries falling into this category, mineral rights are an interest in land – that is, a *profit à prendre*. This is because mineral rights, such as a mining lease, grant the holder a right to enter land belonging to the government for the purpose of exploiting minerals situated in the land.

5 Security of mineral tenure

The term 'security of mineral tenure' has both narrow and broad meanings. In the narrow sense, security of mineral tenure means the mineral right holder's 'right to mine' – that is, the right to proceed from the exploration phase to the mining stage – and it encompasses "the notion that the investor has to be provided with the assurance of being able to develop a successful discovery prior to committing sizeable resources to exploration".[87] As noted by the World Bank, "[t]he investor needs to be assured of the right to the minerals and of the right to proceed from exploration to mining, provided pre-defined criteria are met".[88] In addition, a mineral right "must be of sufficient duration and security to make the [investor's] exploration and development commitment worthwhile".[89] This narrow conception of security of mineral tenure focuses on the host state's assurances and guarantees that mineral right holders will be afforded unhindered opportunity to prospect for minerals and subsequently convert their prospecting right into a mining right. In this narrow sense, then, security of mineral tenure is the legal entitlement of the holder to transit from the discovery of minerals to the mining of minerals.[90]

In the broader sense, security of mineral tenure refers to the stability, continuity, and protection of mineral rights during the three phases of mining – that is, exploration, development, and mining of minerals.[91] James Otto, a principal proponent of the broad conception of security of mineral tenure, identifies three critical questions in relation to the security of mineral tenure. First, are the mineral rights secure for a period long enough to determine whether the allotted ground contains commercially interesting ore bodies? Second, once an ore body is discovered, is the mineral right holder able to convert their prospecting licence into a mining lease and what is the time limit during the transition phase between

---

87 Elizabeth Bastida, "A Review of the Concept of Security of Mineral Tenure: Issues and Challenges" (2001) 19 *Journal of Energy & Natural Resources Law* 31 at 32.
88 World Bank, *Strategy for African Mining* (World Bank Technical Paper Number 181) at 23, http://documents.worldbank.org/curated/en/722101468204567891/pdf/multi-page.pdf
89 *Ibid*. at 23.
90 Badenhorst, *supra* note 66 at 12; Bastida, *supra* note 87 at 35.
91 James Otto, "Security of Mineral Tenure: Time-Limits" in Bastida *et al*., eds., *supra* note 84 at 362.

exploration and mining? Third, having been granted a mining right, how may the mining right be lost through government action or expiry of the grant?[92] Answers to these questions provide a complete picture of the security of mineral tenure in any given country. Underlying Otto's questions is the notion that mineral tenure should be guaranteed during the exploration, production, and mining phases of a mining project. In effect, the host state should provide reasonable assurances and guarantees regarding the continuity of operations over the life of a mining project.[93] In that regard, Omalu and Zamora identify two phases of security of mineral tenure. The first phase requires "that there should be reasonable legal entitlement for extraction rights after successful completion of the exploration phase", while the second phase involves certainty of mineral rights as well as the conditions under which the rights may be revoked or lost.[94]

In sum, the broad conception of security of mineral tenure consists of several component parts, including the duration of mineral rights; renewal of mineral rights; convertibility of exploration rights into production rights; transferability of mineral rights; guarantees against expropriation and nationalization; and the specificity and preciseness of the rules governing termination of mineral rights, including the grounds for the revocation of mineral rights. This book adopts the broad conception of security of mineral tenure.

Security of tenure ranks high among the factors that mining companies take into account when deciding whether to invest in a country.[95] Investors need to be assured that their investment is safe in terms of their physical assets and the duration of their mineral rights. Mining companies desire security of mineral rights for a period long enough to explore for minerals and, in the event of a commercial find, produce the minerals.[96] In fact, "few investors will be prepared to incur significant exploration expenditures without assurance that they can proceed to mine if they prove a deposit" of minerals.[97] Security of tenure provides an incentive for investors to invest in capital-intensive and highly risky mining projects, particularly in developing countries with unstable and unpredictable political climates. As the World Bank has observed, "[s]ecurity and continuity of tenure of mineral rights is essential if there is to be sufficient incentive to undertake high-risk exploration with substantial work commitments and then marshal the large sums necessary for mine development".[98]

African countries have enacted statutory provisions guaranteeing security of mineral tenure, albeit to varying degrees. A common feature of modern mining statutes in Africa is the enhancement of security of tenure for mineral right

92 Otto, *ibid*. at 362–3.
93 Mirian K. Omalu & Armando Zamora, "Key Issues in Mining Policy: A Brief Comparative Survey on the Reform of Mining Law" (1999) 17(1) *Journal of Energy & Natural Resources Law* 13 at 27.
94 *Ibid*. at 27.
95 Otto, *supra* note 91 at 363.
96 *Ibid*. at 363.
97 World Bank, *Strategy for African Mining*, *supra* note 88 at 24.
98 *Ibid*. at 23.

holders. For example, as discussed in Chapter 2, mineral right holders in Africa are statutorily empowered to transfer, sell, assign, or otherwise encumber their mineral rights, provided that the prescribed conditions precedent are fulfilled. Moreover, in countries such as South Africa, a principal objective of the mining regime is to "provide for security of tenure in respect of prospecting, exploration, mining and production operations".[99] In the ensuing pages, I discuss the various manifestations of security of mineral tenure in Africa, including the long-term duration of mineral rights, transition of mineral rights, statutory and contractual guarantees, and the constitutional protections accorded mineral right holders.

5.1 Curtailment of ministerial power and discretion

The degree to which discretion is vested in the Minister and rights-granting agencies directly impacts the security of mineral tenure in any country. The broader the discretion vested in the Minister, the greater the likelihood that mineral tenure would be adversely impacted by the improper use of ministerial discretion. This is particularly so in developing countries lacking institutional and administrative expertise. Conversely, the curtailment of ministerial discretion enhances the security of mineral tenure.[100] In much of Africa, mineral rights are granted, renewed, or terminated by the Minister based on precise statutory rules. These rules oblige the Minister to make informed decisions, including issuing their decisions in writing; supporting their decisions with written reasons; making decisions within the stipulated timeframe; and acting in accordance with the principles of lawfulness, reasonableness, and procedural fairness.[101]

The preciseness of the rules and procedure governing the grant and renewal of mineral rights has a direct impact on the security of mineral rights. The more precise the rules are, the greater the likelihood that security of tenure is enhanced in a given country. Given the long gestation period between the exploration and production stages, investors need assurances regarding the ability to renew their mineral rights, including the conditions under which renewals are granted. Investors want to know whether the renewal of mineral rights is automatic upon the fulfilment of prescribed conditions, or whether administrative discretion is vested in the Minister regarding the renewal of mineral rights.

Modern mining statutes in Africa severely curtail the discretionary power of the Minister and rights-granting agencies by providing clear rules for administrative decisions. For example, mineral rights may be renewed by way of an automatic right of renewal vested in the holder under the mining statute or mining agreement.[102] In some African countries, the Minister is obliged to renew mineral

---

99 *MPRDA*, s. 2.(g).
100 Badenhorst, *supra* note 66 at 15.
101 *Ibid.* at 22.
102 Otto, *supra* note 91 at 368.

rights if the mineral right holder has satisfied the statutory criteria for the renewal of mineral rights. Thus, for the most part, the Minister has little discretion regarding the renewal of mineral rights. In Botswana, for example, the holder of a prospecting licence is entitled to the renewal of their licence provided they are not in default of the terms of the licence and the proposed programme of prospecting operations is adequate.[103] Similarly, "the Minister shall grant an application for renewal" of a mining licence if they are satisfied that the holder has complied with the terms of the licence.[104] In Ghana, the Minister 'shall' grant a renewal of a prospecting licence if the holder has complied with statutory requirements.[105] Likewise, in South Africa the "Minister must grant the renewal of" a prospecting right or a mining right if the holder has satisfied the statutory requirements.[106] Moreover, in some African countries applications for the renewal of mineral rights must be determined within the time stipulated in the statute or regulations. For example, in South Africa, any administrative process or decision regarding mineral rights must be undertaken "within a reasonable time and in accordance with the principles of lawfulness, reasonableness and procedural fairness".[107] Similarly, Zambia requires the Director of Mining Cadastre to approve an application for deferment of mining operations within 60 days of the receipt of the application.[108] The specificity of the timeframe for the renewal of mineral rights improves the security of mineral tenure in Africa.[109]

Moreover, in some countries the duration of a mineral right is automatically extended in situations where there is a delay by the Minister in granting a renewal of the mineral right. In effect, mineral tenure is preserved where the Minister fails to act on a renewal application within a reasonable time. In Ghana, for example, prospecting licences and mining leases are deemed automatically extended until the merits of an application for their extension is determined.[110] Thus,

> [w]here the holder has made an application for an extension of the term of the lease, and the term of the lease would but for this subsection, expire, the lease shall continue in force in respect of the land the subject of the application until the application is determined.[111]

Likewise, in Nigeria, Tanzania, and Zambia the duration of a mineral right may be extended by operation of law. In these countries, upon the submission of an application for the renewal of a mineral right, the extant mineral right is deemed

---

103 See *Mines and Minerals Act, 1999* (Botswana), s. 17.(3).
104 *Ibid*. s. 42.(4).
105 *Minerals and Mining Act, 2006* (Ghana), s. 35.(2).
106 *MPRDA*, ss.18.(3) & 24.(3)
107 *Ibid*. s. 6.(1).
108 *Mines and Minerals Development Act, 2015* (Zambia), s. 33.(3).
109 Badenhorst, *supra* note 66 at 14.
110 *Minerals and Mining Act, 2006* (Ghana), ss. 35.(4) & 44.(4).
111 *Ibid*. s. 44.(4).

134 Security of mineral tenure

to continue in force until the date the renewal is granted or until the application is rejected.[112]

The security of mineral tenure is also apparent in the mandatory language of mining statutes in Africa with regard to the duration of mineral rights. Mining statutes in Africa enhance security of tenure through provisions prescribing long durations for mineral rights. The duration of mining leases in Africa ranges from ten years (for small-scale mining) to 30 years (for large-scale mining).[113] In the vast majority of African countries, mineral rights are granted for a fixed term and, as discussed earlier, these rights are renewable under clear criteria. In some instances, the duration of a mineral right is not predetermined by law. Rather, the duration coincides with the lifespan of the mining operations. In Tanzania, for example, the duration of a special mining licence coincides with the estimated life of the mineral deposit indicated in the applicant's feasibility study report, or such period as requested by the applicant, whichever period is shorter.[114]

The mandatory language of these statutory provisions not only prevents administrative delays within rights-granting agencies, but also constrains arbitrary exercise of discretion, thus ensuring security of tenure.[115] Moreover, the clarity and preciseness of the rules and conditions for renewal curtail ministerial discretion across the continent. These modern statutory provisions contrast sharply with provisions under old and repealed mining statutes in Africa which conferred overly broad and unhindered discretion in the Minister with the result that ministerial discretion was often abused to the detriment of mineral right holders.[116]

That said, some mining statutes in Africa vest a significant level of discretion in the Minister. In Tanzania, for example, the Minister possesses broad discretion to reject an application for the renewal of a special mining licence on the basis of administrative and economic factors, including the applicant's failure to satisfy any relevant stipulation in a development agreement previously entered between the applicant and the government of Tanzania,[117] or the applicant's failure to conduct development of the mining area with reasonable diligence.[118] A more unsatisfactory position is found in Zambia where the *Mines and Mineral Development Act, 2015* does not stipulate any specific conditions for the renewal of an exploration licence, thus leaving room for broad and open-ended discretion on the part of the licensing authority.[119] While it is reasonable to speculate that a renewal would be granted where the holder of the licence is not in breach of the conditions of the

---

112 *Mines and Minerals Development Act, 2015* (Zambia), s. 76.(3); *Nigerian Minerals and Mining Act, 2007*, s. 148; *Mining Act, 2010* (Tanzania), s. 67.
113 See Table 2.1 in Chapter 2.
114 *Mining Act, 2010* (Tanzania), s. 43.
115 See Michael Dale, "Comparative International and African Mineral Law as Applied in the Formation of the New South African Mineral Development Legislation", in Bastida, *et al.*, eds, *supra* note 84, 823 at 833.
116 See Ayisi, "Ghana's New Mining Law", *supra* note 86 at 78.
117 *Mining Act, 2010* (Tanzania), s. 45.(5)(a).
118 *Ibid.* s. 45.(5)(b).
119 See *Mines and Minerals Development Act, 2015* (Zambia), s. 24.

licence or the provisions of the Act, this broad discretion is susceptible to abuse and, in fact, may encourage corruption within the Zambian mining industry.

In order to forestall abuse of ministerial discretion, some countries require the Minister to afford the mineral right holder an opportunity to rectify any default prior to the refusal of an application for renewal of mineral rights. For example, Tanzania requires that, prior to rejecting an application for renewal, the Minister must issue a written notice to the applicant giving the particulars of the default and requiring the applicant to remedy the default within a specified period.[120] In cases where it is impossible to remedy a default, the Minister may, in lieu of remedial action by the applicant, require the applicant to pay reasonable compensation for the default.[121] Perhaps more significantly, some modern mining statutes enable mineral right holders to challenge discretionary decisions reached by rights-granting authorities through judicial review. In Nigeria, for example, the Minister's rejection of an application for a mining lease, or the Minister's failure to determine such application within the prescribed period of 45 days, may be challenged at the Federal High Court.[122] The Federal High Court can compel the Minister to grant the lease within seven days of the court's decision if it determines that the Minister erroneously exercised their discretion.[123] Moreover, mineral rights holders in Nigeria and Tanzania are statutorily empowered to institute legal proceedings challenging the revocation of a mineral right.[124] Relatedly, some statutes allow mineral right holders to curtail ministerial discretion through contractual provisions. For example, in Ghana an investor who proposes an investment exceeding US$500 million may enter into a mineral development agreement containing provisions "relating to the circumstance or manner in which the Minister will exercise a discretion conferred by or under this Act".[125]

Although there is potential for abuse of ministerial discretion, it is appropriate for African countries to vest modest discretionary powers in the Minister or rights-granting authorities. A modest level of administrative discretion is required in order to ensure the attainment of the policy objectives informing modern mining statutes in Africa. As discussed in Chapter 1, these statutes were enacted for the purpose of harnessing Africa's mineral resources for the economic and social advancement of African citizens. These statutes aim to promote equitable access to mineral resources on the part of citizens, including the creation of linkages between the mining industry and other sectors of the economy. These policy objectives may not be achieved if mining companies are given a blank cheque to operate as they please. Thus, African countries must retain a modest level of control over the activities of mining companies and other economic actors within their territories. Such control manifests in the form of administrative discretion vested in

---

120 *Mining Act, 2010* (Tanzania), s. 45.(5)(a).
121 *Ibid.* s. 45.(5)(a).
122 *Nigerian Minerals and Mining Regulations 2011*, reg. 57.(20).
123 *Ibid.* reg. 57.(21)
124 *Nigerian Minerals and Mining Act, 2007*, s. 156 read with the *Nigerian Minerals and Mining Regulations 2011*, reg. 97.(6); *Mining Act, 2010* (Tanzania), s. 65.(2).
125 *Minerals and Mining Act, 2006* (Ghana), s. 49.(2b).

136 *Security of mineral tenure*

the Minister and rights-granting agencies. However, ministerial discretion must be exercised reasonably and in a transparent and accountable manner if Africa is to achieve the stated objective of economic advancement through the exploitation of mineral resources.

5.2 Convertibility and transition of mineral rights

Mining statutes in Africa guarantee the conversion and transition of mineral rights from the exploration phase to the production phase. Akin to the right to automatic conversion of a prospecting right into a mining right in Latin American countries, some African countries vest a statutory right in holders of exploration licences to convert their right into a mining right once they discover commercially viable ore. More specifically, the holder of an exploration licence who complies with statutory and regulatory requirements is guaranteed, upon application, the issuance of a mining lease. For example, in South Africa the holder of a prospecting right has "the exclusive right to apply for and be granted a mining right in respect of the mineral and prospecting area in question".[126] In Tanzania, mineral right holders classified as "entitled applicants" have a statutory right regarding the convertibility of their mineral rights. Holders of prospecting licences and retention licences in Tanzania are "entitled", on application to the Minister, to the grant of a special mining licence in relation to the area covered by their licences.[127] In Nigeria, the holder of an exploration licence who has complied with the conditions attached to the licence, as well as the requisite statutory and regulatory requirements, is "entitled to the grant of a mining lease for any mineral for which he was authorised to explore" under the licence.[128] Furthermore, the holder of an exploration licence who has complied with their obligations under the Act "has the exclusive right to apply for, and be granted" a mining lease in respect of the exploration area or parts of the exploration area.[129] In fact, Angola's mining code anticipates the fusion of the three phases of mining into one investment contract by requiring that the rules, rights, and obligations governing the three phases of mining (that is, prospecting, exploration, and production) "shall be provided for in a single investment contract".[130] In effect, an investor can be granted a prospecting right, an exploration right, and a mineral lease at once, thus ensuring automatic progression through the three phases.

5.3 Retention of mineral rights

A critical component of the concept of security of mineral tenure is the ability of mineral rights holders to retain their mineral rights even where they are temporarily

---

126 *MPRDA*, s. 19.(1b).
127 *Mining Act, 2010* (Tanzania), s. 39.(1).
128 *Nigerian Minerals and Mining Act, 2007*, s. 60.(2).
129 *Ibid.* s. 61.(4).
130 *Mining Code* (Angola), art. 113.(2).

unable to produce minerals from the lease area. Mineral production may be impracticable under certain circumstances, including the holder's inability to raise requisite capital, civil unrest and strife in the lease area, and unfavourable market conditions, such as a sharp drop in commodity prices rendering mining operations uneconomical. To cater to such situations, several African countries, including Botswana, Ethiopia, Kenya, Namibia, South Africa, South Sudan, and Tanzania, provide for a retention licence to enable mineral right holders to retain their mineral rights in the circumstances mentioned above.[131]

Other African countries, though not expressly providing for a retention licence, have enacted liberal provisions permitting mineral right holders to delay commencement of production or suspend production of minerals due to temporary adverse market or economic conditions. In Zambia, for example, mining operations may be deferred on grounds that "the mineral deposit cannot be developed immediately due to adverse economic conditions or technological constraints, or both, which are, or may be, of a temporary nature".[132] Similarly, in Ghana mineral right holders can suspend production of minerals for reasons beyond their control, including unfavourable market conditions, provided they give three months' written notice to the Minister and obtain the Minister's approval.[133] These provisions regarding retention licence and suspension of production provide statutory assurances that mineral tenure is secure even where prevailing market conditions make it impracticable to produce minerals. These retention provisions ensure continuity of mineral tenure in that even though the mineral right holder is unable to produce minerals over an extended period, they retain their mineral right over the area covered by the retention licence until the emergence of market circumstances suitable for the production of minerals.

5.4 Statutory and contractual guarantees regarding security of tenure

The security of mineral tenure may be expressly preserved through assurances and guarantees contained in mining statutes, state-investor contracts, and mining leases. These guarantees cover a broad range of subjects, including the stability of the legal and fiscal regimes governing mining projects, safety of investments, investors' right to peaceful enjoyment of their mineral rights, guarantees against expropriation and nationalization of mineral rights, and guarantees regarding investors' right to regulatory approvals.

The security of mineral tenure is strengthened by stabilization clauses in mining statutes and contracts in Africa. As discussed in Chapter 6, mining contracts in Africa usually contain 'stabilization clauses' designed to protect and preserve the rights of mineral right holders. Stabilization clauses ensure the stability of the legal regimes governing a project, thus circumscribing arbitrary changes to the law. In addition to stabilization clauses, African governments provide broad guarantees

---

131 See Chapter 2, Table 2.1.
132 *Mines and Minerals Development Act, 2015* (Zambia), s. 33.(1)(b).
133 *Minerals and Mining Act, 2006* (Ghana), s. 51.

138 *Security of mineral tenure*

regarding the safety of investments. In Mozambique, for example, the *Mining Law No. 20/2014* states:

> The State assures the safety and legal protection of the ownership of goods and rights, including industrial property rights covered by the authorized and carried out investments under the mining permit issued in accordance with this Law and other applicable legislation.[134]

Likewise, Angola provides an elaborate and wide-ranging guarantee scheme to investors in the mining industry, including statutory guarantees that mineral right holders shall be allowed to exploit "mineral resources revealed during prospecting, without any restrictions, except those that expressly arise out of the regulations in this Code or complementary legislation".[135]

In some African countries, the security of mineral tenure is further guaranteed through investment statutes. In Nigeria, for example, the *Nigerian Investment Promotion Commission Act* guarantees against expropriation of an enterprise or interest in an enterprise. More specifically, the Act provides that "no enterprise shall be nationalized or expropriated by any Government of the Federation" and that

> [t]here shall be no acquisition of an enterprise to which this Act applies by the Federal Government, unless the acquisition is in the national interest or for a public purpose and under a law which makes provision for (a) payment of fair and adequate compensation; and (b) a right of access to the courts for the determination of the investor's interest or right and the amount of compensation to which he is entitled.[136]

The security of mineral tenure in Africa is equally enhanced through contractual guarantees regarding the peaceful enjoyment of mineral rights. To this effect, a mineral development agreement in Liberia provides that:

> The Government hereby warrants and defends the Company's title to, possession and peaceful enjoyment of, all rights granted to it by the Government under this Agreement, including its right to all land and property in Liberia in accordance with applicable law.[137]

Similarly, a mining contract in Guinea provides that "[s]ubject to what is provided for in this Convention, the Investor and Sub contractors have the exclusive right

---

134 *Mining Law No. 20/2014, dated 18 August* (Mozambique), art. 66.(1).
135 *Mining Code* (Angola), art. 91.
136 *Nigerian Investment Promotion Commission Act*, Chapter N117, LFN 2004, s. 25.(1) & (2).
137 *Mineral Development Agreement between the Government of the Republic of Liberia, Potu Iron Ore Mining, Inc., and Mano River Iron Ore Ltd* (Dated September 2, 2010), s. 18.6 ["*Potu Agreement*"].

and the complete freedom to possess, maintain, enjoy, use and dispose of all their assets, rights, titles and interests" in the mining project.[138] The contract provides further that "[t]he State undertakes not to interfere with the full enjoyment by the Investor and Sub contractors, of the legal rights that they possess over their assets, rights, titles and interests".[139] The right to peaceful enjoyment of mineral rights is reinforced by contractual provisions enjoining the host government not to derogate from its obligations under the contract. A mining contract in Liberia, for example, provides that:

> The Government undertakes and affirms that at no time shall the rights (and the full and peaceful enjoyment thereof) granted by it under this Agreement be derogated from or otherwise prejudiced by any law or by the action or inaction of the Government, the Minister or any other official of the Government, or any other Person whose actions or inactions are subject to the control of the Government.[140]

Relatedly, in recognition of the fact that delays in the administrative process could adversely impact the security of mineral tenure, some mining contracts in Africa contain provisions regarding a right to regulatory approvals. These provisions oblige host governments to be diligent in granting approvals for project implementation and execution. For example, a mining contract in Liberia provides that:

> The Company shall have the right to receive all approvals and consents that may be required from the Government (including any local municipalities and state owned or controlled corporations or entities) pursuant to this Agreement or its Operations without delay, and such approvals and consents shall not be unreasonably withheld.[141]

Similarly, a mining contract in Guinea requires the state "to provide all the authorisations necessary for the exercise of the rights and the guarantees provided for by the Convention".[142] A 'right to approval' clause may be couched in the form of an undertaking on the part of the host government. For example, an extant agreement in Liberia provides that the "[g]overnment undertakes and affirms that it shall issue all licences, permits, mining titles, easements, and other

---

138 *Basic Convention between the Republic of Guinea, Bellzone Mining Plc and Bellzone Holdings S.A. for the Development, Processing, Treatment, Transformation, Transport and Commercialization of Kalia Ore Deposits and Related Infrastructure* (4 August 2010), art. 44.1 ["*Bellzone Agreement*"].
139 *Ibid.* art. 44.3.
140 *Mineral Development Agreement between the Republic of Liberia and Amlib United Minerals Incorporated* (March 14, 2002), art. 18.9 ["*Amlib Agreement*"].
141 *Mineral Development Agreement between the Government of the Republic of Liberia, BHP Billiton (Liberia) Inc. and BHP Billiton Iron Ore Holdings Pty Ltd* (dated September 16, 2010), art. 19.10.
142 *Bellzone Agreement, supra* note 138, art. 41.4.

140 *Security of mineral tenure*

authorizations" which are or may be necessary for the mineral right holder to conduct mining operations.[143] These contractual provisions regarding the investor's right to regulatory approvals could prevent administrative delays during the project's implementation phase, thus enhancing the security of mineral tenure.

5.5 *Constitutional guarantees regarding protection of property*

The security of mineral tenure is significantly enhanced in African countries that have enacted statutory provisions characterizing mineral rights as proprietary rights, as well as in countries where the constitution defines property broadly enough to include mineral rights. Constitutional provisions in Africa protect against arbitrary deprivation of property; prohibit enactment of laws that permit the state to deprive investors of property in an arbitrary manner, or laws that limit or restrict the full enjoyment of property rights; guarantee that the state will observe its obligations under investments contracts; and protect against unlawful termination or expropriation of mineral rights. For example, Tanzania's Constitution protects the right to property and makes it "unlawful for any person to be deprived of his property for the purposes of nationalization or any other purposes without the authority of law which makes provision for fair and adequate compensation".[144] Likewise, the Constitution of Zambia provides that:

> Except as provided in this Article, no property of any description shall be compulsorily taken possession of, and no interest in or right over property of any description shall be compulsorily acquired, unless by or under the authority of an Act of Parliament which provides for payment of adequate compensation for the property or interest or right to be taken possession of or acquired.[145]

Similarly, the *Constitution of the Republic of South Africa, 1996* provides that:

> 25.(1) No one may be deprived of property except in terms of law of general application, and no law may permit arbitrary deprivation of property.
> (2) Property may be expropriated only in terms of law of general application
>   (a). for a public purpose or in the public interest; and
>   (b). subject to compensation, the amount of which and the time and manner of payment of which have either been agreed to by those affected or decided or approved by a court.[146]

---

143 *Amlib Agreement, supra* note 140, art. 18.6.
144 *The Constitution of the United Republic of Tanzania, 2005*, art. 24. See also *The Constitution of Kenya, 2010*, art. 40; *The Constitution of Zambia, 1991*, art. 16.
145 *The Constitution of Zambia, 1991*, art. 16. This section is not affected by the *Constitution of Zambia (Amendment) Act, No. 2 of 2016*.
146 *The Constitution of the Republic of South Africa, 1996*, art. 25.

These constitutional provisions not only protect against undue termination or expropriation of mineral rights, but they ensure that due process is observed by host governments in the course of terminating or expropriating mineral rights. Thus, where the host government fails to observe due process, an aggrieved mineral right holder can invoke these constitutional provisions to challenge the validity of the termination or expropriation of their mineral right.

5.6 Legal safeguards against arbitrary revocation of mineral rights

Legal safeguards against unlawful revocation of mineral rights are essential to the security and continuity of mineral rights. Such legal safeguards are provided through multiple avenues, including mining statutes, investment contracts, and the bills of rights under African constitutions. We have previously discussed the constitutional protection accorded to mineral right holders against unlawful deprivation and expropriation in some African countries. In addition, mining statutes and state-investor contracts in Africa contain specific provisions regarding the grounds on which mineral rights may be suspended or revoked by host governments. These provisions impact not only the security of mineral tenure but the ability of mining companies to finance mining projects through loans. If the grounds for revocation of mineral rights are unspecified, "there is a danger that the investor will have no asset by way of mining rights to pledge as security for finance".[147]

The statutory grounds for revoking mineral rights are fairly uniform across Africa and, in some cases, mining statutes in African countries adopt similar wording and phrasing. As discussed in Chapter 7, a mineral right may be revoked if the holder contravenes the mining statute; breaches any material term or condition of the mineral right; fails to meet their financial obligations such as payment of royalties and taxes; or becomes ineligible to hold the mineral right.[148] The lawful revocation of a mineral right extinguishes the rights of the holder, but such revocation does not extinguish the liabilities and obligations incurred by the holder prior to the revocation.[149]

The statutory provisions regarding revocation of mineral rights vests discretionary power in the Minister, given that the Minister is ultimately responsible for determining whether any of the grounds for revocation applies in any given instance. However, unfettered discretionary power offers a "scope for corruption" and could hinder the preservation and continuity of mineral tenure in Africa.[150] In recognition of this possibility, some mining statutes in Africa provide specific legal

---

147 World Bank, *Strategy for African Mining*, supra note 88 at 25.
148 See *MPRDA*, s. 47.(1); *Mines and Minerals Act, 1999* (Botswana), s. 76.(1); *Minerals and Mining Act, 2006* (Ghana), s. 68.(1); *Mines and Minerals Development Act, 2015* (Zambia), s. 72.(1); *Nigerian Minerals and Mining Act*, 2007, s.151; *Mining Act, 2010* (Tanzania), s. 63; *Mining Act, 2016* (Kenya), s. 147.(1).
149 *Mines and Minerals Development Act, 2015* (Zambia), s. 72.(7) & (10); *Nigerian Minerals and Mining Act*, 2007, s.155; *Mining Act, 2010* (Tanzania), s. 63.(7); *Minerals and Mining Act, 2006* (Ghana), s. 68.(3); *Mining Act, 2016* (Kenya), s. 148.
150 Dale, "Security of Tenure", *supra* note 64 at 307.

safeguards against abuse of ministerial discretion. The first safeguard is the requirement of notice prior to the exercise of the Minister's power of revocation. Mining statutes in Africa require the Minister to give a written notice to the mineral right holder specifying the specific failures, breaches, or contraventions prompting the notice, as well as giving the mineral right holder a reasonable time or opportunity (at least 30 days in Botswana and Nigeria, 60 days in Zambia, and not less than 120 days in the case of a mining lease or 60 days in the case of other mineral rights in Ghana) to remedy the failures or contraventions.[151] The Minister cannot suspend or revoke a mineral right where the holder takes remedial action within the period specified in the written notice. The Minister's power to suspend or revoke a mineral right is confined to situations where the mineral right holder fails to remedy the breach specified in the Minister's notice. Even in such cases, some countries afford the Minister some flexibility in exercising their discretion. In Ghana, for example, in situations where the breach cannot be remedied, the holder may yet retain the mineral right if they "show cause to the reasonable satisfaction of the Minister why the mineral right should not be suspended or cancelled".[152]

Second, some countries afford mineral right holders a right of appeal against the Minister's decision to suspend or revoke a mineral right. In Nigeria and Tanzania, the revocation of a mineral right by the Minister can be appealed to the High Court within the stipulated timeframe – that is, within 30 days of receipt of the Minister's decision in Nigeria and within 60 days of the Minister's decision in Tanzania.[153] Third, some countries allow mineral right holders to circumvent the statutory provisions regarding revocation of mineral rights, including the Minister's power of revocation. In Ghana, for example, a development agreement may contain provisions "relating to the circumstance or manner in which the Minister will exercise a discretion conferred by or under this Act".[154] Presumably, then, mining companies that are signatories to development agreements in Ghana can attenuate the Minister's discretionary power by inserting clauses that circumscribe the Minister's power to revoke their mineral right.

Despite the legal safeguards against arbitrary termination of mineral rights and despite the statutory guarantees regarding security of tenure, some African governments have terminated mineral rights during the last few years. As discussed in Chapter 7, in recent years mining concessions were terminated in the DRC, Guinea, Kenya, Tanzania, and Zambia. In 2015, for example, Kenya cancelled 65

---

151 *Mines and Minerals Act, 1999* (Botswana), s. 76.(2); *MPRDA*, s. 47.(2); *Minerals and Mining Act, 2006* (Ghana), ss. 68(2) & 69.(2); *Mines and Minerals Development Act, 2015* (Zambia), s. 72.(2); *Nigerian Minerals and Mining Act,* 2007, s.151(2); *Mining Act, 2010* (Tanzania), s. 63.(2); *Mining Act, 2016* (Kenya), s. 147.(3).
152 *Minerals and Mining Act, 2006* (Ghana), ss. 68.(2) & 69.(2).
153 *Nigerian Minerals and Mining Act, 2007,* s. 156 read with the *Nigerian Minerals and Mining Regulations 2011*, reg. 97.(6); *Mining Act, 2010* (Tanzania), s. 65.(2).
154 *Minerals and Mining Act, 2006* (Ghana), s. 49.(1)(b).

prospecting and mining licences,[155] while, in 2018, Tanzania revoked 11 retention licences.[156] While some of these licences were revoked on justifiable grounds, such as the non-fulfilment of the terms and conditions of the licences, others were revoked on spurious and questionable grounds. This serves as a reminder that the broad discretionary power vested in the Minister in some African countries is susceptible to abuse.

The potential for abuse of ministerial discretion is particularly acute in Tanzania where the Minister may compel a mineral right holder to transfer their interest without compensation. As noted above, a mineral right may be revoked if the holder becomes ineligible to hold a mineral right. Such is the case where the holder becomes an undischarged bankrupt or, in the case of a body corporate, the body corporate is wound up or dissolved. Tanzania takes this provision further than other African countries by requiring compulsory relinquishment and assignment of mineral rights in situations where a joint owner of a mineral right becomes ineligible to hold the mineral right. In that country, a joint owner of a mineral right may be compelled to relinquish their interest to the other joint owner(s) if an event occurs that makes them ineligible to hold the mineral right.[157] Compulsory assignment can also occur where one or more of the joint owners "fails to comply with an obligation which, under the terms and conditions of the licence or a relevant development agreement, is a several obligation".[158] In such instances, the Minister is obliged to serve on such person

> a notice of compulsory assignment requiring the affected person unconditionally, without consideration and free from any encumbrance, to assign to the licence holders who are not affected persons (in this section referred to as "unaffected persons") the entire interest in the licence held by the affected person.[159]

This provision appears harsh as it borders on expropriation of mineral rights without compensation. Presumably, the person to whom a mineral right is compulsory assigned under this provision would pay adequate compensation for the right, but it is preferable to afford the joint owner an opportunity to sell their interest to the other joint owners or third parties. That said, the harshness of this provision appears to be cushioned because the Minister's order regarding compulsory assignment of a mineral right is appealable to the High Court of Tanzania

---

155 Library of Congress, "Kenya: Many Prospecting and Mining Licences Cancelled; New Mining Law Considered", www.loc.gov/law/foreign-news/article/kenya-many-prospecting-and-mining-licenses-cancelled-new-mining-law-considered
156 Fumbuka Ng'wanakilala, "Tanzania Cancels License of Barrick, Glencore Nickel Project" (Reuters, May 12, 2018), www.reuters.com/article/us-tanzania-mining/tanzania-cancels-license-of-barrick-glencore-nickel-project-idUSKCN1ID0O7
157 *Mining Act, 2010* (Tanzania), s. 63.(4).
158 Ibid. s. 63.(4).
159 Ibid. s. 63.(4).

144 *Security of mineral tenure*

within 60 days of the order.[160] The Tanzanian provision contrasts sharply with the provisions in other African countries. In Nigeria, for example, the government cannot compel the ineligible joint owner to assign their interest because "no person who owns, whether wholly or in part, the capital of any enterprise shall be compelled by law to surrender his interest in the capital to any other person".[161] Thus, in Nigeria, if one of the joint holders of a mineral right becomes ineligible to hold the mineral right, they may sell their interest to the remaining eligible joint holders or other third parties.

6 Conclusion

This chapter analyzes the degree to which Africa ensures the security and protection of mineral rights. It also analyzes the source and nature of mineral rights because the degree to which the tenure of mineral rights is assured in a country depends partly on the legal characterization of mineral rights. It argues that in many African countries the source of mineral rights is not monolithic; rather, mineral rights may be both contractual and statutory. This appears to be the case in countries where mineral rights are granted in the form of contracts encompassing terms and conditions consensually agreed upon by the state and the investor, but some of the terms of the contracts are mandated by the enabling statute. Because the enabling statute dictates some of the terms of the contract, the mineral rights could be said to flow partly from the statute. With regard to the nature of mineral rights, this chapter argues that, with the exception of countries where mineral rights are classified as 'property' and countries where 'property' is defined broadly enough to include mineral rights, mineral rights in Africa are a *profit à prendre* because they grant the holder a right to enter land belonging to the government for the purpose of exploiting the minerals situated in the land.

African countries have attempted to preserve the security of mineral tenure through statutory, contractual and constitution provisions. Statutory regimes in Africa guarantee security of tenure during prospecting, exploration, and production of minerals. In some countries, the security of mineral tenure also manifests in the form of an automatic right to renew mineral rights, as well as in the ability of mineral right holders to retain their mineral rights even though they are temporarily unable to produce minerals. More significantly, African countries offer investors certain constitutional and contractual guarantees regarding the security of mineral tenure. In countries such as South Africa and the DRC, mineral rights are expressly characterized as 'property', while in Kenya and Zambia, 'property' is defined broadly to include mineral rights. The classification of mineral rights as 'property' is significant because it confers constitutional protection on mineral right holders. Even in African countries where mineral rights are not expressly

160 *Ibid.* s. 65.(2).
161 *Nigerian Investment Promotion Commission Act*, Chapter N117, LFN 2004, s.25.(1)(b).

classified as 'property', legal protection is nonetheless accorded to mineral right holders because mineral rights qualify as an interest in land. The faithful implementation of these statutory and constitutional schemes could ensure that mineral rights are secured throughout the life circle of mining projects in Africa.

5 Fiscal regimes for mineral exploitation

1 Introduction

The fiscal regimes governing hard-rock minerals in Africa include royalties, taxes, concession and licence fees, rents, custom duties on imports, and the various fiscal incentives offered to companies engaged in mineral exploitation. The overarching goal of the fiscal regimes is to encourage the inflow of FDI into Africa, which, in turn, could aid the economic development aspirations of resource-rich African countries. For example, South Africa's fiscal regimes are informed in part by "the need to promote local and rural development and the social upliftment of communities affected by mining".[1] Africa desires to use mineral resources as catalysts for broad-based growth and development, including the optimization of linkages into Africa's domestic economies.[2] In the view of resource-rich African countries, economic development can be achieved through two interrelated policy objectives: the enhancement of revenues generated from resource exploitation and the enhancement of the economic competitiveness of individual countries.[3] This chapter assesses the underlying features of the fiscal regimes for mineral extraction in Africa against the backdrop of the policy objectives informing the regimes. The chapter focuses on the fiscal regimes in Botswana, Ghana, Nigeria, South Africa, and Zambia, but occasional references are made to other African countries. While noting that the fiscal regimes have encouraged FDI inflow as well as an increase in mineral production in Africa, it argues that Africa has yet to optimize mining revenues due partly to the abuse of the fiscal regimes by mining companies.

2 Royalty and tax regimes

The word 'royalty' is not susceptible to a precise definition, but it is universally acknowledged to possess three attributes. First, the royalty system for mineral

1 *MPRDA*, Preamble.
2 African Union, *Africa Mining Vision* (February 2009) at 13, www.africanminingvision. org/amv_resources/AMV/Africa_Mining_Vision_English.pdf
3 See, for example, the Preamble to South Africa's *MPRDA* which speaks of "the need to create an internationally competitive and efficient administrative and regulatory regime" for mining.

resources is created legally through statutory enactment. Second, the purpose of a royalty is to compensate the owner of minerals for transferring to the mineral right holder the right to exploit and sell minerals. Third, royalty is unique to minerals and other natural resources as it is usually not imposed on other industries.[4] Royalties and taxes are predicated on the concept of economic rent – that is, "the difference between the value of production and the costs" of extracting the resources.[5] As owners of mineral resources, African governments participate in the sharing of "the surplus revenue" earned by resource extractors "after accounting for the costs of all capital and labour inputs".[6] The sharing of mineral resource economic rent in Africa is achieved primarily by way of royalties and taxes. Thus, African states usually require resource extractors to pay a royalty as compensation for the exploitation of mineral resources. A royalty is, in effect, "the means by which the mineral owner shares in the production of the substances from [its] land".[7] Royalties are viewed favourably by most governments because they ensure "a share of revenue for the government as soon as production [of mineral resources] commences".[8]

2.1 Royalties

Royalties for mineral exploitation are usually set out in statutory and quasi-statutory instruments and in concession contracts between African governments and mining companies. For the most part, mining statutes in Africa anticipate the payment of royalties once a mineral right holder begins commercial production of minerals. Aside from production, however, a mining contract may prescribe other triggering events for the payment of royalties. For example, the payment of royalties may be tied to the actual sale of minerals by the mineral right holder, particularly where the rate of royalty is based on income or profits. Royalties are paid at the time or period prescribed in the mining statute or concession contract. In some instances, royalties are paid annually while in other instances royalties are paid several times a year (usually monthly or quarterly), depending on stipulations in the governing statute, regulation, or contract.

2.1.1 Types of royalties

Royalties for minerals may be based on the market value of minerals (that is, *ad valorem* value-based royalty), on the metric weight of each unit of minerals (that

---

4 James Otto et al., *Mining Royalties: A Global Study of Their Impact on Investors, Government, and Civil Society* (Washington DC: The World Bank, 2006) at 50.
5 Daniel Johnston, *International Petroleum Fiscal Systems and Production Sharing Contracts* (Tulsa, OK: PennWell Books, 1994) at 6.
6 Carole Nakhle, *Petroleum Taxation – Sharing the Oil Wealth: A Study of Petroleum Taxation Yesterday, Today and Tomorrow* (London & New York: Routledge, 2008) at 17.
7 John Bishop Ballem, *The Oil and Gas Lease in Canada*, 4th Edition (Toronto: University of Toronto Press, 2008) at 178.
8 Nakhle, *supra* note 6 at 23.

is, unit-based royalty), or on the profits made by a company (that is, profit-based royalty).[9] Royalty may also be hybrid where it is based on a combination of the value of minerals, the metric weight of each unit of minerals, and profits arising from sale of the minerals. The unit-based system is rarely used for precious stones and metals, but it is appropriate for minerals that are sold in bulk (such as coal, iron ore, and phosphate) and minerals that are homogenous in terms of quality, such as gravel and limestone.[10] Although the unit-based royalty system is said to protect the government from a downturn in the value of commodities, it is disadvantageous because it precludes an increase in government revenue in times of rising commodity prices.[11] The unit-based system is also susceptible to manipulation by companies, particularly in developing countries where government officials are usually unable to physically monitor the actual production of minerals, relying instead on the record of production submitted by mining companies.[12]

The profit- or income-based method is disadvantageous because, under that method, royalties are payable only when a company makes a profit or generates income from the sale of minerals. The mere production of minerals would not trigger the payment of royalties under the profit-based method. Moreover, because most mining projects are not profitable in the early stages, profit-based royalties "are almost never paid in the early years of a project".[13] Thus, the profit- or income-based royalty system is uncommon in Africa. Besides, given the administrative deficiencies in Africa's public service, African tax administrators are often unable to determine the actual income and profits generated by mining companies. There is also the issue of royalty- and tax-avoidance practices, such as round-tripping, which enable mining MNCs to understate their income and profits. That said, profit- or income-based royalties could be advantageous because "they can be applied to any type and scale of mineral operation without the need to differentiate between the types of minerals being produced", and "[b]ecause they are based simply on revenues and costs, calculation procedures can be similar for all mine types and sizes".[14]

The particular type of royalty adopted by a country could implicate not only the steadiness of mineral resource revenue but also the amount of revenue generated by the government. It could also impact adversely on mining projects, particularly marginal projects. Hence, policy-makers in each country must carefully consider and balance the economic aspirations of the country with the reasonable expectations of investors, including the need to motivate investors and attract FDI.

---

9 Otto *et al.*, *supra* note 4 at 50–4.
10 *Ibid.* at 50.
11 Mirian K. Omalu & Armando Zamora, "Key Issues in Mining Policy: A Brief Comparative Survey on the Reform of Mining Law" (1999) 17 *Journal of Energy & Natural Resources Law* 13 at 29.
12 *Ibid.* at 29.
13 Otto *et al.*, *supra* note 4 at 64.
14 *Ibid.* at 44.

Policy-makers must also take into account the merits and demerits of each type of royalty. As James Otto et al. explain:

> Unit-based and ad valorem-type royalties are certain to be paid in all years when production takes place, whereas profit- and income-based royalties will be paid in years with profits or income. Unit-based and ad valorem royalties also satisfy the objective of providing revenue in the early years of a project, whereas a profit or income type probably will not yield a return. Unit-based and ad valorem royalties are also transparent and easy to administer compared with profit- or income-based royalty taxes.[15]

The choice of any particular model of royalty depends on a number of factors, including the steadiness of revenue generation, the size and diversity of mining operations in a country, the institutional competence and expertise of mining regulators, the cost of administering the royalty regime, and the price volatility of the mineral.[16] However, irrespective of the type of royalty, a country may choose to impose a single royalty rate for all minerals or prescribe a different rate for different types of minerals.

The foremost consideration in designing royalty policy in Africa appears to be the ability to generate steady revenue for the government. Where a government desires to generate revenue from inception of production, the government may want to adopt value-based royalty or unit-based royalty because royalties are payable under these methods "irrespective of whether the mine is making a profit or losing money".[17] While a few African countries have adopted the profit-based royalty regime, the most common type of mineral royalty in Africa is the value-based royalty. The prevalence of value-based royalty regimes in Africa is premised on Africa's desire to use mineral resources as catalysts for economic development. It is spurred partly by Africa's desire to capture a greater amount of mineral revenues during times of commodity boom. Under the value-based system, the government's mineral revenue increases proportionately with the increase in commodity prices.

Given the pre-eminence of revenue generation, it is hardly surprising that most African countries impose *ad valorem* royalty based on the value of minerals determined at the mine mouth, or on the gross revenues arising from the sale of minerals, or on the prevailing commodity prices on the international market. In Tanzania, for example:

> Every authorised miner shall pay to the Government of the United Republic a royalty on the gross value of minerals produced under his licence at the rate –
> (a) in the case of uranium, of five per centum;
> (b) in the case of gemstone and diamond, of six per centum;

15 *Ibid.* at 64.
16 *Ibid.* at 66.
17 *Ibid.* at 52.

(c) in the case of metallic minerals such as copper, gold, silver, and platinum group minerals, of six per centum;
(d) in the case of gem, of one per centum; and
(e) in the case of other minerals, including building materials, salt, all minerals within the industrial minerals group, of three per centum.[18]

2.1.2 Valuation methods for royalties

A country may choose to impose a uniform rate of royalty for all minerals or adopt rate differentials for groups of minerals. A uniform rate of royalty may be prescribed for each group or class of minerals (such as precious stones and precious metals), or a country may choose to create a distinct royalty rate for each type or group of mineral, as is apparent from the Tanzanian provision above. The latter approach is common in situations where some types of minerals attract higher prices than others on the international commodity market.

The valuation method for royalties vary between the types of royalties, and even within a particular type of royalty, there may be a combination of valuation methods. For example, the *ad valorem* value-based royalty may be determined based on the gross value of minerals as determined at the mine gate or on the value of minerals at the time of sale. The value of minerals may be determined based on a reference price, as is the case in Zambia where the value of minerals is based on prices prevailing on the London Metal Exchange.[19] In addition, the *ad valorem* value-based royalty may be predicated on a sliding scale based on cumulative sales of minerals or on the price of minerals on the international market. Under the sliding-scale method, the rate of royalty increases with the volume of sales or price, so that the greater the volume of sales or the higher the price, the higher the rate of royalty. In Zambia, for example, the royalty rate for copper increases when the price of copper on the international market is US$4,500 or above.[20]

The valuation method commonly adopted for the profit-based royalty regime is the gross revenue or net revenue arising from mining operations minus the cost of producing the minerals, while the unit-based regime is calculated on the basis of the weight of a unit of the minerals, such that the rate of royalty increases with the number of units of minerals produced by a company.[21]

2.2 Taxes

In addition to royalties, mineral-rich African countries participate in sharing the economic rent accruing from mineral extraction by imposing taxes on mineral

18 *Mining Act, 2010* (Tanzania), s. 87 (as amended by *The Written Laws (Miscellaneous Amendments) Act, 2017* (No. 7), s. 23).
19 *Mines and Minerals Development Act, 2015* (Zambia), s. 89.(5) (as amended by the *Mines and Minerals Development (Amendment) Act*, No. 14 of 2016).
20 *Mines and Minerals Development Act, 2015* (Zambia), s. 89.(2).
21 Otto *et al.*, *supra* note 4 at 56.

extractors. The taxes imposed on mineral extractors are, for the most part, profit-based or value-based and they include corporate income tax, capital gains tax, withholding tax on dividends accruing to non-residents, resource rent tax, and value-added tax.[22] In exceptional circumstances, a windfall tax may be imposed on mineral extractors to capture an additional share of unexpectedly large profits, particularly in times of boom in commodity prices.[23] Relatedly, a country may impose additional taxes on "excess profits" to cater to rising commodity prices. For example, in Nigeria, in addition to the general tax rate of 30%, a special levy of 15% is imposed on "excess profits" made by companies. In this regard, "excess profits" means "the difference between total profits as computed in accordance with section 31 of [the *Companies Income Tax Act*] and standard profits as calculated in accordance with" section 40.(3) of the Act.[24]

The most common type of mineral tax in Africa is the corporate income tax that is levied on the revenues generated by mineral extractors after a deduction of the costs of generating the revenues. Quite often, the corporate income tax is levied at a predetermined percentage rate of a company's revenues, less the costs of generating the revenues. However, the predetermined tax rate system may be disadvantageous to mineral-rich countries with regard to their share of the economic rent in circumstances where a mining project becomes highly profitable or where prices of minerals rise sharply on the international market.[25] Hence, some countries adopt a progressive profit tax system "by having a stepped tax rate schedule linked to higher product prices, production volume, sales turnover, or the profit-to-sales ratio".[26]

3 Case studies of mineral royalties and taxes

This section examines the royalty and tax regimes in five African countries namely Botswana, Ghana, Nigeria, South Africa, and Zambia. The focus of the analysis is the types and rates of royalties and taxes in each country, the valuation method used in determining royalty and tax liability, and the fiscal incentives offered to mining companies operating in these countries.

3.1 Botswana

Botswana operates an *ad valorem* value-based royalty system predicated on the gross market value of minerals. In Botswana, the royalty rate for precious stones

---

22 See Thomas Baunsgaard, "A Primer on Mineral Taxation" (IMF Working Paper WP/01/139) at 6–11, available at www.imf.org/external/pubs/ft/wp/2001/wp01139.pdf
23 Omalu & Zamora, *supra* note 11 at 28.
24 *Companies Income Tax Act*, CAP. C21, LFN 2004, s. 40.(2).
25 See Baunsgaard, *supra* note 22 at 7.
26 *Ibid.* at 7–8.

Table 5.1 Royalty and tax rates in selected African countries

Country	Royalty rate	Type and method of valuation of royalty	Corporate income tax rate
Angola	5%: Strategic minerals; precious metallic minerals and precious stones 4%: Semi-precious stones 3%: Non-precious metallic minerals 3%: Artisanal diamonds 2%: Other minerals & construction materials	Ad valorem (Market value of minerals)	25% (mining industry) 30% (other industries)
Botswana	10%: Precious stones 5%: Precious metals 3%: Other minerals	Ad valorem (Gross market value of minerals)	25% (minimum)
Burkina Faso	8%: Uranium 7%: Diamond and other precious stones 4%: Precious metals 3%: Base metals and other mineral substances 3%–5%: Gold The sliding scale of 3%–5% for gold is based on the market price as follows: • 3% when price is less than US$1,000 • 4% when price is between US$1,000 and US$1,300, and • 5% when price exceeds US$1,300	Ad valorem (Market value of minerals)	30%
Democratic Republic of Congo	10%: Strategic substances 6%: Precious and coloured stones 3.5%: Non-ferrous metals and precious metals 1%: Industrial minerals, solid hydrocarbons, iron and ferrous metals	Ad valorem (Sales value of minerals)	30%

Country	Royalty rate	Type and method of valuation of royalty	Corporate income tax rate
Ghana	5% of gross revenue However, the Ghana Chamber of Mines reports that large-scale gold producers that have signed development agreements with the government pay sliding-scale rates as follows: • 3% when price is less than US$1,300 • 3.5% when price is between US$1,300 and US$1,449.99 • 4% when price is between US$1,450 and US$2,299.99 • 5% when price is US$2,300 and above	*Ad valorem* (Gross sales value of minerals)	35%
Guinea	5%: Gold 3%: Iron ore 0.075%: Bauxite	*Ad valorem* (Market value of minerals)	35%
Kenya	12%: Diamond 10%: Rare earth minerals, niobium, titanium ores, and zircon 8%: Coal; metallic ores, iron, manganese, chromium, nickel, bauxite, and other ores; 5%: Gold, silver, platinum, platinoid group metals, gemstones, fluorspar, diatomite, and all other minerals 1%: Industrial minerals	*Ad valorem* (Gross sales value of minerals)	30%
Mozambique	8%: Diamond 6%: Precious metals, precious and semi-precious stones 3%: Base minerals, coal, and ornamental rocks	*Ad valorem* (Market value of minerals)	32% (profit-based)
Namibia	10%: Rough diamonds; rough emeralds, rubies, and sapphires; 3%: Gold, copper, zinc and other base metals; 2%: Semi-precious stones 2%: Industrial minerals including fluorspar and salt	*Ad valorem* (Market value of minerals)	55% (diamond-mining companies) 37.5% (non-diamond-mining companies)

(continued)

Table 5.1 (continued)

Country	Royalty rate	Type and method of valuation of royalty	Corporate income tax rate
Nigeria	5%: Barite, gypsum, limestone, kaoline, sapphire, phosphate, dolomite, emerald, quartz, mica, etc. 3%: Columbite, gold, iron ore, tantalite, magnesite, etc.	Ad valorem (Market value of minerals)	30%
Sierra Leone	15%: Special stones having a market value above US$500,000 6.5%: Precious stones 5%: Precious metals 3%: All other minerals	Ad valorem (Market value of minerals)	30%
South Africa	7%: Unrefined minerals 5%: Refined minerals	Ad valorem (Gross sales value of minerals)	28% (rate is set through annual budget)
Tanzania	6%: Gemstone and diamond 6%: Metallic minerals such as copper, gold, silver, and platinum group minerals 5%: Uranium 3%: Industrial minerals and other minerals 1%: Gem	Ad valorem (Gross market value of minerals)	30%
Zambia	Sliding-scale rates of 4%–6% of norm/gross value depending on price of minerals as follows: • 5% of norm value of base metals other than copper • 5% of gross value of energy and industrial minerals • 6% of gross value of gemstones • 6% of norm value of precious metals	Ad valorem (Norm or Gross value of minerals)	30%
Zimbabwe	10%: Precious stones 3.5%: Precious metals 2%: Base metals & industrial minerals 1%: Coal	Ad valorem (Gross market value of minerals)	15%

Sources: Mining statutes, mining regulations, and income tax statutes.

(such as diamonds) is 10% of the gross market value of the minerals, while the royalty rate for precious metals (such as gold and platinum) is 5% of the gross market value of the minerals.[27] Other minerals attract royalty at the rate of 3% of the gross market value of the minerals.[28] In that regard, 'gross market value' means "the sale value receivable at the mine gate in an arms-length transaction without discounts, commissions or deductions for the mineral or mineral product on disposal".[29] However, where a mineral is disposed of through a transaction that is not an arm's-length transaction, the Minister may impose a royalty rate on the basis of the prices prevailing in the industry.[30]

With regard to taxes, mining companies in Botswana are liable to pay income taxes and withholding taxes. The tax rate for diamonds is not predetermined but subject to negotiation based on financial, technical, and commercials factors. Thus, in Botswana the tax rate for diamonds varies between projects. However, regarding other minerals, the income tax of mining companies in Botswana is determined in accordance with the formula specified in the Twelfth Schedule to the *Income Tax Act*.[31] Under that schedule, mining profits, other than profits from diamond mining, are taxed according to the following formula:

$$\text{Annual tax rate} = 70 - \frac{1500}{X}$$

where X is the profitability ratio, given by taxable income as a percentage of gross income:

> Provided that the rate shall not be less than the company rate (current 25 per cent) made up of 15 per cent basic company tax and 10 per cent additional company tax rate.[32]

For this purpose, gross income from mining operations include:

(a) ... all amounts accrued to [the mining company] during the tax year from all mining and prospecting operations carried on by [the company];
(b) ... all amounts, whether in cash or otherwise, accrued to [the company] or to any associated person, during the tax year from processing, marketing, servicing, financial or administrative operations whether
  (i) any such operation is carried on in or out of Botswana;
  (ii) the source of any such amount is situate in or out of Botswana, to the extent to which the Commissioner General is of the opinion that such amounts are related to mining operations;

---

27 *Mines and Minerals Act, 1999* (Botswana), s. 66.(2).
28 *Ibid.* s. 66.(2).
29 *Ibid.* s. 66.(3).
30 *Ibid.* s. 66.(5).
31 *Income Tax Act*, Chapter 52:01 (Botswana), s. 55(d).
32 *Ibid.* 12th Schedule, para. 4.

(c) ... the amount of any royalty-
  (i) remitted, exempted or repaid during that tax year under section 66(5) or 67 of the Mines and Minerals Act where such royalty has been allowed as a deduction under section 43(1)(c);
  (ii) ... ;
(d) ... the amount of any excess of disposal value property, disposed of in that tax year and included in mining capital expenditure incurred in that tax year or any previous tax year.[33]

However, the income tax rate under this formula cannot be less than 22% of the profit of the company.

Gross income also includes all amounts, whether in cash or otherwise, accruing

by way of royalty, premium or other consideration, howsoever described, for the right to extract minerals from land situate in Botswana; or in respect of the disposal of any share or interest in the capital or income of a company holding mineral rights over land situate in Botswana.[34]

Thus, mining companies in Botswana are liable to pay taxes on income generated by way of sale, transfer, or assignment of their mineral rights, as well as income accruing from the sale of their interest in other mining companies. In addition, 'gross income' includes "any commercial royalty, entertainment fee or management or consultancy fee accrued or deemed to have accrued" to a company "from a source situated or deemed to be situated in Botswana".[35]

In addition to income taxes, mining companies in Botswana are charged withholding taxes at the rate of 15% on each payment of dividend made to a resident; 15% on each payment of dividend, interest, commercial royalty, management, or consultancy fee made to a non-resident; and 10% on each payment of entertainment fee made to a non-resident.[36]

3.2 Ghana

Until March 2010, Ghana adopted a sliding-scale royalty regime that required holders of mining leases and licences to pay royalties in respect of minerals obtained from their mining operations at a rate that is not more than 6% or less than 3% of the total revenue of minerals obtained by the holders.[37] The sliding-scale regime was replaced in 2010 with a provision to the effect that

---

33 *Ibid.* s. 31.(1).
34 *Ibid.* s. 31.(2).
35 *Ibid.* s. 33.(1).
36 *Ibid.* 12th Schedule, para. 9.
37 *Minerals and Mining Act, 2006* (Ghana), s. 25. Section 25 was repealed and replaced by the *Minerals and Mining (Amendment) Act, 2010 (Act 794)*.

[a] holder of a mining lease, restricted mining leases or small scale mining license shall pay royalty in respect of minerals obtained from its mining operations to the Republic at the rate of 5% of the total revenue earned from minerals obtained by the holder.[38]

However, in 2015 Ghana abolished the flat rate of 5% of total revenues through the *Minerals and Mining (Amendment) Act, 2015 (Act 900)*. This statute does not provide for a particular rate of royalty, but it stipulates that mineral right holders shall pay royalty to the Republic at the rate and in the manner that may be prescribed. This amendment promotes flexibility within Ghana's royalty regime, but curiously, even though the amendment was enacted in 2015, the government of Ghana has yet to publish the rate of royalty. However, while some companies still pay the flat rate of 5%, the government has reportedly reintroduced a sliding-scale rate ranging between 3% and 5% of the gross revenues realized from large-scale mining operations.[39] According to the Ghana Chamber of Mines, the royalty rate for large-scale gold producers in Ghana is 3% of gross revenue when the price of minerals is US$1,300 per ounce; 3.5% when the price is between US$1,300 and US$1,449.99 per ounce; 4% when the price is between US$1,450 and US$2,299.99 per ounce; and 5% when the price is US$2,300 and above.[40]

The royalty regime in Ghana is value-based, calculated on the basis of the gross value of minerals produced by a company. Under Ghana's sliding-scale system, the value of minerals is determined based on the price of minerals on the international market. In addition to the royalty payment, holders of mineral rights are also required to pay an annual ground rent to the owners of the land over which a mineral right is granted.[41] In the case of mineral rights over Stool Lands, the annual ground rent is paid to the Office of the Administrator of Stool Lands.[42]

In terms of taxes, a 'mineral income tax' is imposed on mining companies operating in Ghana. For each year of assessment the rate of mineral income tax payable by mining companies is 35% of the chargeable income of the companies.[43] For this purpose, 'chargeable income' means the income derived from mining operations in each year of assessment less the total amount of deductions allowed under the *Income Tax Act*.[44] The assessable income is the income derived by companies from any business or investment in the year of assessment.[45] In determining the assessable income, each separate mineral operation is regarded as an independent business and the tax liability for the business is calculated

---

38 *Minerals and Mining (Amendment) Act, 2010 (Act 794)*.
39 Ghana Chamber of Commerce, "Performance of the Mining Industry in 2017" at 27, https://ghanachamberofmines.org/wp-content/uploads/2016/11/Performance-of-the-Industry-2017.pdf
40 *Ibid*. at 27.
41 *Minerals and Mining Act, 2006* (Ghana), s. 23.
42 *Ibid*. s. 23.(2).
43 *Income Tax Act, 2015*, Act 896 (Ghana), s. 77 (read with the 1st Schedule, para. 6).
44 *Ibid*. s. 2.(1).
45 *Ibid*. s. 3.

independently for each year of assessment.[46] Similarly, where a mineral right is held jointly by two or more companies under an arrangement other than a partnership, the assessable income of each of the companies is calculated separately and the companies are treated as if they are in a 'controlled relationship'. In this regard, a 'controlled relationship' exists where a person or company controls an entity (such as a company, partnership, or trust) or benefits directly or indirectly from 50% or more of the voting power or rights to income or capital of the entity.[47]

Ghana's *Income Tax Act* casts a wide net regarding the scope of the 'mineral income tax'. Income from mineral operations includes any amount derived from the disposal of minerals through arm's-length transactions; any compensation or payment received in respect of loss or destruction of minerals whether derived under an insurance policy or otherwise; any amount derived in respect of the sale of information pertaining to mineral operations or mineral reserves; any monetary gain from the assignment or disposal of an interest in a mineral right; any amount in respect of a surplus in an approved rehabilitation fund; and any other amount derived from or incidental to mineral operations.[48]

In addition, companies operating in Ghana are obliged to remit a 'withholding tax' to the government based on a prescribed percentage of employee income and investment returns. For example, the withholding tax for non-resident employees is 20%, while a withholding tax of 8% applies to dividends.[49] In addition, payments made by mineral right holders in connection with mining operations are subject to a withholding tax at the rate of 15%.[50]

3.3 Nigeria

A number of statutes require mining companies to pay royalties, taxes, fees, and rents to the Nigerian government.[51] Nigeria charges value-based or *ad valorem* royalties for minerals, but the percentage of royalty differs depending on the type mineral in question. A 3% *ad valorem* royalty rate is imposed on gold, iron ore, lead, zinc, tantalite, bitumen or tar sand, coal, magnesite, and tin ore; while a royalty rate of 5% is payable for minerals such as barite, laterite, limestone, marble aggregates, kaoline, sapphire, phosphate, crystal quartz, and emerald.[52] However, the royalty may be reduced or waived where the Minister is satisfied that the mineral is being exported solely for the purpose of analysis or experiment or as a

---

46 *Ibid.* s. 77.(4).
47 *Ibid.* s. 128.
48 *Ibid.* s. 80.
49 *Ibid.* ss. 85, 114–5 (read with the 1st Schedule, para. 8).
50 *Ibid.*
51 See *Nigerian Minerals and Mining Act, 2007*, s. 33; *Companies Income Tax Act*, CAP. C21, LFN 2004, s. 40.
52 *Nigerian Minerals and Mining Regulations 2011*, reg. 99 & Schedule 4.

scientific specimen, provided that the quantity sought to be exported is not greater than is reasonably necessary for such purposes.[53]

The income tax rate for all companies operating in Nigeria, including mining companies, is 30%.[54] In addition, companies operating in Nigeria are charged an 'annual education tax' at the rate of 2% of their assessable profits.[55] Holders of mineral titles other than a reconnaissance permit are obliged to pay an annual service fee "equal to the number of cadaster units that comprise the title area multiplied by the fee per cadaster unit for [the] type of title" held by the titleholder.[56] In addition, the holder of a mining lease, small-scale mining lease, quarry lease, and water use permit is required to pay an annual surface rent as determined by the Minister.[57] Mining companies are also liable to pay sundry fees, such as mining fees, milling fees, and quarry fees.[58]

3.4 South Africa

The South African government requires mineral resource extractors to pay a royalty in respect of the transfer of mineral resources, although different rates of royalty apply to "refined" mineral resources and "unrefined" mineral resources.[59] South Africa's value-based royalty regime is unique in the sense that it consists of a variable-scale regime based on the gross sales of a company. Under this regime, royalty rates for "refined" and "unrefined" minerals have a maximum cap of 5% and 7%, respectively.[60] The rates are determined by multiplying the gross sales of the extractor in respect of the mineral resource during the year of assessment by the percentage determined in accordance with the formula prescribed in section 4 of the *Mineral and Petroleum Resources Royalty Act, 2008*.[61] Under this formula, royalties for refined mineral resources are ascertained by multiplying the extractor's gross sales for the year of assessment by "0.5 + [earnings before interest and taxes/(gross sales in respect of refined mineral resources x 12.5)] x 100".[62] However, the percentage of royalty for refined mineral resources must not exceed 5%.[63] In the case of unrefined mineral resources, royalties are determined by multiplying the gross sales of the extractor during the year of assessment by "0.5 + [earnings before interest and taxes/(gross sales in respect of unrefined mineral

---

53 *Nigerian Minerals and Mining Act, 2007*, s. 33.(2).
54 *Companies Income Tax Act*, CAP. C21, LFN. 2004, s. 40.
55 *Tertiary Education Trust Fund (Establishment, Etc.) Act, 2011*, s. 1.(2).
56 *Nigerian Minerals and Mining Regulations 2011*, reg. 98.(1).
57 Ibid. reg. 100.
58 See the Schedule to the *Taxes and Levies (Approved List for Collection) Act (Amendment) Order 2015*.
59 *Mineral and Petroleum Resources Royalty Act, 2008* (Act No. 28, 2008), ss. 2 & 3.
60 Ibid. s. 4.
61 See ibid. ss. 3 & 4.
62 Ibid. s. 4.(1).
63 Ibid. s. 4.(3)(a).

160 *Fiscal regimes for mineral exploitation*

resources x 9)] x 100", provided, however, that the percentage determined under this formula does not exceed 7%.[64]

For this purpose, "gross sales" means "the amount received or accrued during the year of assessment in respect of the transfer of" mineral resources,[65] while "earnings before interest and taxes" means the aggregate of the gross sales of unrefined mineral resources by the extractor during the year of assessment and the amount allowed as deduction from the extractor's income under the *Income Tax Act*, whether in the year of assessment or a previous year, in respect of the assets used or expenditure incurred directly to win, recover, and develop those mineral resources.[66] However, "earnings before interest and taxes" does not include any "amount that is received or accrued from the disposal of assets the cost of which has in whole or in part been included in capital expenditure taken into account" under the *Income Tax Act*.[67]

Mineral extractors who qualify as a 'small business' are exempt from the payment of royalty in certain circumstances. A mineral extractor is exempt from the payment of royalty in respect of a year of assessment if:

> (a) gross sales of that extractor in respect of all mineral resources transferred does not exceed R10 million during that year; (b) the royalty in respect of all mineral resources transferred that would be imposed on the extractor for that year does not exceed R100,000; (c) the extractor is a resident as defined in section 1 of the Income Tax Act throughout that year; and (d) the extractor is registered for that year pursuant to section 2 of the Administration Act.[68]

In addition, a mineral extractor is exempted from payment of royalty in respect of mineral resources won or recovered for the purposes of testing, identification, analysis, and sampling "if the gross sales in respect of those mineral resources does not exceed R100,000 during a year of assessment".[69] However, a small business extractor is not exempt from the payment of royalty if:

> (a) the extractor at any time during that year holds the right to participate (directly or indirectly) in more than 50 per cent of the share capital, share premium, current or accumulated profits or reserves of, or is entitled to exercise more than 50 per cent of the voting rights in, any other extractor;
> (b) any other extractor at any time during that year holds the right to participate (directly or indirectly) in more than 50 per cent of the current or accumulated profits of the extractor;
> (c) any other person at any time during that year holds the right to participate (directly or indirectly) in more than 50 per cent of the profits of the

64 Ibid. s. 4.(2) & (3)(b).
65 Ibid. s. 6(1)(a) & (2)(a).
66 Ibid. s. 5.(1) & (2).
67 Ibid. s. 5.(1)(b) & (2)(b).
68 Ibid. s. 7.(1).
69 Ibid. s. 8.

extractor and more than 50 per cent of the current or accumulated profits of any other extractor; or

(d) the extractor is a registered person mentioned in section 4 of the Administration Act.[70]

South Africa provides financial relief to companies disposing of their mineral resources. For the purpose of royalty payment,

> a disposal of a mineral resource by an extractor that forms part of the disposal of a going concern, or of a part of a going concern which is capable of separate operation, by that extractor to any other extractor is deemed not to be a disposal.[71]

In effect, a mineral extractor disposing of their mineral resources in the course of a sale or acquisition of the company (or a part thereof) by another company is exempt from paying royalty on such disposed mineral resources. However, this exemption does not mean that government will forgo payment of royalty on the mineral resources so disposed because the company acquiring the mineral resources from the extractor in those circumstances would eventually be obliged to pay a royalty on the mineral resources.[72]

With regard to taxes, South Africa adopts a flexible tax regime for resource extraction in the sense that the *Income Tax Act* does not prescribe any particular tax rates for resource extraction. Rather, the Act empowers the Minister of Finance to determine, on an annual basis, the rate(s) of taxes chargeable in respect of income and announce same through the national annual budget.[73] Currently, the tax rate for companies operating in South Africa (other than small business corporations) is 28% of taxable income – that is, gross income less exemptions and allowable deductions.[74] Over the last decade, the company income tax rate in South Africa has remained steady at 28%.[75]

3.5 Zambia

Zambia operates a value-based royalty regime but the royalty rates for mineral resources vary depending on the type of mineral operations undertaken by the

---

70 *Ibid.* s. 7.(2).
71 *Ibid.* s. 9.(1).
72 *Ibid.* s. 9.(2).
73 *Income Tax Act, 1962*, s. 5.(2) (as amended by *Taxation Laws Amendment Act, No 15 of 2016*, s. 6).
74 South African Revenue Service, *Tax Statistics 2019*, at 147, www.sars.gov.za/AllDocs/Documents/Tax%20Stats/Tax%20Stats%202019/Tax%20Stats%202019%20Full%20doc.pdf
75 *Ibid.* at 159.

162 Fiscal regimes for mineral exploitation

titleholder and the nature of the mineral resource in question. In this regard, the *Mines and Minerals Development Act, 2015* provides that:

> A holder of a mining licence shall pay a mineral royalty at the rate of –
> (a) five percent of the norm value of the base metals produced or recoverable under the licence, except when the base metal is copper;
> (b) five percent of the gross value of the energy and industrial minerals produced or recoverable under the licence;
> (c) six percent of the gross value of the gemstones produced or recoverable under the licence; and
> (d) six percent of the norm value of precious metals produced or recoverable under the licence.[76]

In Zambia, the royalty rate for copper fluctuates depending on the price of copper on the international market. Currently, copper attracts a royalty rate of:

> (a) four percent of the norm value when the norm price of copper is less than four thousand five hundred United States dollars per tonne;
> (b) five percent of the norm value, when the norm price of copper is four thousand five hundred United States dollars per tonne or greater but less than six thousand United States dollars per tonne; and
> (c) six percent of the norm value, when the norm price of copper is six thousand United States dollars per tonne or greater.[77]

For the purpose of calculating royalty, "norm value" means the monthly average of the cash price per metric tonne on the London Metal Exchange multiplied by the quantity of the metal sold, while "gross value" means "the realised price for a sale, free-on-board, at the point of export from Zambia or point of delivery within Zambia".[78] The "norm price" is the monthly average cash price per tonne on the London Metal Exchange or any other exchange market approved by the Commissioner-General.[79]

Royalties are due and payable in Zambia within 14 days after the end of the month in which the sale of minerals occurred.[80] In order to ensure the proper collection of royalties, all mining companies are required to submit monthly mineral royalty returns in the prescribed form within 14 days after the end of the month in which the sale of minerals occurred.[81] However, a provisional assessment of mineral royalty may be made where it is impractical to assess the amount of any mineral royalty due, but, in such a case, the mineral right holder is obliged

---

76 *Mines and Minerals Development Act, 2015* (Zambia), s. 89.(1) (as amended by the *Mines and Minerals Development (Amendment) Act*, No. 14 of 2016).
77 *Mines and Minerals Development Act, 2015* (Zambia), s. 89.(2) (as amended).
78 *Ibid.* s. 89.(5).
79 *Ibid.* s. 89.(5).
80 *Ibid.* s. 90.
81 *Ibid.* s. 91.

to pay the balance when the actual amount of royalty is eventually ascertained.[82] To ascertain the quantity and value of minerals produced by companies in Zambia, the government can compel mineral right holders to produce for inspection any books, accounts, vouchers, documents, or records of any kind relating to the mineral right.[83] In addition, the government may compel witnesses to attend before a designated officer to "answer questions relating to minerals obtained or the value of the minerals obtained".[84]

Failure by a mining licence holder to pay royalty on the due date may cause the Commissioner-General (appointed under the *Zambia Revenue Authority Act*) to issue an order prohibiting the licence holder from disposing of any mineral from the mining area until an arrangement satisfactory to the Commissioner-General is made with regard to the payment of royalty.[85] The contravention of a prohibition order issued by the Commissioner-General is an offence punishable in the case of an individual with a fine "not exceeding five hundred thousand penalty units or imprisonment for a term not exceeding five years or both"; and in the case of a body corporate, "a fine not exceeding one million penalty units".[86] However, the payment of royalty may be waived or deferred where "the cash operating margin of the holder in respect of mining operations in the mining area falls below zero".[87]

As regards taxes, income from mining operations is taxed at the rate of 30%.[88] In addition, the holder of a mining right or a mineral-processing licence in Zambia is required to pay an annual area charge as may be determined by the Minister.[89] The annual area charge is payable on the grant of a mining right or mineral-processing licence and thereafter annually on the anniversary date of the right or licence until the termination of the right or licence.[90]

4 Fiscal incentives and allowances

African countries offer generous fiscal incentives to mining companies in order to encourage foreign investment in the mining industry. These fiscal incentives are either statute-based (mining statute and income tax act) or contract-based. In some African countries, mining companies are afforded both statute-based and contract-based incentives. Contract-based incentives may be personalized and tailored to the peculiar needs of the mining companies that are signatories to the

---

82 *Ibid*. s. 92.
83 *Ibid*. s. 78.(1).
84 *Ibid*. s. 78.(2)(b).
85 *Ibid*. s. 95.(1).
86 *Ibid*. s. 95.(2).
87 *Ibid*. s. 94.(1).
88 *Income Tax Act*, Chapter 323, Statutes of Zambia, s. 14.(1) read with para. 3 of the Charging Schedule to the *Income Tax Act* (as amended by the *Income Tax (Amendment) Act*, No. 11 of 2016).
89 *Mines and Minerals Development Act, 2015* (Zambia), s. 77.(1).
90 *Ibid*. s. 77.(2).

contracts containing such incentives. Generally speaking, fiscal incentives in Africa include capital allowances and deductions; cost-based incentives; tax relief and tax holidays; investment allowances; investment guarantees regarding transfer of capital, profits, and dividends; and exemption from import and export duties. This section discusses some of the fiscal incentives offered to mining companies in Africa.

4.1 Investment allowances, tax credits, and loss deductions

Mining companies in Africa are afforded certain capital allowances and deductions in assessing their chargeable income, thus reducing their tax liability to African governments. Capital allowances enable mining companies to deduct the cost of the assets used to produce minerals from their taxable income. These allowances may take the form of investment allowances and investment tax credits. An investment allowance confers on "the taxpayer the right to offset a percentage of its capital expenditure against its taxable income in the year the expenditure is made, rather than spread over time through depreciation", while "an investment tax credit enables a taxpayer to reduce the amount of tax payable by a portion of its investment expenditure in the first year, rather than reduce its taxable income, as with investment allowances".[91] However, as discussed below, deductible expenditure must be incurred exclusively for mining purposes.

Botswana provides for a 'mining capital allowance' which is "computed in accordance with 100 per cent of the mining capital expenditure made in the year in which such expenditure was incurred with unlimited carry forward of losses".[92] For this purpose, 'mining capital expenditure' means expenditure on the acquisition of mineral rights; exploratory work; buildings and structures, including plant, machinery, and equipment; provision of water, light, or power; provision of residential accommodation and welfare facilities for employees; and expenditure on general administration and management, including the interest payable on any loan to finance mining operations.[93] Mining companies in Botswana are also allowed to deduct from their chargeable income expenditure incurred in the production of their income, including expenditure on prospecting operations.[94] Furthermore, any amount paid or owed to the government as royalties is deductible from a company's chargeable income.[95] In addition, mining companies can deduct any assessed loss determined by the Commissioner General of Taxes as arising from mining operations, provided the deduction does "not exceed the amount of the chargeable income of the next subsequent tax year".[96] Other deductible

---

91 OECD, *Tax Incentives in Mining: Minimising Risks to Revenue*, at 28, www.oecd.org/tax/beps/tax-incentives-in-mining-minimising-risks-to-revenue-oecd-igf.pdf [*Tax Incentives in Mining*].
92 *Income Tax Act, Chapter 52:01* (Botswana), 12th Schedule, para. 1.
93 *Ibid.* s. 2.
94 *Ibid.* s. 43.
95 *Ibid.* s. 43.(1)(c) & 12th Schedule, para. 8.
96 *Ibid.* s. 46.

amounts in Botswana include expenditure exclusively incurred during the tax year on the acquisition of mineral rights; head office expenses; and expenditure on the training of personnel, provided such training is approved by the Commissioner General of Taxes.[97]

In Ghana, holders of mining leases are entitled to the capitalization of expenditure on reconnaissance and prospecting operations where they commence development of commercial quantities of minerals.[98] Companies that incur revenue expenditure or capital expenditure during reconnaissance or prospecting operations are required to place the expenditure in a single pool, and the balance of the expenditure in the pool is carried forward from year to year until the company commences production of minerals.[99] In addition,

> [Where the company] commences production of a commercial find with respect to a separate mineral operation, the balance in the pool of reconnaissance and prospecting expenditure at the time of production shall be capitalized and capital allowances shall be granted by the Commissioner-General in respect of those expenditures.[100]

Ghana also grants capital allowances in respect of depreciable assets owned and used in the production of income.[101] In addition, mining companies in Ghana can deduct from their chargeable income royalties and ground rents paid with respect to mineral operations; any other expenses incurred during the year of assessment in connection with mineral operations as may be deductible under the *Income Tax Act, 2015*.[102] However, financial costs incurred in relation to mineral operations in Ghana are deducted from chargeable income "only to the extent that a relevant financial gain has been included in calculating the income of" the company.[103]

Likewise, in Nigeria, mining companies are entitled to deduct from their "assessable profits a capital allowance of ninety-five percent of the Qualifying Capital Expenditure incurred in the year in which the investment is incurred", including all certified exploration, development, and processing expenditure; feasibility study and sample assaying costs; and "all infrastructure costs incurred regardless of ownership and replacement".[104] These deductions are made "as far as it is possible from the assessable profits of the first year of assessment" after the year in which the loss or expenditure was incurred.[105] However, where it is not feasible to make the deductions during the first year of assessment, then

---

97 *Ibid.* ss. 43.(2), 44 & 12th Schedule, para 5.
98 *Minerals and Mining Act, 2006* (Ghana), s. 28.
99 *Income Tax Act, 2015*, Act 896 (Ghana), s. 79.(2) & (5).
100 *Ibid.* s. 79.(7).
101 *Ibid.* ss. 14 & 81.(1).
102 *Ibid.* s. 81.(1).
103 *Ibid.* s. 81.(4).
104 *Nigerian Minerals and Mining Act, 2007*, s. 24.(1). See also *Companies Income Tax Act*, CAP. C21, LFN 2004, 2nd Schedule, para. 6 read with Table 1.
105 *Nigerian Minerals and Mining Act, 2007*, s. 24.(2).

deductions are made "from assessable profits of the next year of assessment, and so on up to a limit of four years after which period any unrelieved loss shall become lapse".[106] 'Qualifying capital expenditure' includes expenditure on plant, machinery, and fixtures; capital expenditure on the construction of buildings, structures, or works of a permanent nature; capital expenditure incurred in connection with, or in preparation for, the working of a mine or other source of mineral deposits of a wasting nature (otherwise called 'qualifying mining expenditure'); capital expenditure incurred in connection with the acquisition of mineral rights or on the purchase of information relating to the existence and extent of mineral deposits; and capital expenditure relating to searching, testing, and winning of mineral deposits.[107]

Nigeria's *Companies Income Tax Act* allows mining companies to deduct from their taxable profits any sum payable as interest on any money borrowed and employed as capital in acquiring the profits; rent in respect of land or building occupied for the purposes of acquiring the profits; expenses incurred in respect of staff salary, wages, and benefits; expenses incurred for the repair of premises, plant, machinery, or fixtures employed in acquiring the profits; bad debts incurred in the course of business, provided the debts are "proved to have become bad during the period for which the profits are being ascertained"; and any contribution to a pension, provident, or other retirement benefits fund.[108] In addition, Nigeria grants a 'rural investment allowance' to companies operating in rural areas.[109] The 'rural investment allowance' enables companies that provide infrastructural facilities to rural communities to deduct the costs of such infrastructure from their taxable profits. The 'rural investment allowance' applies

> where a company incurs capital expenditure on the provisions of facilities such as electricity, water or tarred road for the purpose of a trade or business which is located at least 20 kilometres away from such facilities provided by the government.[110]

The 'rural investment allowance' varies depending on the degree to which infrastructural facilities previously existed in the community and the type of infrastructure in question. Where no facilities previously existed in a community, a company providing infrastructural facilities is entitled to deduct 100% of the costs

---

106 *Ibid.* s. 24.(2).
107 *Companies Income Tax Act*, CAP. C21, LFN 2004, 2nd Schedule, para. 1
108 *Ibid.* s. 24.
109 *Ibid.* s. 34. This statute equally grants a 'reconstruction investment allowance' which allows companies to deduct 10% of the expenditure on plant and equipment from their taxable income. *Ibid.* s. 32. However, the 'reconstruction investment allowance' and the 'rural investment allowance' are mutually exclusive in the sense that both allowances cannot be granted in relation to the same assets. See *Companies Income Tax Act*, CAP. C21, LFN 2004, s. 34.(1).
110 *Companies Income Tax Act*, CAP. C21, LFN 2004, s. 34.(1).

of the facilities from its taxable profits.[111] However, the rates of deduction for electricity, water, and road facilities are 50%, 30%, and 15%, respectively, where these facilities did not previously exist.[112]

The 'rural investment allowance' is particularly relevant to the mining industry given that mining activities in Nigeria occur predominantly in rural communities. Mining companies operating in rural communities sometimes provide facilities for these communities in order to foster a cordial relationship with the communities. Thus, the 'rural investment allowance' complements the legal obligations imposed on mining companies in Nigeria to negotiate and sign CDAs with local host communities.[113] As discussed in Chapter 8, CDAs often contain provisions on the construction of infrastructural facilities in host communities. The result, then, is that companies providing infrastructural facilities for rural communities under a CDA are able to claim the 'rural investment allowance' in relation to such facilities.

In South Africa, assets used or expenditures incurred directly to win, recover, and develop mineral resources are deductible from the taxable income of mineral extractors.[114] The *Income Tax Act* allows general deductions in determining taxable income, including deduction of "expenditure and losses actually incurred in the production of the income, provided such expenditure and losses are not of a capital nature".[115] Also deductible are expenditures incurred during the year of assessment on prospecting operations (including surveys, boreholes, trenches, pits, and other exploratory work preliminary to the establishment of a mine), as well as any other expenditure which is incidental to such mining operations.[116] In addition, companies may set-off against their taxable income any balance of assessed loss incurred in any previous year, which was carried forward from the preceding year of assessment.[117]

Similarly, in Zambia mining companies are entitled to tax deductions on their losses and expenditures, but expenditures of a capital nature are non-deductible.[118] In Zambia, prospecting expenditure incurred by mining companies is deductible from the income of the companies for tax purposes.[119] Parent companies engaged in mining activities in Zambia can renounce their deductible prospecting expenditures to a new subsidiary company incorporated for the specific purpose of continuing the prospecting operations of the parent company or carrying on mining operations in Zambia.[120] In such a case, the parent company is

111 *Ibid.* s. 34.(2).
112 *Ibid.* s. 34.(2).
113 *Nigerian Minerals and Mining Act, 2007*, s. 116.
114 See *Income Tax Act, 1962* (Act 58) (South Africa), s. 11.(hA) (as amended by s. 12 of the *Income Tax Act, 1997* (Act No. 28 of 1997)).
115 *Ibid.* s. 11(a).
116 *Ibid.* s. 15.
117 *Ibid.* s. 20.
118 *Income Tax Act*, Chapter 323, Statutes of Zambia, s. 29.(1).
119 *Ibid.* s. 33 read with paras 20–3 of the 5th Schedule.
120 *Ibid.* 5th Schedule, para. 21.(3).

required to give a written notice to the Commissioner-General within 12 months after the incorporation of the new subsidiary company indicating that the parent company has irrevocably renounced their prospecting expenditure to the subsidiary.[121] These prospecting expenditures, which are deductible from the income of mining companies in Zambia, are deemed as losses occurring in the charge year in which the expenditures were incurred.[122] Furthermore, mining companies are allowed to deduct any interest on loans, but any such deduction in a charge year "shall not exceed the interest on any borrowings in excess of a loan-to-equity ratio of 2:1".[123] However, deduction of losses is allowed only where a company owns two or more mines, one of which is in regular production while the other mine incurs a loss. To that end, the *Income Tax Act* provides that:

> Where a person is carrying on mining operations in a mine which is in regular production and is also the owner of, or has the right to work, a mine which is not contiguous with the producing mine and from which the person has a loss in the charge year, the amount of such loss may be deducted in ascertaining the gains or profits from his mining operations in that charge year:
> Provided that the amount of tax which would otherwise be payable by such person in such charge year is not reduced by more than twenty per centum as a result of this deduction.[124]

That said, a recent amendment introduced some restrictions with regard to deductions of losses from the assessable income of mining companies in Zambia. Thus, a loss incurred by a mining company from their mining operation in a charge year is deductible from 50% of the income of the company accruing from the mining operation.[125] Where a loss arising from a mining operation exceeds 50% of the company's income for a charge year, "the excess shall, as far as possible, be deducted from fifty percent of that [company's] income from the mining operation in the following charge year".[126] However, a loss incurred by a mining company cannot be carried forward beyond ten subsequent years after the charge year in which the loss was incurred.[127]

4.2 Tax holidays, deferment of royalty, and other exemptions

Some African countries offer tax-free periods (usually referred to as tax holidays or tax relief) to mining companies, particularly during the early stages of mineral production. A tax holiday may consist of an exemption from all taxes over a specified period, or an exemption from a specific tax such as income tax over a

---

121 *Ibid*. 5th Schedule, para. 21.(3).
122 *Ibid*. 5th Schedule, para. 21.(4).
123 *Ibid*. 5th Schedule, para. 22.(7).
124 *Ibid*. 5th Schedule, para. 23.
125 *Ibid*. s. 30.(1)(a) (as amended by the *Income Tax (Amendment) Act*, No. 6 of 2015).
126 *Ibid*. s. 30.(2)(b) (as amended).
127 *Ibid*. s. 30.(3)(a) (as amended).

specified period, or a reduction in the rate of taxes over a specified period. In some instances, a tax holiday is based on the volume of production, such that where a company exceeds a specified level of production, the tax holiday lapses. In other instances, tax holidays may be tied to a prescribed minimum investment threshold or the number of new jobs created by the investor. In some cases, tax relief may take the form of a reduction in the rate of income tax for a given period. For example, in 2018 Mali granted Randgold Resources Limited a 50% corporate tax reduction for four years in relation to the Gounkoto mine which, upon completion, will be one of the largest opencast gold mines in Africa.[128]

However, the more common practice is to attach tax holidays to a specified period. For example, Nigeria grants an income tax relief period of three years commencing on the date of commencement of mining operations.[129] This tax relief period may be extended by the Minister for one further period of two years if the Minister is satisfied with

> (a) the rate of expansion, standard of efficiency and level of development of the company in mineral operations for which the mineral title was granted; (b) the implementation of any conditions upon which the lease was granted; and (c) the training and development of Nigerian personnel in the operation of the mineral concerned.[130]

The *Companies Income Tax Act* of Nigeria reiterates this tax exemption by providing that "a new company going into the mining of solid minerals shall be exempt from tax for the first three years of its operation".[131] Mining companies in Nigeria may also qualify for the 'pioneer status' tax holiday which allows companies to operate in their formative years without paying taxes.[132] However, to be eligible for the 'pioneer status' a company must demonstrate the tangible impact its business activity will have on Nigeria's economy including diversification of the economy and transfer of skills and technology to Nigerian citizens.[133] As of the time of writing, 39 companies had 'pioneer status' in Nigeria while eight other companies were provisionally approved for 'pioneer status'.[134]

---

128 Reuters, "Randgold Resources' Mali mine granted four-year, 50% tax concession", www.mining.com/web/randgold-resources-mali-mine-granted-four-year-50-tax-concession
129 *Nigerian Minerals and Mining Act, 2007*, s. 28.(1).
130 *Ibid*. s. 28.(2) & (3).
131 *Companies Income Tax Act*, CAP. C21, LFN 2004, s. 36.
132 *Industrial Development (Income Tax Relief) Act*, CAP. 17, LFN 2004s. 10.(1).
133 Nigerian Investment Promotion Commission & Federal Inland Revenue Service, *Compendium of Investment Incentives in Nigeria*, at 8, https://nipc.gov.ng/ViewerJS/?#../wp-content/uploads/2019/01/Compendium-of-Investment-Incentives-in-Nigeria-final.pdf
134 Nigerian Investment Promotion Commission, "Report on Pioneer Status Incentive Applications Q1 2020: 01 January to 31 March 2020", https://nipc.gov.ng/wp-content/uploads/2020/06/Q1-2020-PSI-Report-09062020.pdf

170 *Fiscal regimes for mineral exploitation*

Tax holidays are said to be inefficient and ineffective incentives for mining not only because they are susceptible to abuse but also because mining companies place low priority on tax holidays in making investment decisions.[135] The Organisation for Economic Co-operation and Development (OECD) reports that because mining investors usually have two or more operations within a country, they tend to allocate profits from their operations to the mining operation that is subject to a tax holiday, thereby reducing the investor's taxable income.[136] This is particularly so in countries where mining operations are not ring-fenced. Moreover, mining companies that are granted tax holidays sometimes engage in 'high-grading' practices, particularly where the tax holiday is time-limited. "High-grading involves companies increasing the rate of extraction or preferentially extracting high-grade ore compared to what they would otherwise do absent tax considerations", with the result that "the amount of tax relief is well above that originally envisioned by government".[137] The abuse of tax holidays deprives governments of mining revenues; hence, the OECD recommends that governments should avoid granting tax holidays to mining companies.[138]

Some African countries offer deferment of royalty payment, while other countries provide for the waiver of royalty. In Zambia, the payment of mineral royalty may be waived or deferred in very limited circumstances. A company may be exempted from royalty payment with regard to any samples of minerals acquired for the specific purposes of assay, analysis, or other examination, provided the quantity or value of the minerals is within the prescribed limit.[139] With regard to the deferment of royalty payment, a holder of a mining licence in Zambia is eligible for deferment of royalties if its cash operating margin – that is, the amount derived by deducting operating costs from revenue – in respect of mining operations in the mining area falls below zero.[140] In other words, the payment of royalty is deferred where the costs of operations exceed the revenues from a particular project. The deferred royalty becomes payable during a royalty payment period in which, after the deduction of the royalty then due, the cash operating margin of the mineral right holder is positive, provided that the deferred sum does not exceed a sum which would reduce the mineral right holder's cash operating margin for the relevant royalty payment period below zero.[141] Any amount deferred as royalty payment may be accumulated with any other deferred payment of mineral royalty that is outstanding.[142] Zambia's provisions on the deferment of

135 OECD, *Tax Incentives in Mining*, supra note 91 at 21–2.
136 *Ibid*. at 21.
137 *Ibid*. at 20.
138 *Ibid*. at 21–2.
139 *Mines and Minerals Development Act, 2015* (Zambia), s. 93.(2).
140 *Ibid*. s. 94.(1) & (3).
141 *Ibid*. s. 94.(2).
142 *Ibid*. s. 94.(2)(a).

royalty appear to be aimed at ensuring that royalties are paid only when a company makes a profit from its mining operations.

Unlike many African countries, South Africa exempts companies that qualify as "small business" from payment of mineral royalty. In South Africa, a mineral extractor is exempt from royalty in respect of a year of assessment if:

(a) gross sales of that extractor in respect of all mineral resources transferred does not exceed R10 million during that year;
(b) the royalty in respect of all mineral resources transferred that would be imposed on the extractor for that year does not exceed R100 000;
(c) the extractor is a resident as defined in section 1 of the Income Tax Act throughout that year; and
(d) the extractor is registered for that year pursuant to section 2 of the Administration Act.[143]

However, in South Africa a mineral extractor is not exempt from royalty in four enumerated circumstances; namely: (a) where the mineral extractor at any time during the year of assessment holds the right to participate (directly or indirectly) in more than 50% of the share capital, share premium, current or accumulated profits or reserves of, or is entitled to exercise more than 50% of the voting rights in any other mineral extractor; (b) where any other mineral extractor at any time during the year of assessment holds the right to participate (directly or indirectly) in more than 50% of the current or accumulated profits of the mineral extractor; (c) where any other person at any time during the year of assessment holds the right to participate (directly or indirectly) in more than 50% of the profits of the mineral extractor and more than 50% of the current or accumulated profits of any other mineral extractor; or (d) where the mineral extractor is a registered person mentioned in section 4 of the Administration Act.[144]

In addition, South Africa exempts from royalty minerals won or recovered by a mineral extractor for purposes of testing, identification, analysis, and sampling pursuant to a prospecting right or an exploration right "if the gross sales in respect of those mineral resources does not exceed R100 000 during a year of assessment".[145]

4.3 Import duty exemptions and currency transfer guarantees

Holders of mineral rights in Africa are granted additional fiscal incentives, such as exemptions from custom duties and foreign exchange rules, as well as guarantees regarding repatriation of profits and dividends. Ghana exempts mining companies "from payment of customs import duty in respect of plant, machinery, equipment

---

143 *Mineral and Petroleum Resources Royalty Act, 2008* (Act No. 28, 2008), s. 7.(1).
144 *Ibid.* s. 7.(2).
145 *Ibid.* s. 8.

172 *Fiscal regimes for mineral exploitation*

and accessories imported specifically and exclusively for the mineral operations".[146] It also grants exemptions for the staff of mining companies regarding the payment of income tax on furnished accommodations at the mine site, immigration quotas in respect of the approved number of expatriate personnel, and personal remittance quotas for expatriate personnel, free from tax imposed by an enactment regulating the transfer of money out of Ghana.[147] Likewise, mining companies in Nigeria are exempt from the payment of customs and import duties in respect of plant, machinery, equipment, and accessories imported specifically and exclusively for mining operations.[148] However, prior to such importation, mining companies must obtain the approval of the Mines Inspectorate Department.[149] Mining companies are equally entitled to expatriate quotas and resident permits for their foreign personnel, as well as personal remittance quotas for expatriate personnel, free from any tax imposed by any law in Nigeria for the transfer of foreign currency out of Nigeria.[150]

Mining companies in Africa are equally granted broad exemptions from foreign exchange laws, including laws prohibiting unauthorized transfer of foreign currencies. For example, holders of mineral rights in Ghana are allowed to retain in a bank account "an amount not less than twenty five percent of the foreign exchange" earned from their mining operations for purposes of acquiring spare parts, raw materials, machinery, and equipment for their operations; servicing of debts and payment of dividends to shareholders; remittance in respect of quotas for their expatriate staff; and transfer of capital in the event of a sale or liquidation of mining operations.[151] Similarly, Nigeria permits mining companies to retain in a domiciliary account a portion of any foreign exchange earned from the sale of minerals, but such foreign exchange must be spent on "acquiring spare parts and other inputs required for the mining operations which would otherwise not be readily available without the use of such earning".[152] Perhaps more significantly, mining companies

> [are] guaranteed free transferability through the Central Bank [of Nigeria] in convertible currency of – (a) payments in respect of loan servicing where a certified foreign loan has been obtained by the holder for his mining operations; and (b) the remittance of foreign capital in the event of sale or liquidation of the mining operations or any interest therein attributable to foreign investment.[153]

---

146 *Minerals and Mining Act, 2006* (Ghana), s. 29.
147 *Ibid.* s. 29.
148 *Nigerian Minerals and Mining Act, 2007*, s. 25.(1)(a).
149 *Ibid.* s. 25.(2).
150 *Ibid.* s. 25.(1)(b) & (c).
151 *Minerals and Mining Act, 2006* (Ghana), s. 30.(1) & (2).
152 *Nigerian Minerals and Mining Act, 2007*, s. 26.
153 *Ibid.* s. 27. See also *Minerals and Mining Act, 2006* (Ghana), s. 30.(4).

4.4 Allowances regarding environmental rehabilitation

Mining companies are sometimes able to deduct expenditures on environmental and waste treatment plant, as well as pollution control and monitoring equipment, provided the equipment is utilized in the course of the company's operations. Expenditures incurred in conserving or maintaining land covered by a mineral right are deemed to be incurred in the production of income and thus deductible from the taxable income of mining companies. Some African countries grant tax exemptions to mining companies in relation to monies paid into the environmental rehabilitation fund, as well as the cost of rehabilitating the area covered by a mineral right.[154] For example, mining companies in Ghana can deduct from their chargeable income expenditures incurred in the course of reclamation, rehabilitation, and closure of mineral operations.[155] In Nigeria, mining companies are entitled to a tax-deductible reserve for costs incurred in relation to environmental protection, mine rehabilitation, reclamation, and mine closure.[156] However, the appropriateness of such a tax-deductible reserve must be verified and certified by an independent qualified person who must be satisfied that (a) the reserve is appropriately recorded in the audited financial statements of the company; (b) tax deductibility is restricted to the actual amount incurred for the purpose of the reclamation; and (c) a sum equivalent to the reserve amount is set aside every year and invested in a dedicated account or trust fund managed by independent trustees.[157]

5 An appraisal of the fiscal regimes

African countries believe that financial incentives for mineral extraction are necessary to attract FDI; hence, they offer generous concessions to investors in the mining industry.[158] Africa hopes that the infusion of FDI into the mining industry will enhance economic development across the continent. While, as noted below, the incentive schemes may not be beneficial to Africa when viewed in the context of the various costs associated with the schemes, there is evidence suggesting that the schemes are attracting FDI to Africa. For example, the United Nations Conference on Trade and Development (UNCTAD) has reported that the increase in FDI in Africa's mining industry is due in part to the "increased incentives" offered to investors by African countries.[159] According to the UNCTAD, the incentive

---

154 See *Income Tax Act, 2015*, Act 896 (Ghana), s. 84.
155 *Ibid*. s. 81.(1).
156 *Nigerian Minerals and Mining Act, 2007*, s. 30.
157 *Ibid*. s. 30.
158 Ousman Gajigo, Emelly Mutambatsere, & Guirane Ndiaye, "Gold Mining in Africa: Maximizing Economic Returns for Countries" (African Development Bank Working Paper No. 147, March 2012) at 14, www.afdb.org/fileadmin/uploads/afdb/Documents/Publications/WPS%20No%20147%20Gold%20Mining%20in%20Africa%20Maximizing%20Economic%20Returns%20for%20Countries%20120329.pdf
159 UNCTAD, *Economic Development in Africa: Rethinking the Role of Foreign Direct Investment* (Geneva: United Nations, 2005) at 39 [*Economic Development in Africa*].

regimes, coupled with the deregulation of the mining sector, has made Africa "much more attractive as a location for mining FDI".[160] Statistics in the last two decades appear to bear out UNCTAD's assertion. For example, investments in natural resource exploitation accounted for most of the $36 billion FDI inflows to Africa in 2006.[161] In addition, new mining projects that came into full production by the end of 2018 accounted for about US$18 billion in investment across the continent.[162] However, the low commodity prices of the last few years have negatively impacted the inflow of FDI to Africa. In 2015, for example, FDI inflow to West Africa was US$9.9 billion, representing a decline of 18% over the previous year, while FDI inflow to Central Africa declined by 36% to US$5.8 billion.[163]

Although the fiscal and tax incentives have contributed to the increase in FDI in Africa, it would be naïve to assume that the surge in FDI in Africa's mining industry is attributable solely to the fiscal incentives offered by African governments. As Magnus Blomstrom has rightly noted, "international investment incentives play only a limited role in determining the international pattern of [FDI]".[164] Admittedly, investors consider the fiscal incentives available in a country in deciding whether to invest in that country. But that is not all. Investors also consider other factors, such as the political stability of a country, the degree of legal protection accorded mineral right holders, and "market characteristics, relative production costs and resource availability".[165] Resource availability plays a prominent role in FDI in the extractive industries. Because commercial quantities of mineral resources are available in a limited number of countries, prospective investors in the extractive industries have a constricted pool from which to choose the location of their investments. Given the limited availability of mineral resources, coupled with the intense global competition for access to these resources, it is arguable that some of the FDIs in Africa's extractive industries would have been attracted to Africa in the absence of the fiscal incentives offered by African countries.

The increase in FDI has also enhanced production of mineral resources in Africa by enabling resource extractors to expand their exploration and production capacity, including "the development of new mines".[166] In Tanzania, for example, gold exports accounted for less than 1% of the country's total export earnings in

---

160 *Ibid.* at 43.
161 UNCTAD, *World Investment Report 2007: Transnational Corporations, Extractive Industries and Development* (New York & Geneva: United Nations, 2007) at 34.
162 Deloitte, "State of Mining in Africa: In the Spotlight" at 7, www2.deloitte.com/content/dam/Deloitte/tr/Documents/energy-resources/za-state-of-mining-africa-09022015.pdf
163 UNCTAD, *World Investment Report 2016: Investor Nationality: Policy Challenges* (New York & Geneva: United Nations, 2016) at 40.
164 Magnus Blomström, "The Economics of International Investment Incentives", in *International Investment Perspectives* (OECD, 2002) 165 at 169, www.oecd.org/daf/inv/investment-policy/2487874.pdf
165 *Ibid.* See also Shandre M. Thangavelu, Yik Wei Yong, & Aekapol Chongvilaivan, "FDI, Growth and the Asian Financial Crisis: The Experience of Selected Asian Countries" (2009) *The World Economy* 1461 at 1462.
166 UNCTAD, *Economic Development in Africa*, supra note 159 at 43.

the late 1990s, but by 2003 gold exports accounted for over 40% of the export earnings of that country.[167] Copper production in Zambia increased significantly following the introduction of generous fiscal incentives under the privatization exercise undertaken by the Zambian government in the 1990s.[168] More specifically, in 2006 copper production in Zambia increased by about 7.1% due to an increase in FDI, while the contribution of the mining industry to Zambia's GDP increased from 6.2% in 2000 to 11.8% in 2005.[169]

The enhancement of mineral production has had a consequential positive impact on the revenues accruing to mineral-rich African countries. In fact, mining revenues in Africa have increased modestly in the last two decades. For example, between 1990 and 2004, Ghana's gold exports increased three-fold, resulting in the mining industry accounting for about 11% of the total revenues collected by the government of Ghana.[170] Mining revenues in Zambia increased by 99% between 2010 and 2011, partly due to the new tax regimes introduced in 2008 and partly as a result of increase in the production of copper.[171]

That said, it is worth noting that recent modest increases in mining revenues in Africa may not be due solely to the increase in FDI in the mining industry. Mining revenues have also been enhanced by the adoption of value-based and profit-based royalties and taxes in Africa. As opposed to the unit-based system that fixes royalty and tax rates at a predetermined figure or price, the value-based and the profit-based systems assess royalties on the basis of the ascertained value of minerals and the profitability of the mining operation, respectively.[172] As is apparent in the analysis above, royalties and taxes for mineral resources in Botswana, Ghana, Nigeria, South Africa, and Zambia are for the most part value-based, although some countries have a hybrid of the value-based and the profit-based systems. The unit-based system has been rightly abandoned by African countries not only because it is rigid, but also because its "posted price" regime may be unable to cater to changing economic circumstances in the global economy. It may in fact be disadvantageous financially to a host country in circumstances where the value or market price of a mineral resource increases significantly after the enactment of the law stipulating a posted price for the mineral.[173]

---

167 Ibid. at 48.
168 Neo Simutanyi, "Copper Mining in Zambia: The Developmental Legacy of Privatization" (Institute for Security Studies Occasional Paper 165, July 2008), https://issafrica.org/research/papers/copper-mining-in-zambia-the-developmental-legacy-of-privatisation
169 Ibid.
170 UNCTAD, Economic Development in Africa, supra note 159 at 50.
171 Zambia Extractive Industries Transparency Initiative, Reconciliation Report for the Year 2011 (February 2014) at 12, https://eiti.org/sites/default/files/migrated_files/Zambia-2011-EITI-Report.pdf
172 See Otto et al., supra note 4 at 50–5.
173 See James Otto, "Legal Approaches to Assessing Mineral Royalties" in James Otto, ed., The Taxation of Mineral Enterprises (London: Graham & Trotman/Martinus Nijhoff, 1995) 131 at 133 (noting that the unit-based royalty system "precludes an increase in revenue in times of rising prices").

5.1 Disconnect between mining revenues and the value of minerals

Although Africa's generous fiscal regimes have attracted FDI to the mining industry and while mining-related revenues have trended upwards in recent years, the modest increase in mining revenues is not commensurate with the volume and value of minerals produced in Africa. As John Jacobs observes, "the success in attracting investment has not been matched by proportionate increases in government revenue".[174] For example, "Sierra Leone received about $10 million from $179 million exports of minerals in 2007".[175] Likewise, in Tanzania "six major mining companies earned total export revenues of about $890 million (between 1997 and 2002), from which the government received $86.9 million".[176] Ghana suffered a similar fate in 2003 when it received about $46.7 million out of a total mineral export value of $893.6 million.[177]

The sub-optimal revenues generated by African countries from mineral exploitation is partly attributable to the generous fiscal and tax regimes which pervade much of Africa. As Francis Botchway notes, mining companies in many African countries pay neither value-added tax nor import and export duties.[178] Moreover, very few African countries impose a withholding tax on mining companies and even the "few countries that levy withholding tax do so at rates between 10 and 15 percent, compared with the world-wide industry average of about 30 percent".[179] Furthermore, African countries do not impose a windfall tax on excessive profits earned by mining companies, particularly during periods of commodity boom. Zambia introduced a windfall profits tax in 2008, but the government repealed the tax in 2010 partly due to the global financial crisis that impacted adversely on mining investments, and partly due to the immense pressure exerted on the government by mining MNCs.[180]

In some instances, the investment allowances and tax holidays offered to mining companies enable these companies to mine minerals without paying taxes. For example, in Ghana where most mining companies operate open-pit mines with relatively short lifespans, the generous tax incentives and capital allowances ensure that the "companies benefit from a virtual tax holiday throughout their operations and most often run out of [mineral] reserves and close before they are expected to

---

174 John Jacobs, "An Overview of Revenue Flows from the Mining Sector: Impacts, Debates and Policy Recommendations", in Bonnie Campbell, ed. *Modes of Governance and Revenue Flows in African Mining* (Basingstoke: Palgrave Macmillan, 2013) at 20.
175 Francis N. Botchway, "Mergers and Acquisitions in Resource Industry: Implications for Africa" in Botchway, ed., *Natural Resource Investment and Africa's Development* (Cheltenham: Edward Elgar, 2011) 159 at 182.
176 UNCTAD, *Economic Development in Africa, supra* note 159 at 48–50.
177 *Ibid.* at 50.
178 Botchway, *supra* note 175 at 182.
179 *Ibid.*
180 Anthony Simpasa *et al.*, "Capturing Mineral Revenues in Zambia: Past Trends and Future Prospects", at 12 (UNDP Discussion Paper, August 2013), www.un.org/en/land-natural-resources-conflict/pdfs/capturing-mineral-revenues-zambia.pdf

pay corporate income tax".[181] Until recently, mining companies in Tanzania were exempted from paying income tax for the first ten years of operations, hence the paucity of mining revenues in Tanzania. For example, "AngloGold Ashanti paid $1 million tax in 2007, more than ten years after it acquired the Geita property in Tanzania".[182] While some African countries have increased the royalty and tax rates applicable to mining in recent years, rampant corruption and the numerous loopholes in the fiscal regimes enable mining companies to diminish their tax liability to African countries and, in some cases, evade payment of taxes altogether.

The period preceding the developmental statutes of the current era witnessed widespread deregulation of the African mining industry, including privatization of state-owned enterprises and the relaxing of regulation. During this period, some African countries deliberately set royalty and tax rates for mineral resources below industry standards in order to attract foreign investors. While African countries have recently attempted to reassert control over the mining industry through fiscal reforms, foreign investors continue to exert pressure on African governments to enact royalty and tax regimes that are overly generous towards investors. Zambia apparently succumbed to such pressure when it amended its mining statute in 2016 and reduced the royalty rates for minerals from 9% to 6%.[183] Prior to the 2016 amendment, the royalty rates for minerals in Zambia where 9% of the norm value of base metals and precious metals produced through opencast mining operations; 6% of the gross value of gemstones or energy minerals produced through underground mining operations; and 6% of the gross value of the industrial minerals produced or recoverable under a mining licence.[184] Similarly, Ghana reduced the royalty rate for minerals from an upper limit of 12% (under the *Mineral and Mining Law, 1986*) to 6% in 2006 and 5% in 2010.[185]

5.2 Abuse of the fiscal regimes through tax-avoidance schemes

The fiscal regimes and incentive schemes for mineral extraction in Africa are often manipulated by MNCs for their financial benefit. MNCs manipulate Africa's fiscal regimes through tax evasion and unethical practices such as transfer pricing. Zambia is reportedly losing about US$3 billion a year due to tax dodging by mining companies.[186] Mining companies intentionally inflate the cost of production of minerals, such as cost of equipment, machinery, building, and

---

181 Thomas Akabzaa, "Mining Legislation and Net Returns from Mining in Ghana" in Bonnie Campbell, ed., *Regulating Mining in Africa: For Whose Benefit* (Nordiska Afrikainstitutet Discussion Paper 26) 25 at 29.
182 Botchway, *supra* note 175 at 182.
183 See *Mines and Minerals Development (Amendment) Act, No. 14 of 2016* (Zambia).
184 *Mines and Minerals Development Act, 2015* (Zambia), s. 89 (repealed).
185 *Minerals and Mining (Amendment) Act, 2010* (Act 794). The *Minerals and Mining (Amendment) Act*, 2010 (Act 794) was itself repealed in 2015 through the *Minerals and Mining (Amendment) Act*, 2015 (Act 900).
186 NIPSA, "Extracting Minerals, Extracting Wealth: How Zambia is losing $3 billion a year from corporate tax dodging", www.waronwant.org/sites/default/files/WarOnWant_ZambiaTaxReport_web.pdf

management, thus increasing their tax deductions and allowances. For example, an audit of the Mopani Copper Mines Plc in Zambia indicates that the company inflated the cost of labour, fuel, insurance, security, and safety, apparently with a view to increasing its tax deductions.[187] Mopani Copper Mines Plc equally engaged in non-arm's-length transactions with an affiliate company, Glencore UK Ltd, to whom it apparently sold copper at prices that are incompatible with prevailing market prices.[188]

In addition, mining companies in Africa intentionally engage in high levels of borrowing from parent or affiliate companies based in foreign countries.[189] They do so in order to claim the interest on such loans as tax deductions. As discussed above, interest on foreign loans is tax-deductible in many African countries. However, because these loans are obtained from parent companies, subsidiary companies in Africa are able to transfer parts of their income to the parent company as interest on the loans, thus reducing their tax liability in Africa. In extreme cases, the interest rates on loans provided to subsidiary companies by parent companies are deliberately set high in order to reduce the chargeable income of the subsidiary companies. In other instances, mining companies manipulate the sliding-scale royalty system to reduce royalty payment to the host government. Unless the price of minerals is tied to a prescribed metric or index, mining companies can easily abuse the sliding-scale royalty system by adopting tax-planning strategies to avoid falling into a higher royalty bracket.[190] In Ghana, some mining companies undermine the sliding-scale royalty system by quoting "different prices for gold sold on the same day, leading to different royalty payment calculations".[191] The sliding-scale royalty system is particularly prone to abuse where the sliding-scale system is aggregate in nature. The aggregate sliding-scale system incentivizes companies to sell minerals at prices slightly below the price boundary because "the tax rate in the aggregate structure jumps at each price boundary".[192] Under the aggregate sliding-scale structure, "the seller has an incentive to price up to the boundary change but not beyond this, where extra revenue gained from selling at a higher price is offset by the additional royalty due from paying the higher rate".[193]

The value-based and profit-based royalty and tax systems in Africa are equally manipulated by mining companies when they under-report the volume and value of mineral ore produced in a country, thus reducing their royalty and tax liability.

187 Grant Thornton, "Pilot Audit Report – Mopani Copper Mines Plc", www.facing-finance.org/wp-content/blogs.dir/16/files/2012/03/2010_Report_audit_Mopani.pdf
188 *Ibid.* at 8 & 18.
189 Pietro Guj *et al.*, *Transfer Pricing in Mining with a Focus on Africa: A Reference Guide for Practitioners* (World Bank, 2017) at 96.
190 OECD, *Tax Incentives in Mining, supra* note 91 at 39.
191 Open Society Institute of South Africa *et al.*, "Breaking the Curse: How Transparent Taxation and Fair Taxes can Turn Africa's Mineral Wealth into Development", at 28, https://elaw.org/system/files/breaking-the-curse-march2009.pdf?_ga=2.90897050.1455428545.1579812908-917430816.1579812908 ["Breaking the Curse"].
192 OECD, *Tax Incentives in Mining, supra* note 91 at 40.
193 *Ibid.*

In Tanzania, for example, Acacia Mining Plc under-reported the volume and value of its mineral production and exports.[194] In fact, between 2010 and 2015 Acacia did not pay corporate income tax to Tanzania even though it earned significant profits from its mining operations in Tanzania.[195] Ironically, during the same period Acacia distributed US$444 million in dividends to its shareholders.[196] A subsidiary of Barrick Gold Corp. in Tanzania has similarly evaded paying more than US$40 million in taxes through a sophisticated tax-evasion scheme.[197] In addition, mining companies abuse the import duty exemption by importing dual-use goods and goods not directly related to mining (such as office supplies) without paying import duty. The import duty exemption is particularly prone to abuse because, in many African countries, the scope of the exemption is broad and open-ended. Tax administrators in Africa are often unable to detect the abuse and manipulation of the fiscal regimes due to their incapacity, coupled with the complex nature of the accounting practices adopted by mining MNCs.

MNCs are usually able to manipulate Africa's fiscal regimes through the complex nature of their ownership structure. MNCs adopt a multi-layered ownership structure not only as a risk management strategy but also as a tax-avoidance mechanism. MNCs usually have an intricate web of ownership involving shareholders, parent companies, holding companies, and subsidiary companies spread across multiple countries and continents.[198] This complex ownership arrangement is widespread in the African mining industry where many companies have an ownership structure that is 'layered' and linked to offshore affiliates.[199] For example, Sierra Leone Hard Rock (SL) Limited, a mining company, operates "through three separate offshore holding companies (two registered in Guernsey and one in Bermuda) with a primary owner registered in Bermuda, owned in turn by three separate holding companies (two of which were registered in London and one in China)".[200]

This complex ownership structure not only allows MNCs to evade domestic regulation but also enables them to engage in unethical practices such as tax

194 Thabit Jacob & Rasmus H. Pedersen, "New Resource Nationalism? Continuity and Change in Tanzania's Extractive Industries" (2018) 5 *The Extractive Industries and Society* 287 at 287–8.
195 Maya Forstater & Alexandra Readhead, "A brutal lesson for multinationals: Golden tax deals can come back and bite you", *The Guardian*, Thursday 6 July 2017, www.theguardian.com/global-development-professionals-network/2017/jul/06/a-brutal-lesson-for-multinationals-golden-tax-deals-can-come-back-and-bite-you
196 *Ibid.*
197 Geoffrey York, "Barrick Gold subsidiary evaded Tanzanian taxes, tribunal rules", www.theglobeandmail.com/report-on-business/international-business/african-and-mideast-business/barrick-gold-subsidiary-evaded-tanzanian-taxes-tribunal-rules/article29533858
198 See UNCTAD, *World Investment Report 2016*, supra note 163 at xiii.
199 Africa Progress Panel, *Equity in Extractives: Stewarding Africa's Natural Resources for All*, at 60–1, https://static1.squarespace.com/static/5728c7b18259b5e0087689a6/t/57ab29519de4bb90f53f9fff/1470835029000/2013_African+Progress+Panel+APR_Equity_in_Extractives_25062013_ENG_HR.pdf
200 *Ibid.* at 61.

evasion, price transfer practices, royalty payments to subsidiaries, and round-tripping. Through price transfer practices, mining MNCs are able to transfer goods, services, knowledge, and intellectual property between their subsidiaries and affiliates, thereby reducing the overall costs of production and enhancing their profits.[201] In the mining industry, for example, MNCs engage in transfer pricing practices such as the provision of debt finance by parents companies to subsidiary companies at above-market interest rates; the provision by parent companies of capital goods and machinery to subsidiary companies at above-market costs; and the claiming of excessive management fees, headquarter costs, and consultancy charges by parent companies for services rendered to subsidiary companies in Africa.[202] Mining companies can maximize their deductible capital expenditures and capital investment allowances by procuring "capital goods and machinery in leasing arrangements at above-market costs charged by a related-party lessor".[203] MNCs abuse Africa's fiscal regimes by artificially "increasing the cost of imported equipment and material procured from related parties to reduce taxable income in the mining country".[204] Mining MNCs artificially inflate costs either by "paying the retail price for older equipment and machinery that has been used by an affiliate company in operations elsewhere" or by "paying a higher markup on the cost of equipment and machinery purchased through a corporate services hub located in a low- or zero-tax jurisdiction".[205] Subsidiary companies are able to reduce their tax liability to African countries by inflating such expenditures because most African countries allow mining companies to deduct their capital expenditures from their earnings.

Transfer pricing also allows MNCs to reduce their tax obligations by manipulating their accounts to show losses or low profits in African countries and large profits in foreign tax havens. Transfer pricing is a simple technique through which MNCs "ship goods and services whose unit value, decided by the [MNC] itself, is often biased by tax considerations".[206] Through transfer pricing, MNCs are able to "pay higher amounts to affiliates [based in countries] where taxes are lower, and show lower values [in countries] where taxes and/or tariffs are higher".[207] In more extreme cases, mining companies engage in transfer pricing in Africa by undertaking non-arm's-length transactions such as selling minerals to their foreign subsidiaries at below-the-market prices. The World Bank captures the transfer pricing practices of a mining MNC in South Africa in the following terms:

---

201 Peter T. Muchlinski, *Multinational Enterprises and the Law*, 2nd Edition (Oxford: Oxford University Press, 2007) at 269.
202 Baunsgaard, *supra* note 22 at 21.
203 *Ibid.* at 21.
204 OECD, *Tax Incentives in Mining*, *supra* note 91 at 35.
205 *Ibid.* at 35.
206 Farok J. Contractor, "Tax Avoidance by Multinational Companies: Methods, Policies, and Ethics" (2016) 1(1) *Rutgers Business Review* 27 at 29.
207 *Ibid.* at 29.

[A] foreign investor (Head Co.) controls two PGMs mines in South Africa through a local subsidiary (Holding Co.). This subsidiary holds 70% and 60% equity in each of them with the remainder held by a BEE partner (BEE Co.) which purchased shares in the mines. The foreign investor also registered and holds 100% equity in an offshore trading company (Trader Co.) to market the commodities from the two mines. BEE Co., however, is not a shareholder in Trading Co. The minerals produced are priced by international buyers/traders and sellers at a South African harbour on a FOB basis, determining their "Spot Export Market Price." Under the influence of the holding Head Co. the mines entered into a marketing agreement to transfer their product to the foreign-owned Trading Co. at a 7% discount to the "Spot Export Market Price," which the Trading Co. claims is an appropriate margin for the marketing and trading services rendered. In addition, occasionally Holding Co. derives a further margin by selling the commodity to end users at a price higher than the "Spot Market Price."[208]

This marketing arrangement, particularly the sale of minerals to the foreign affiliate of the MNC at the discounted rate of 7%, minimized the profits made by the MNC's South African subsidiary, thus reducing both its tax liability to the government as well as dividends to its South African shareholders. An audit of this transaction found that a discount of between 2% and 3% should have been sufficient to cover all fixed and variable costs, including a reasonable return associated with the services provided by the foreign affiliate.[209] In effect, the transaction did not meet the arm's-length requirements in South Africa.

The value-based royalty system in some African countries appear to incentivize MNCs to sell minerals to subsidiary companies at below-the-market prices. Because African states lack the capacity to determine the appropriate value of minerals, tax administrators often adopt the low prices (invoiced by mining companies) as the basis for calculating the companies' royalty and tax liabilities. This problem can be avoided by tying the value of minerals to the market value of minerals as determined by international commodity exchanges such as the London Metal Exchange. Zambia adopted this strategy in 2008. In that country, the value of base metals and precious metals produced under a mining licence are determined based on the monthly average cash prices per metric tonne of the metals on the London Metal Exchange.[210]

The multi-layered and offshore ownership structure of some mining MNCs not only aids tax evasion but also serves as a conduit for corruption, money laundering, and bribery in Africa.[211] The secrecy shrouding business transactions in the "offshore world creates an impenetrable barrier behind which" Africa's corrupt

208 Guj et al., supra note 189 at 115.
209 Ibid. at 115.
210 Mines and Minerals Development Act, 2015 (Zambia), s. 89 (as amended by the Mines and Minerals Development (Amendment) Act, No. 14 of 2016).
211 Africa Progress Panel, Equity in Extractives, supra note 199 at 61.

public officials and unscrupulous political leaders hide the diversion of mineral wealth.[212] Moreover, the rampant corruption pervading public agencies in Africa enables MNCs to manipulate the fiscal regimes by gratifying greedy public officers. Companies can avoid paying taxes by bribing corrupt public officials, as was the case in Nigeria where Halliburton admitted that it bribed Nigerian officials to evade the payment of taxes.[213] Admittedly, MNCs sometimes create a complex web of ownership in order to advance their legitimate business interest, but it is also the case that MNCs "will endeavour to affect any change in ownership structures in the most advantageous manner possible, especially from a fiscal and risk management perspective".[214]

5.3 Revenue costs of the generous fiscal regimes

The fiscal and tax incentives in the African mining industry may have contributed to the increase in mineral production and revenues across Africa, but it is not altogether certain that these incentive schemes are economically beneficial to Africa. This is especially so when we consider that the incentive schemes have certain costs that are borne by individual African countries, including "revenue costs" and costs of implementation of the schemes.[215] Fiscal and tax incentives, by their nature, require countries to forgo certain revenues that would otherwise accrue to them. For example, the exemption from import duty imposes a direct financial cost on host African countries because it permits mining companies to use port facilities and services without paying for such. More significantly, by exempting mining companies from import duty African countries forgo the income they would otherwise have generated from the payment of import duty.

The exclusion of certain items from the taxable profits of mining companies and the exemption of these companies from the payment of customs duty constrict the tax base available to African countries. The constriction of the tax base not only reduces the amount of taxes collected by African countries, but also has a detrimental effect on the overall development of resource-dependent countries. Alexander Klemm captures this sentiment best when he notes that tax incentives impose wide-ranging costs on host countries, including "distortions to the economy as a result of preferential treatment of investment qualifying for incentives, administrative costs from running and preventing fraudulent use of incentive schemes, and social costs of rent-seeking behavior, including possibly an increase

---

212 *Ibid.* at 61.
213 Securities and Exchange Commission, "SEC Charges KBR and Halliburton for FCPA Violations", www.sec.gov/news/press/2009/2009-23.htm; Jim Cason, "Nigeria: U. S. Firm Halliburton Acknowledges Bribe to Nigerian Official", https://allafrica.com/stories/200305090511.html
214 UNCTAD, *World Investment Report 2016, supra* note 163 at 124.
215 Alexander Klemm, "Causes, Benefits, and Risks of Business Tax Incentives" (IMF Working Paper WP/09/21) at 11, www.imf.org/en/Publications/WP/Issues/2016/12/31/Causes-Benefits-and-Risks-of-Business-Tax-Incentives-22628

in corruption".[216] These costs are particularly detrimental to mineral-dependent African countries because, as argued earlier, some of the foreign investments attracted by the fiscal incentives would, in the absence of the incentives, still have been attracted to Africa. This being so, "the entire tax revenue waived" under these incentive schemes constitutes direct revenue cost to African countries.[217]

In other cases, the fiscal policy choices made by African countries impose additional revenue costs on the government, thus diminishing mining revenues. As discussed in Chapter 6, African countries enter into fiscal stabilization agreements some of which prescribe royalty rates that are lower than the statutory rates. These stabilization agreements compromise the ability of African states to optimize mining revenues. For example, until 2008 when Zambia revised mining royalties upwards, Zambia gained minimal revenues from the mining industry because of the low rate of royalty prescribed under the development agreements (DAs) that governed the operations of mining MNCs.[218] The low rate of royalty prescribed in the DAs ensured that the copper price boom of the early to mid-2000s yielded virtually no revenue to the government of Zambia.[219] Moreover, stabilization agreements are sometimes deliberately designed to supersede the fiscal regimes specified in domestic African statutes, thus ensuring that the low royalty rates in the stabilization agreements take precedence over the statutory rates.[220]

In addition, some African countries are signatories to bilateral agreements against double taxation. For example, the President of Zambia is authorized to

> enter into an agreement, which may have retrospective effect, with the government of any other country or territory with a view to the prevention, mitigation or discontinuance of the levying, under the laws of the Republic [of Zambia] and of such other country or territory, of taxes in respect of the same income.[221]

Double taxation agreements may not prevent host African governments from imposing taxes on companies but, in some cases, they restrict the ability of host countries to tax the income of mining companies.

Furthermore, the rapid expansion in mineral exploration and production in Africa, which is aided largely by the generous fiscal regimes for mineral extraction, may be exacerbating the social costs of mineral exploitation, including environmental pollution, human rights violations, and conflicts between mining companies and host communities. Thus, unless the increase in mineral exploitation is matched by a corresponding increase in the enforcement of regulatory standards,

---

216 *Ibid.*
217 *Ibid.*
218 See Christopher Adam & Anthony M. Simpasa, "Harnessing Resource Revenues for Prosperity in Zambia", at 36–7, www.economics.ox.ac.uk/materials/working_papers/4930/oxcarrerp201036.pdf
219 *Ibid.* at 37.
220 Otto *et al.*, *supra* note 4 at 248.
221 *Income Tax Act*, Chapter 323, Statutes of Zambia, s. 74.(1).

the social instability prevalent in some mineral-rich African countries is likely to be sustained, if not exacerbated. As this author has noted elsewhere, certain strategies, such as the provision of adequate funds, equipment, and personnel, may ameliorate the problem of regulatory enforcement in Africa.[222]

5.4 Lack of institutional capacity to implement royalty and tax regimes

Mining revenues are not currently optimized in Africa not only because of the overly generous nature of the fiscal regimes, but also because of the lack of effective implementation of royalty and tax policies. Africa lacks the institutional capacity to implement complex royalty and tax regimes. As mentioned earlier, mining royalties and taxes in many African countries are value-based or profit-based.[223] The effective administration of value-based or profit-based royalties and taxes "requires the collection, presentation, and audit of prices, volumes, and a sometimes bewildering variety of costs on a timely basis".[224] In countries lacking the ability to detect tax-avoidance schemes, mining companies are often able to reduce revenues or inflate expenditure deductions in order to minimize their tax liability.[225] Such is the case in Africa where tax administrators lack the capacity to analyze and audit mineral prices and volumes, thus offering mining companies "considerable scope for tax evasion through misrepresentation of revenues or cost inflation".[226] Quite often, institutions and agencies responsible for administering the fiscal regimes for mineral extraction in Africa lack adequately trained staff, adequate funding, and the infrastructure necessary for effective implementation of the fiscal regimes.[227] They are thus often unable to assess the accuracy or reasonableness of financial returns submitted by mining companies for royalty and tax purposes.[228] Africa's institutional weaknesses incentivize mining MNCs to manipulate the fiscal regimes and engage incessantly in tax avoidance.

Moreover, the ability of MNCs to manipulate Africa's fiscal regimes is aided by the incompetence and lack of expertise on the part of African tax administrators. For example, it has been reported that under Ghana's erstwhile sliding-scale royalty system, which consisted of royalty rates fluctuating between 3% and 6% of the gross value of minerals, no company "ever paid more than 3% in royalties, partly because of the high capital allowances, and partly because Ghana's tax collection authorities do not know how to use the formulae" for determining royalties.[229]

---

222 See Evaristus Oshionebo, *Regulating Transnational Corporations in Domestic and International Regimes: An African Case Study* (Toronto: University of Toronto Press, 2010) at 244.
223 See Otto *et al.*, *supra* note 4 at 83–4.
224 World Bank, *Taxation and State Participation in Nigeria's Oil and Gas Sector* (Washington, DC: The World Bank, 2004) at 12 [*Taxation and State Participation*].
225 See Baunsgaard, *supra* note 22 at 6.
226 *Ibid.* at 12.
227 World Bank, *Taxation and State Participation*, *supra* note 224 at 49.
228 *Ibid.*
229 "Breaking the Curse", *supra* note 191 at 29.

The inability of tax administrators in Ghana to assess and collect royalties and taxes has reportedly led to that country losing at least US$387.74 million between 1990 and 2007.[230] Although Ghana abolished the sliding-scale royalty system and replaced it with a flat rate of 5% in 2010,[231] Ghana reverted to the sliding-scale system in 2015 due partly to the pressure exerted on the government by mining companies.[232]

6 Anti-tax-avoidance rules

As discussed above, the fiscal regimes for mining in Africa offer generous incentives to investors. Despite the generosity of the fiscal regimes, mining companies in Africa routinely abuse these regimes through unethical practices designed to evade or limit their financial liability to African countries. Africa's desire to use mining revenues as a catalyst for economic development may not be realized unless Africa devises strategies to optimize mining revenues. Revenue-maximizing strategies must include mechanisms that prevent mining companies from avoiding or limiting their financial obligations in host African countries. One strategy would be to limit the scope of the financial allowances, deductions, and exemptions afforded to mining companies. In particular, as opposed to the current situation in many African countries where the list of items exempted from import duties is expansive, this list should be constricted to a few items. Even then, exemption from import duties should not apply throughout the duration of a mine; rather, the exemption should cover a specified period, such as the early stages of a mine.

A prominent strategy for addressing tax avoidance is the enactment of anti-tax-avoidance legislation with clear implementation guidelines. Such legislation should not only provide for the adoption of the 'arm's-length principle' regarding the sale of mineral resources, but must also impose disclosure obligations on mining companies. Legislation may also 'ring-fence' mining operations for tax purposes such that where a mining company owns several mining operations in a country, each mining operation is segregated and treated as a separate asset for tax purposes. In such cases, each mining operation is insulated from any loss arising from other mining operations owned by the company. This section considers anti-tax-avoidance legislation in Africa focusing on Botswana, Ghana, and South Africa.

6.1 Suspicious and fictitious transactions

Anti-tax-avoidance legislation empowers a designated government official to disregard any suspicious transaction or scheme by mining companies to avoid or limit

---

230 Ibid.
231 *Minerals and Mining Act, 2006* (Ghana), s. 25 (as amended by the *Minerals and Mining (Amendment) Act, 2010* (Act 794)).
232 *Minerals and Mining (Amendment) Act, 2015* (Act 900).

payment of royalties and taxes. Hence, in practice, transactions that are fictitious or artificial and transactions intended to reduce, postpone, or avoid the tax liability of mining companies are disregarded by tax administrators when determining the royalty and tax liability of companies.[233] In Botswana, for example, the Commissioner General of Taxes has the power to disregard certain expenditures, including (1) head office expenses that exceed the prescribed limit of 1.5% of the gross income of the company for the year of assessment, and (2) excessive or 'above-market' interest rates on foreign loans.[234] In regard to the latter, where a foreign company grants a loan to its subsidiary in Botswana at a rate that is in excess of the market rate, the Commissioner is obliged to disallow that part of the interest which exceeds the market rate, and the amount of interest so disallowed is treated and taxed as a dividend.[235]

In Ghana, a transaction is disregarded for tax purposes if it is entered into or carried out as part of a tax-avoidance scheme, or if it is fictitious, or lacks a substantial economic effect, or its form is not reflective of its substance.[236] In addition, mineral right holders in Ghana are prohibited from engaging in income splitting and thin capitalization practices if the purpose is to reduce their tax liability.[237] With regard to thin capitalization, where a mining entity

> in which fifty percent or more of the underlying ownership or control is held [by a non-resident person] either alone or together with an associate has a debt-to-equity ratio in excess of three-to-one at any time during a basis period, a deduction is disallowed for any interest paid or foreign currency exchange loss incurred by that entity during that period on that part of the debt which exceeds the three-to-one ratio.[238]

Similarly, in South Africa, the Tax Commissioner has the power to disregard any transaction (such as a disposal, transfer, operation, scheme, or understanding) that is designed to avoid, postpone, or limit a resource extractor's royalty liability; transactions that do not serve a bona fide business purpose; and transactions that are undertaken solely or mainly for the purpose of obtaining a royalty benefit.[239] Furthermore, a transaction is disregarded if it "created rights or obligations which would not normally be created between persons dealing at arm's length".[240]

---

233 *Income Tax Act, Chapter 52:01* (Botswana), s. 36 (as amended by the *Income Tax (Amendment) Act, 2018* (No. 38), s. 2); *Income Tax Act, 2015, Act 896* (Ghana), s. 34; *Mineral and Petroleum Resources Royalty Act, 2008* (Act No. 28, 2008), s. 12.
234 *Income Tax Act, Chapter 52:01* (Botswana), 12th Schedule, paras 5 & 7.
235 *Ibid.* 12th Schedule, para. 7.
236 *Income Tax Act, 2015,* Act 896 (Ghana), s. 34.
237 *Ibid.* ss. 32 & 33.
238 *Ibid.* s. 33.
239 *Mineral and Petroleum Resources Royalty Act, 2008* (Act No. 28, 2008), s. 12.
240 *Ibid.* s. 12.(1)(ii).

6.2 The arm's-length standard

In order to ensure that resource extractors pay the appropriate rate of taxes, some African countries require mining companies to engage in arm's-length transactions, particularly with regard to the sale, purchase and disposal of mineral resources and assets. Failure to do so could result in tax administrators adjusting or re-characterizing a transaction for tax-liability purposes. The arm's-length standard discourages tax-avoidance practices such as transfer pricing between affiliated companies by ensuring that transactions are undertaken based on prevailing market conditions. In Botswana, for example, where a person engages directly or indirectly in any transaction, operation or scheme "with a connected person, the amount of each person's taxable income derived from the transaction shall be consistent with the arm's length principle".[241] In that country, transactions that fail to comply with the arm's-length principle are disregarded in calculating the tax liability of the parties. More specifically,

> Where a transaction between connected persons is not consistent with the arm's length principle in accordance with subsection (1), then –
> (a) any amount of income that would have accrued to either of the connected persons; and
> (b) any amount of income taxable in Botswana if the conditions of the transactions had been consistent with the arm's length principle, but have not accrued to that person due to the non-arm's length conditions, shall be included in the taxable income of that person and taxed accordingly.[242]

However, a transaction is considered to be consistent with the arm's-length principle if "the conditions of the transaction do not differ from the conditions that would have applied between independent persons in comparable transactions carried out under comparable circumstances".[243]

Likewise, Ghana imposes an arm's-length standard on mining transactions, particularly where the parties to the transaction are in a controlled relationship. Parties in a controlled relationship are required "to quantify, characterize, apportion and allocate amounts to be included in or deducted from income to reflect an arrangement that would have been made between independent persons".[244] Tax administrators in Ghana are empowered to adjust any transaction that does not comply with the arm's-length standard, including the power to re-characterize a debt financing arrangement as equity financing; power to re-characterize the source and type of income, loss, amount or payment; and power to apportion and allocate expenditure.[245] Similarly, in South Africa, where the earnings or gross

---

241 *Income Tax Act, Chapter 52:01* (Botswana), s. 36A.(1) (incorporated into the *Income Tax Act* by the *Income Tax (Amendment) Act, 2018* (No. 38), s. 3).
242 *Ibid.* s. 36A.(3).
243 *Ibid.* s. 36A.(2).
244 *Income Tax Act, 2015, Act 896* (Ghana), s. 31.(2).
245 *Ibid.* s. 31.(5).

188 *Fiscal regimes for mineral exploitation*

sales reported by a resource extractor in a year of assessment differ from the earnings or gross sales that would have been derived if the extractor had engaged in arm's-length transactions, the Commissioner may, for the purpose of royalty, adjust the earnings or gross sales to reflect the earnings or gross sales that would have been derived if the extractor had engaged in arm's-length transactions.[246]

6.3 Exclusivity standard

Although African countries permit deduction of mining-related expenditures from taxable income, such deduction is allowed only if the expenditures satisfy the exclusivity requirement. To be deductible, an expenditure must be incurred exclusively for mining purposes. In Botswana, for example, mining companies can deduct from their chargeable income expenditure that is "wholly, exclusively and necessarily incurred" in the production of their assessable income.[247] In Ghana, a deductible expenditure is an expenditure incurred solely for the purpose of mineral operations. Thus, an expenditure is not deductible unless it:

> (i) is wholly, exclusively and necessarily incurred in the acquisition or improvement of a valuable asset used in the marginal operation; or (ii) is wholly, exclusively and necessarily incurred in acquiring services or facilities for the mineral operation; and is income of the recipient which has a source in Ghana.[248]

Similarly, Nigeria allows mining companies to deduct from their taxable profits all expenses "wholly, exclusively, necessarily and reasonably incurred in the production of those profits", including "any sum payable by way of interest on any money borrowed and employed as capital in acquiring the profits" and rent and premium incurred during the assessment period in respect of land or building occupied for the purposes of acquiring the profits.[249] South Africa permits deduction of "expenditure and losses actually incurred" in the production of the income earned by mining companies,[250] while, in Zambia, mining companies can deduct "losses and expenditure, other than of a capital nature, incurred in [the charge] year wholly and exclusively for the purposes" of their mining business.[251]

6.4 Ring-fencing rules

In some African countries, mining operations are ring-fenced for tax purposes. 'Ring-fencing' refers to the practice of segregating each mining operation or asset, so that the owner of two or more mining operations is not able to consolidate the

246 *Mineral and Petroleum Resources Royalty Act, 2008*(Act No. 28, 2008), s. 11.
247 *Income Tax Act, Chapter 52:01* (Botswana), s. 43.
248 *Income Tax Act, 2015, Act 896* (Ghana), s. 81.(2)(b).
249 *Companies Income Tax Act*, CAP. C21, LFN 2004, s. 24.
250 *Income Tax Act* (Act 58 of 1962) (South Africa), s. 11(a) (as amended).
251 *Income Tax Act*, Chapter 323, Statutes of Zambia, s. 29.(1).

income of the mining operations for tax purposes. In Botswana and Ghana, for example, a mineral operation pertaining to each mine is regarded as a separate mineral operation for tax-related purposes.[252] More specifically, in Botswana,

> Where separate and distinct mining operations are carried on in mines which are not contiguous, the deduction to be allowed shall be calculated separately and shall not be transferable between such operations, except expenditure on a license or lease which has been relinquished by the mining company.[253]

In addition, in Ghana a mineral operation with a shared processing facility is regarded as a separate mineral operation.[254] Thus, mining companies that own two or more mineral operations in Ghana cannot consolidate their operations for tax purposes. In effect, a loss incurred in one mining operation cannot be deducted from the profits earned in another mining operation, even though both mining operations are owned by the same company. In that regard, the *Income Tax Act* of Ghana stipulates that unrelieved "losses from a separate mineral operation may be deducted only in calculating future income from that operation and not income from any other activity".[255]

Unlike the situation in Botswana and Ghana, mining companies owning two or more mining operations in Zambia are allowed to consolidate their mining operations for the purpose of tax deductions, provided that the mining operations are not contiguous. For example, where a mining company

> is carrying on mining operations in a mine which is in regular production and is also the owner of, or has the right to work, a mine which is not contiguous with the producing mine and from which the [company] has a loss in the charge year, the amount of such loss may be deducted in ascertaining the gains or profits from [their] mining operations in that charge year.[256]

However, such consolidation of deductions is allowed only if "the amount of tax which would otherwise be payable by such a person in such charge year is not reduced by more than twenty per centum as a result of this deduction".[257]

6.5 Contractual guarantees regarding payment of royalties and taxes

In addition to anti-tax-avoidance provisions, mining companies may be required to execute binding guarantees or a 'performance bond' regarding the discharge of their financial obligations, including payment of royalties and taxes to the host

---

252 *Income Tax Act, Chapter 52:01* (Botswana), 12th Schedule, para. 2; *Income Tax Act, 2015*, Act 896 (Ghana), s. 78.
253 *Income Tax Act, Chapter 52:01* (Botswana), 12th Schedule, para. 2
254 *Income Tax Act, 2015*, Act 896 (Ghana), s. 78.(1).
255 *Ibid*. s. 82.
256 *Income Tax Act* (Zambia), 5th Schedule, para. 23.
257 *Ibid*.

government. In Botswana, parent companies are required to sign contractual guarantees regarding the discharge of the subsidiary's financial obligations to the government.[258] In countries such as Tanzania, mining companies are required to execute an "integrity pledge" which enjoins companies to desist from engaging "in any arrangement that undermines or is otherwise prejudicial to Tanzania's tax system".[259] In Tanzania, failure to comply with the integrity pledge amounts to a breach of the terms of the mining licence or permission "and such licence or permission shall be deemed to have been withdrawn or cancelled".[260] It is unlikely that Tanzania will cancel mining licences based solely on non-compliance with the integrity pledge; however, the integrity pledge could be utilized as a shaming mechanism through the publication of the names of mining companies that breach their integrity pledge.

7 Ineffectiveness of anti-tax-avoidance rules in Africa

Much of the anti-tax-avoidance legislation in Africa is defective because it does not include appropriate documentation and disclosure requirements.[261] Moreover, tax administrators in Africa lack requisite knowledge of the mining industry and, consequently, they tend not to fully identify or address the specificities of the mining sector.[262] A further consequence of the institutional incapacity in Africa is that tax administrators rarely audit mining companies to determine compliance with anti-tax-avoidance legislation. In addition, the complexity of the audit processes, "coupled with the high cost of implementing a transfer pricing audit function in general", impedes the ability of tax administrators to conduct transfer pricing audits in Africa.[263] Even in the few instances where audits are undertaken, tax administrators may not have access to reliable data on mining transactions such as sales invoices, and there is a paucity of domestic comparables, which are essential to the effective enforcement of transfer pricing legislation.[264]

The result of the institutional weakness in Africa is that anti-tax-avoidance statutes are rarely enforced. In practice, only a relatively small number of African countries enforce anti-tax-avoidance legislation to any appreciable degree.[265] A recent study on transfer pricing in the African mining industry observes that:

> Most jurisdictions see the area of mineral transfers/sales as the main transfer pricing risk; however, only a quarter of respondents [African countries] had

---

258 *Mines and Minerals Act, 1999* (Botswana), 1st Schedule, Form V, Annexure 1, "Parent Company Guarantee".
259 *Mining Act, 2010* (Tanzania), s. 106.(2)(c) (introduced by *The Written Laws (Miscellaneous Amendments) Act, 2017* (No. 7), s. 28)).
260 *Ibid.* s. 106.(4).
261 Guj *et al., supra* note 189 at 94.
262 *Ibid.* at 94.
263 *Ibid.* at 97.
264 *Ibid.* at 95.
265 *Ibid.* at 96.

systems in place to check the degree to which the prices applied to minerals transferred to related parties comply with the arm's length principle, and only two or three do so systematically. Similarly, only a quarter of respondents systematically carry out value-chain analysis to determine whether marketing and hedging fees charged by related parties comply with the arm's length principle. About three quarters of respondents reported high levels of borrowing by the mining industry, largely from related overseas parties with remittance of the related interest expenses to the lenders being subject to various levels of withholding tax. However, in many instances the rate of withholding tax may have been reduced or even waived in line with the terms of stability agreements and/or double taxation agreements.[266]

The successful implementation of anti-tax-avoidance legislation requires strengthening of the capacity of tax administrators, particularly "in the area of transfer pricing", including the enhancing of the knowledge of mining industry processes within tax authorities.[267] Apart from the strengthening of domestic capacity, Africa needs to close the legal loopholes that enable abuse of the fiscal regimes. For example, there is currently no penalty for abuse of the fiscal regimes. This lacuna could be addressed through statutory and contractual provisions prohibiting tax-avoidance practices. More specifically, any transaction designed to evade or limit taxes or royalties should be statutorily declared void and should be regarded as an offence punishable with a fine. For this purpose, the onus must be on mining companies to establish that a transaction is not designed to evade or limit taxes or royalties. Persistent breaches of such statutory provision should be grounds for terminating a mineral right.

8 Conclusion

This chapter analyzes the fiscal and incentive regimes governing mineral mining in several African countries, including Botswana, Ghana, Nigeria, South Africa, and Zambia. The policy objectives informing these fiscal regimes include the promotion of economic development and an increase in mining revenues accruing to African governments. While there is evidence that extant fiscal regimes have made Africa more attractive to foreign investors, and although the increase in FDI in Africa has led to a consequential increase in mining revenues, the economic impacts of the fiscal regimes appear compromised by the generosity of the regimes. As noted in this chapter, the generous nature of the fiscal regimes is detrimental to Africa's desire and ability to optimize mining revenues. However, given the current global economic climate and, in particular, the increasing global demand for mineral resources, it is no longer realistic for Africa to grant overly generous fiscal regimes to resource extractors. Rather, Africa should adopt dynamic and flexible fiscal regimes that can adapt to changing economic and social circumstances.

---

266 *Ibid.* at 96.
267 *Ibid.* at 97.

6 Legal stabilization of mining investments

1 Introduction

Mineral-rich African countries have a number of common features. These countries often lack the technological expertise and the financial power to exploit mineral resources. Thus, for the most part, African countries rely on foreign companies, usually MNCs, for the exploration and exploitation of their mineral resources. In addition, some of these countries are politically unstable, ruled in the main by autocratic and unresponsive governments. Furthermore, the legal and judicial processes in many of these countries are deficient. These conditions pose certain investment risks for foreign investors in the extractive industries, including the risk that the legal and fiscal regimes governing a project could be amended or changed by the host government mid-way through the life of the project. There is also the risk of expropriation and nationalization of investment projects. Investments in the extractive industries are particularly prone to these risks because of the long period of time it takes to accomplish mineral extraction projects.

To forestall these risks, foreign investors usually require host African countries to issue contractual and statutory guarantees on the stability of the legal and fiscal regimes governing investment projects. As was observed a long time ago, foreign investors often seek and obtain "greater guarantees of stability in developing countries than they would dare hope for in similar projects in advanced countries".[1] These guarantees and assurances, known commonly as 'stabilization clauses', help to reduce "the risk of sudden and unpredictable changes being made by the government after the investment takes place".[2] Such clauses insulate foreign investors from future changes to the host country's laws.[3]

---

1 David N. Smith & Louis T. Wells, Jr., "Mineral Agreements in Developing Countries: Structures and Substance" (1975) 69 *American Journal of International Law* 560 at 565.
2 World Bank, *Strategy for African Mining* (World Bank Technical Paper Number 181, August 1992), at 24, http://documents.worldbank.org/curated/en/722101468204567891/pdf/multi-page.pdf
3 Roland Brown, "The Relationship between the State and the Multinational Corporation in the Exploitation of Resources" (1984) 33 *International and Comparative Law Quarterly* 218 at 222–3.

While stabilization clauses may be beneficial to investors in the sense that they assure, at least on paper, a stable and predictable investment climate, there are questions as to whether these clauses are of any benefit to developing countries. As argued in this chapter, stabilization clauses may be counter-productive to the economic development aspirations of African countries because they constrain the fiscal policy options available to African countries. In addition, stabilization clauses impede the ability of host countries to regulate the activities of foreign corporations, particularly as they relate to human rights and environmental protection.

2 Stabilization regimes

Although this book focuses on the mining industry, this chapter draws on stabilization provisions in the extractive industries more broadly, including contractual provisions in Africa's oil and gas industry. An industry-wide examination of stabilization clauses is informed by the author's desire to capture and analyze the various forms and manifestations of stabilization clauses. As discussed below, stabilization clauses in Africa are contained in contracts between African countries and MNCs or provided for in statutory enactments. In addition, investment treaties between African countries and the economically advanced countries often contain provisions designed to stabilize the legal regimes governing investments by nationals of the treaty states. Thus, this section of the book considers the legal stability regimes in the context of investment contracts, statutes, and investment treaties.

2.1 Contract-based stabilization clauses

Three types of stabilization clauses are common in mining leases and concessions. The first type, referred to as a 'freezing' stabilization clause, has the effect of freezing the law governing a project for a specified duration or throughout the life of the project. The second type, the economic equilibrium clause, requires the host state to maintain the economic equilibrium struck by the parties to the investment contract, while the third type is a hybrid clause containing features of the both the freezing stabilization clause and the economic equilibrium clause.

2.1.1 'Freezing' stabilization clause

A freezing stabilization clause usually provides that the law governing an investment project is the law of the host state in force at the date of execution of the investment contract. Freezing stabilization clauses are aimed at preserving "the law of the host country as it applies to the investment at the time the State contract is concluded", thus ensuring that "future changes to the law of the host country are inapplicable to the foreign investment contract".[4] Freezing stabilization clauses may be 'full' in the sense that they freeze all laws applicable to a project, or

---

4 UNCTAD, *State Contracts: UNCTAD Series on Issues in International Investment Agreements* (New York & Geneva: United Nations, 2004) at 26.

194　*Legal stabilization of mining investments*

'partial' where they freeze specific aspects of the laws governing a project. An example of a 'full' freezing stabilization clause is found in the mining agreement between the DRC and Lundin Holdings Ltd. to the effect that:

(a) The State guarantees, for the whole duration of this Agreement, to T.F. M., its shareholders, its consultant(s), its salaried expatriate agents and its lenders, stability of legislation and regulations which are in force on the date of the Original Convention, and in particular with respect to judicial, land, fiscal and customs, commercial, monetary, social, employment, health and Mining legislation matters, and in matters of residence and work conditions for foreigners.
(b) No legal or regulatory provision effective after the date of the Original Convention may entail a restriction or reduction of the special advantages of this Agreement or hinder the exercise or the rights resulting thereof.[5]

Similarly, the mineral concession agreement between the Republic of Guinea and BSG Resources provides that:

The Government warrants the Company from the date of grant of the Concession and throughout its full duration the stabilization of Current Legislation and of all provisions, particularly fiscal and concerning customs and excise, stipulated in this Agreement.

Accordingly, all changes to current legislation, particularly fiscal and/or concerning customs and excise, after the date of grant of the Concession that would as a result increase, whether directly or indirectly, the Company's tax and/or customs and excise charges would not be applicable [to] it.

On the other hand, the Company may validly take advantage of such changes if their effect is to reduce its tax and/or customs and excise charges.[6]

'Freezing' stabilization clauses may prohibit the host state from unilaterally amending the legal and fiscal regimes governing a project.[7] However, a variant of the 'freezing' stabilization clause may permit amendments to the host state's legal regimes with the mutual consent of the contracting parties – that is, the state and the investor. For example, a mining lease may stipulate that the rights vested in

---

5　*Amended and Restated Mining Convention among The Democratic Republic of Congo and La Generale des Carrieres et des Mines and Lundin Holdings Ltd. and Tenke Fungurume Mining S.A.R.L.* (Dated 28 September 2005), art. 30, https://resourcecontra cts.org/contract/ocds-591adf-6935276590/view#/pdf
6　*Basic Agreement between The Republic of Guinea and BSG Resources* (December 2009), Clause 32, http://icsidfiles.worldbank.org/icsid/icsidblobs/OnlineAwards/C6587/DS10326.pdf
7　See Peter D. Cameron, "Stabilisation in Investment Contracts and Changes of Rules in Host Countries: Tools for Oil & Gas Investors" at 28 (Final Report, 5 July 2006), www.international-arbitration-attorney.com/wp-content/uploads/arbitrationlaw4-Stabilisation-Paper.pdf

the parties to the contract shall not be altered except by mutual consent of the parties. The 'freezing' stabilization clause may also provide that amendments to existing law as well as any new law enacted after the mining contract is executed shall not apply to the investor. In this regard, a stability agreement between Ghana and AngloGold Ashanti provides that:

> The Government hereby agrees that neither Ashanti nor any of its Ghanaian subsidiaries shall, for a period of 15 years commencing the Effective time, (a) be adversely affected by any new law, enactment, orders, instruments or other actions made under any new enactment or changes to any enactment, orders or instruments in existence as of the date of this Agreement or other actions taken under these that have the effect or purport to have the effect of imposing obligations upon Ashanti or any of its Ghanaian subsidiaries, or (b) be adversely affected by changes after the date of this Agreement to (i) the level of any payments of any customs or other duties relating to the entry of materials, goods, equipment and any other inputs necessary to the mining operations or project, or (ii) the level of any payments of taxes, fees and other fiscal imports, or (iii) laws relating to exchange control, transfer of capital and dividend remittance .[8]

In effect, this variant of freezing stabilization clauses insulates the investor from all laws enacted after the investment contract is signed.

As opposed to the 'full' freezing stabilization clause that freezes the entirety of the legal regimes governing a project, a stabilization clause may be partially freezing in the sense that it freezes specific aspects of the legal regimes governing a project.[9] For example, the Mineral Development Agreement between Liberia and Mittal Steel provides that:

> The GOVERNMENT hereby undertakes and affirms that at no time shall the rights (and the full and peaceful enjoyment thereof) granted by it under Article 19 (Income Taxation), Article 20 (Royalty) and Article 22 (Other Payments to the GOVERNMENT) of this Agreement be derogated from or otherwise prejudiced by any Law or the action or inaction of the GOVERNMENT, or any official thereof, or any other Person whose actions or inactions are subject to the control of the GOVERNMENT ...[10]

---

8 *Stability Agreement between Ghana and AngloGold Ashanti Limited* (Ratified by Parliament on February 18, 2004), (cited in Martin K. Ayisi, "The Review of Mining Laws and the Renegotiation of Mining Agreements in Africa: Recent Developments from Ghana" (2015) 16(3) *Journal of World Investment & Trade* 467 at 498).
9 Cameron, *supra* note 7 at 30.
10 *An Act Ratifying the Amendment to the Mineral Development Agreement (MDA) Dated August 17, 2005 Between the Government of the Federal Republic of Liberia (The Government) and Mittal Steel Holding A.G. and Mittal Steel (Liberia) Holdings Limited (The Concessionaire)*, art. 16(E), www.leiti.org.lr/uploads/2/1/5/6/21569928/52423536-an-act-ratifying-the-amendment-to-the-mineral-development-agreement-mda-dated-august-17-2005-between-the-government-of-liberia_and_mittal_steel.pdf [*Act Ratifying Mittal MDA*].

Similarly, erstwhile development agreements between the government of Zambia and several mining MNCs provided, in identical terms that, the government "undertakes that it will not for the Stability Period: (a) increase corporate income tax or withholding tax rates applicable to the Company ... from those prevailing at the date hereof ...; (b) otherwise amend the [Value Added Tax] and corporate Tax regimes applicable to the Company ...; (c) impose new taxes or fiscal imposts" on the company; increase the rates of royalty, import duty and export from the levels set out in the agreement.[11] These stabilization clauses provide 'partial stability' because they freeze only the specific matters – for example, taxes and royalties – mentioned in the clauses.

Freezing stabilization clauses were common in the Africa mining industry during the era immediately following attainment of political independence. However, in modern times they are infrequently inserted into mining contracts partly because of their limited utility and partly because they are inimical to the economic and social advancement of African countries. As argued below, while they attempt to prevent host states from making changes to the legal and fiscal regimes governing a project, freezing stabilization clauses cannot, in law, prevent host states from amending or changing their legal and fiscal regimes. Hence, modern mining contracts in Africa are increasingly adopting 'economic equilibrium clauses', although in some cases 'partial freezing' stabilization clauses are adopted by the contracting parties as the mechanism for stabilizing the investment contract.

2.1.2 Economic equilibrium clause

Contract-based stabilization clauses have evolved in recent years, culminating in the second type of stabilization clause that is designed to maintain economic equilibrium between the host state and the investor. Usually referred to as an 'economic equilibrium' or 'economic balancing' clause, this variant of stabilization clauses strives to maintain the economic interests of the contracting parties in the event that the legal or fiscal regimes governing the project are changed or amended.[12] The 'economic equilibrium' clause neither freezes the legal and fiscal regimes nor prohibits amendments or changes to the legal and fiscal regimes governing a project. Rather, it envisages that where such changes occur, the investor shall be restored to the position they occupied prior to the changes.

---

11 See *The Government of the Republic of Zambia and Mopani Copper Mines Plc: Mufulira Mine, Smelter and Refinery and Nkana Mines, Concentrator and Colbalt Plant Development Agreement* (dated 31 March 2000), art. 16, https://landmatrix.org/media/uploads/1683-mopani-copper-mines-plc-mufulira-nkana-concession-2000.pdf [*Mopani Mines Agreement*]; *The Government of the Republic of Zambia and Konkola Copper Mines Plc: Amended and Restated Development Agreement*, art. 15, www.mmdaproject.org/presentations/MMDA%20Zambia%202004%20Amended%20and%20Restated-1.pdf [*Konkola Mines Agreement*].

12 Andrea Shemberg, "Stabilization Clauses and Human Rights" (March 11, 2008) at 7–8, http://documents.worldbank.org/curated/en/502401468157193496/pdf/452340WP0Box331ation1Paper01PUBLIC1.pdf. See also Cameron, *supra* note 7 at 31–8.

The 'economic equilibrium' clause is a modern feature of mining contracts in Africa, and in some countries such as Tanzania, stabilization clauses must "as much as possible, be based on the economic equilibrium principle".[13]

Economic equilibrium clauses usually prescribe the mechanisms for maintaining the economic equilibrium in the mineral extraction contract.[14] These include provisions to the effect that the host state shall compensate the investor in the event that the state amends the legal regimes while the project is ongoing and the amendment adversely affects the economic interest of the investor. Economic equilibrium may also be attained through a provision that the parties shall negotiate in good faith an amendment to the contract in a manner that restores the economic equilibrium envisaged in the original contract.[15] For example, the agreement governing the West African Gas Pipeline Project provides that the governments of Benin, Ghana, Nigeria, and Togo shall, in the event of an amendment to the fiscal regime which adversely affects the economic interest of the West African Gas Pipeline Company, "endeavour in good faith to negotiate a solution which restores the Company and/or its shareholders to the same or an economically equivalent position it was or they were in prior to such change".[16] Tanzania's model production-sharing contract provides for the rebalancing of economic equilibrium thus:

> If at any time or from time to time there should be a change in legislation or regulations which materially affects the commercial and fiscal benefits afforded by the Contractor under this contract, the Parties will consult each other and shall agree to such amendments to this Contract as are necessary to restore as near as practicable such commercial benefits which existed under the Contract as of the Effective Date.[17]

Similarly, the model concession agreement for gas and crude oil exploration and exploitation in Egypt provides in part that:

> In case of changes in existing legislation or regulations applicable to the conduct of Exploration, Development and production of Petroleum, which take place after the Effective Date, and which significantly affect the economic interest of this Agreement to the detriment of CONTRACTOR or which imposes on CONTRACTOR an obligation to remit to the A.R.E. [Arab Republic of Egypt] the proceeds from sales of CONTRACTOR's Petroleum,

---

13 *The Written Laws (Miscellaneous Amendments) Act, 2017* (No. 7), s. 25 (incorporating s. 100E.(3) into the *Mining Act, 2010*).
14 Shemberg, *supra* note 12 at 6; Cameron, *supra* note 7 at 31.
15 Shemberg, *ibid.* at 6; Cameron, *supra* note 7 at 31.
16 *West African Gas Pipeline Agreement: International project Agreement*, 22 May 2003, Clause 36.2(a), cited in Cameron, *supra* note 7 at 44–5.
17 *Tanzania Model Production Sharing Contract* (November 2004) art. 30, reproduced in *Barrows Basic Oil Laws and Concession Contracts*, Supplement 161, South and Central Africa.

CONTRACTOR shall notify EGPC [Egypt General Petroleum Corporation] of the subject legislative or regulatory measure and also the consequent effects upon issuing legislation or regulation which impact on the stabilization. In such case, the Parties shall negotiate possible modifications to this Agreement designed to restore the economic balance thereof which existed on the Effective Date.

The Parties shall use their best efforts to agree on amendments to this Agreement within ninety (90) days from aforesaid notice.[18]

2.1.3 Hybrid stabilization clause

The third type is a hybrid of the 'freezing' clause and the 'economic equilibrium' clause. The 'hybrid' stabilization clause contains elements of the freezing clause and the economic equilibrium clause. It may or may not freeze the fiscal regimes governing a project, but it "protect[s] against the financial implications of all changes of law, by requiring compensation or adjustments to the [agreement], including exemptions from new laws", in order to maintain the economic equilibrium.[19] A 'hybrid' stabilization clause is envisaged under South Africa's *Mineral and Petroleum Resources Royalty Act, 2008* (MPRRA), which provides, first, that the state shall guarantee the fiscal terms and conditions stipulated in stabilization agreements for as long as the extractor holds the mining right (freezing element); and, second, if the state fails to abide by the terms of the stabilization clause and the failure has a material adverse economic impact on the determination of the royalty payable by the mineral extractor, the state shall restore the economic equilibrium by paying to the extractor compensation in respect of the increase in the royalty caused by the failure.[20]

2.2 Statute-based stabilization clauses

In the early years of Africa's political independence, the fiscal stabilization measures adopted by African countries were, for the most part, contract-based. These measures were effected by way of investment contracts, usually long-term, between natural resource extraction companies and host countries. However, because the fiscal stabilization measures were contractual in nature, it was easy for host governments to breach them, thus resulting in adverse financial consequences for resource extraction companies. This was particularly apparent in the 1970s when Libya revoked and nationalized the oil concessions granted to several foreign companies, despite Libya's contractual undertaking that "the contractual rights expressly created by the concession shall not be altered except by the mutual

---

18 *Concession Agreement for Petroleum Exploration and Exploitation between the Arab Republic of Egypt and the Egyptian General Petroleum Corporation and ...*, art. XIX, https://apexintl.com/wp-content/uploads/2018/04/7-2016-EGPC-Model-Agreement.pdf
19 Shemberg, *supra* note 12 at 9.
20 *MPRRA*, s. 14.

consent of the parties".[21] Given this outcome, resource extraction companies have in the recent past exerted pressure on African governments to enact statutory provisions guaranteeing the stability of the legal and fiscal regimes for resource extraction. A number of African countries have in fact succumbed to this pressure by enshrining fiscal stabilization provisions in statutory enactments. Thus, the position in some African countries today is that stabilization clauses in contractual agreements are backed by statutory provisions prohibiting host states from unilaterally amending the fiscal regimes governing specific projects. Usually, statutory stabilization provisions grant power to a designated officer or agency of government to enter into stabilization agreements on behalf of the host government, as is the case in Ghana and South Africa.[22] In the alternative, the statute may provide a direct stabilization guarantee for investors, as is the case under the *Nigeria LNG (Fiscal Incentives, Guarantees and Assurances) Act* (the *Nigeria LNG Act*), which provides that the fiscal regime contained in the Act "shall not be amended in any way, except with the written consent of the Government, the company and the company's shareholders".[23]

Generally speaking, statutory stabilization provisions often provide that while the host state has the power to amend the fiscal regimes governing its extractive industries, such amendments are inapplicable to companies that signed fiscal stabilization agreements with the government prior to the amendment.[24] For example, holders of mineral rights in Ghana can enter into a mutually negotiated "stability agreement" with the government ensuring that they are not adversely affected by future changes to the legal and fiscal regimes for mining. Ghana's *Minerals and Mining Act, 2006* provides:

(1) The Minister may as a part of a mining lease enter into a stability agreement with the holder of the mining lease, to ensure that the holder of the mining lease will not, for a period not exceeding fifteen years from the date of the agreement,
   (a) be adversely affected by a new enactment, order instrument or other action made under a new enactment or changes to an enactment, order, instrument that existed at the time of the stability agreement, or other action taken under these that have the effect or purport to have the effect of imposing obligations upon the holder or applicant of the mining lease, and

---

21 *Libyan American Oil Co. v Government of the Libyan Arab Republic* (1981) 20 I.L.M. 1 at 17 [*LIAMCO v Libya*].
22 See *Minerals and Mining Act, 2006* (Ghana), s. 48; *MPRRA*, s. 13.
23 *Nigeria LNG (Fiscal Incentives, Guarantees and Assurances) Act No. 39 of 1990*, 2nd Schedule, para. 2 (as amended by the *Nigeria LNG (Fiscal Incentives, Guarantees and Assurances) (Amendment) Act No. 113 of 1993*).
24 See Noah Rubins & N. Stephan Kinsella, *International Investment, Political Risk and Dispute Resolution: A Practitioner's Guide* (Oxford: Oxford University Press, 2005) at 51–2.

(b) be adversely affected by subsequent changes to
  (i) the level of and payment of customs or other duties relating to the entry of materials, goods, equipment and any other inputs necessary to the mining operations or project,
  (ii) the level of and payment of royalties, taxes, fees and other fiscal imports, and
  (iii) laws relating to exchange control, transfer of capital and dividend remittance.[25]

Similarly, South Africa's MPRRA empowers the Minister of Finance (Minister) to enter into fiscal "stabilization agreements" with resource extractors that guarantee that the fiscal regimes contemplated in the Act apply to the mineral right of the extractor "for as long as the extractor holds the right (and for all participating interests subsequently held by the extractor in respect of the right)".[26] The Minister may in fact enter into such agreements "in anticipation of the extractor acquiring a mineral resource right".[27] However, a stabilization agreement concluded in anticipation of the acquisition of a mineral resource right has no force and effect unless the mineral resource right is granted within one year after the date on which the agreement was concluded.[28] The resource extractor can assign its rights under a stabilization agreement if the extractor disposes of its mineral resource right.[29] Although a stabilization agreement relating to a prospecting right or exploration right is assignable without preconditions or restrictions on the class of persons to whom the assignment may be made, a stabilization agreement relating to a mining right or a production right is assignable in limited circumstances. The assignment of a stabilization agreement in the latter case can be made only if both the resource extractor (assignor) and the assignee form part of the same group of companies on the date of the disposal of the mining or production right.[30] Also, an extractor that concludes a stabilization agreement with the Minister can "unilaterally terminate the agreement at any time with effect from the day after the last day of the year of assessment during which the extractor terminated the agreement".[31]

The terms and conditions of stabilization agreements in South Africa are spelled out in the MPRRA, which provides as follows:

14. (1) An amendment of section 4 has no force and effect in respect of an extractor that is party to an agreement contemplated in section 13.(1) if the amendment has the effect that the extractor becomes subject to a royalty

---

25 *Minerals and Mining Act, 2006* (Ghana), s. 48.
26 *MPRRA*, s. 13.(1).
27 *Ibid.* s. 13.(1).
28 *Ibid.* s. 13.(2).
29 *Ibid.* s. 13.(3).
30 *Ibid.* s. 13.(4).
31 *Ibid.* s. 13.(5).

which is greater than the royalty to which the extractor would otherwise have been subject.

(2) If the State fails to comply with the terms of an agreement contemplated in section 13.(1) and the failure has a material adverse economic impact on the determination of the royalty payable by the extractor that is party to that agreement, the extractor is entitled to compensation in respect of the increase in the royalty caused by the failure (and interest at the prescribed rate calculated on the compensation from the date of the failure) or to an alternative remedy that eliminates the full impact of the failure.[32]

Section 14 of the MPRRA contemplates the hybrid stabilization clause in the sense that it contains elements of the 'freezing' and 'economic equilibrium' clauses. It does not prohibit the South African government from amending or altering the fiscal regimes governing the extractive industries, but it envisages that the government may amend the fiscal regimes set out in section 4 of the Act – hence the requirement of compensation. However, where an amendment of the fiscal regimes has "a material adverse economic impact" on the royalty payable by resource extractors, the government is obliged to maintain the economic equilibrium in the stabilization agreement by paying compensation to the resource extractors or by granting "an alternative remedy that eliminates the full impact" of the amendment.[33] That said, such an amendment has no legal force in respect of an extractor that is a party to a prior stabilization agreement, if the amendment imposes a royalty that is greater than the royalty to which the extractor would otherwise have been subject. In other words, section 14(1) of the MPRRA 'grandfathers' all stabilizations agreements made prior to any amendments to the royalty regimes in South Africa. The effect of the provision is that any future amendment to the fiscal regimes that increases the royalty rates for mineral rights will have limited force and application. The amendment will apply to extractors other than extractors that are parties to prior stabilization agreements.

Interestingly, while section 14 of the MPRRA grandfathers stabilization clauses against future increases in royalty rates, it enables resource extractors to pay royalty rates that are lower than the rates specified in extant stabilization agreements where an amendment to the MPRRA reduces the royalty rate for minerals. In effect, future amendments to the royalty rates for mineral resources are applicable to resource extractors that are parties to prior stabilization agreements if the amendments introduce royalty rates that are lower than the royalty rates prescribed in the stabilization agreements. By providing expressly that an amendment has no legal force as against an extractor that signed a prior stabilization agreement if the amendment has the effect of subjecting the extractor to a higher rate of royalty, section 14(1) implies that amendments that do not have such effects, including amendments lowering the rate of royalty, would have legal force as

---

32 *Ibid.* s. 14.
33 *Ibid.* s. 14.(2).

against the parties to a prior stabilization agreement. In this sense, the MPRRA offers resource extractors the best of both worlds.

Nigeria offers some investors in the natural gas industry certain statutorily guarantees regarding the stability of the fiscal regimes governing their investments. For example, in recognition of the magnitude of the investments of foreign investors in the Nigeria Liquefied Natural Gas Project (LNG Project), the government issued certain fiscal guarantees and assurances to the investors through the *Nigeria LNG Act*, which provides in part that:

> Without prejudice to any other provision contained herein, neither the company not its shareholders in their capacity as shareholders in the company shall in any way be subject to new laws, regulation and taxes, duties, imports or charges of whatever nature which are not applicable generally to companies incorporated in Nigeria or to shareholders in companies incorporated in Nigeria respectively
>
> ...
>
> The Government shall take such executive, legislative and other actions as may be necessary so as to effectively grant, fulfil and perfect the guarantees, assurances and undertakings contained herein. In order to afford the degrees of security required to enable the company's investment to be made, the government further agrees to ensure that the said guarantees, assurances and undertakings shall not be suspended, modified or revoked during the life of the venture except with the mutual agreement of the government and the shareholders of the company.[34]

2.3 Treaty-based stabilization regimes

Many mineral-rich countries in Africa have signed bilateral investment treaties (BITs) and multilateral investment treaties (MITs) with economically developed countries, which are usually the home countries of the MNCs operating in the African mining industry. Some of these investment treaties are in force while others, though signed, are not yet in force. With regard to BITs that are in force, Canada, a country that is home to a large number of mining companies operating in Africa, has BITs with Benin, Burkina Faso, Cameroon, Côte d'Ivoire, Egypt, Mali, Guinea, Senegal, and Tanzania.[35] Likewise, the United States has BITs with Cameroon, Congo, the DRC, Egypt, Morocco, Mozambique, Rwanda, Senegal,

---

34 *Nigeria LNG (Fiscal Incentives, Guarantees and Assurances) Act No. 39 of 1990*, 2nd Schedule, paras 3 & 6 (as amended by the *Nigeria LNG (Fiscal Incentives, Guarantees and Assurances) (Amendment) Act No. 113 of 1993*).
35 UNCTAD, "Investment Policy Hub – Canada – Bilateral Investment Treaties", https://investmentpolicy.unctad.org/international-investment-agreements/countries/35/canada

and Tunisia,[36] while the United Kingdom has BITs with Benin, Burundi, Cameroon, Congo, Côte d'Ivoire, Egypt, Eswatini, Ghana, Kenya, Mauritius, Morocco, Mozambique, Nigeria, Senegal, Sierra Leone, Tanzania, Tunisia, and Uganda.[37]

BITs and MITs not only reinforce stabilization clauses in mining contracts, but also provide reciprocal investment stability for nationals of the treaty states. These treaties require state parties to protect the legitimate expectations of investors as enshrined in investment contracts. BITs and MITs usually contain investment stabilization provisions regarding 'full protection and security', 'fair and equitable treatment', as well as 'umbrella clauses'. Taken together, these treaty provisions protect investors from political and legal volatility in host countries. In fact, in appropriate cases, these treaty provisions may confer treaty status on stabilization clauses in investment contracts.[38] In such cases, the host state's breach of a stabilization clause in a mining contract may amount to breach of the relevant BIT and MIT.

2.3.1 'Full protection and security' and 'fair and equitable treatment' clauses

The USA–DRC BIT is reflective of the 'full protection and security' and the 'fair and equitable treatment' clauses as follows:

> Investments of nationals and companies of either Party shall at all times be accorded fair and equitable treatment and shall enjoy protection and security in the territory of the other Party. The treatment, protection and security of investment shall be in accordance with applicable national laws, and may not be less than that recognized by international law. Neither Party shall in any way impair by arbitrary and discriminatory measures the management, operation, maintenance, use, enjoyment, acquisition, expansion, or disposal of investment made by nationals or companies of the other Party ...[39]

---

36 UNCTAD, "Investment Policy Hub – United States of America – Bilateral Investment Treaties", https://investmentpolicy.unctad.org/international-investment-agreements/countries/223/united-states-of-america

37 UNCTAD, "Investment Policy Hub - United Kingdom - Bilateral Investment Treaties", https://investmentpolicy.unctad.org/international-investment-agreements/countries/221/united-kingdom

38 Thomas J. Pate, "Evaluating Stabilization Clauses in Venezuela's Strategic Association Agreements for Heavy-Crude Extraction in the Orinoco Belt: The Return of a Forgotten Contractual Risk Reduction Mechanism for the Petroleum Industry" (2009) 40 *University of Miami Inter-American Law Review* 347 at 363–4.

39 *Treaty between the United States of America and the Republic of Zaire (Democratic Republic of Congo) Concerning the Reciprocal Encouragement and Protection of Investment*, art. II(4), (signed August 3, 1984; entered into force July 28, 1989), https://investmentpolicy.unctad.org/international-investment-agreements/treaty-files/828/download [*USA–DRC Treaty*].

Similarly, the Canada–Guinea BIT stipulates that "[e]ach Party shall accord to a covered investment treatment in accordance with the customary international law minimum standard of treatment of aliens, including fair and equitable treatment and full protection and security".[40]

Treaty provisions regarding 'full protection and security' and 'fair and equitable treatment' oblige state parties to protect foreign investors from legal and administrative actions capable of diminishing the value of investments. However, the scope of the protection afforded by these provisions remains unsettled. Regarding the 'full protection and security' clause, two lines of interpretation have been adopted by arbitral tribunals. The first line of interpretation, which I will refer to simply as the conventional view, is that 'full protection and security' imposes an obligation on states to exercise due diligence to protect against physical damage or physical injury to foreign investment projects.[41] The alternative interpretation adopts an expansive view of the 'full protection and security' clause by holding that the clause imposes obligations on states to protect foreign investors against both physical injury and non-physical injury, including protection from legal and administrative measures capable of depriving investors of their investments. The 'full protection and security' clause under a BIT, according to one tribunal, implies that state parties to the BIT guarantee the stability of investment "in a secure environment, both physical, commercial and legal".[42] Thus the "host State is obligated to ensure that neither by amendment of its laws nor by actions of its administrative bodies is the agreed and approved security and protection of the foreign investor's investment withdrawn or devalued".[43] If this latter view is correct, the host state's amendment of the legal regimes governing a mining project would amount to a breach of the BIT where the amendment changes in some material respect the legal regimes prescribed in the mining contract.

More significantly, such amendment could infringe on the 'fair and equitable treatment' clause under the BIT. The scope of the 'fair and equitable treatment' principle has been the subject of intense debate in international investment law,[44] but it is generally thought that the principle enjoins host states to protect the

---

40 *Agreement for the Promotion and Reciprocal Protection of Investments between Canada and the Republic of Guinea*, art. 6(1), https://investmentpolicy.unctad.org/international-investment-agreements/treaty-files/5095/download
41 See *Suez v Argentine Republic* (ICSID Case No. ARB/03/17), Decision on Liability (30 July 2010) at paras 167 & 173; *Saluka Investments B.V. v Czech Republic* (UNCITRAL), Partial Award (17 March 2006) at para. 484; *Lauder v Czech Republic* (UNCITRAL), Final Award (3 September 2001) at para. 308.
42 *Biwater Gauff (Tanzania) Ltd v United Republic of Tanzania* (ICSID Case No. ARB/05.22), Award (24 July 2008) at para. 729.
43 *CME Czech Republic BV (The Netherlands) v Czech Republic* (UNCITRAL), Partial Award (13 September 2001) at para. 613. See also *Azurix Corp. v Argentine Republic* (ICSID Case No. ARB/01/12), Award (14 July 2006) at paras 406–8.
44 See Peter Muchlinski, "'Caveat Investor'?: The Relevance of the Conduct of the Investor Under the Fair and Equitable Treatment Standard" (2006) 55(3) *International & Comparative Law Quarterly* 527 at 530 ("[t]he fair and equitable treatment standard is still shrouded with considerable uncertainty").

reasonable expectations of foreign investors arising primarily from the terms of the investor-state contract, including "the conditions offered by the host State at the time of the investment".[45] As noted earlier, such conditions usually include contractual guarantees regarding the stability of the legal and fiscal regimes governing the investment. Thus, where the host state grants a stabilization clause relating to the legal regimes governing the investment project and the state subsequently amends the legal regimes, such an amendment not only breaches the stabilization clause, but, more importantly, it also breaches the 'fair and equitable treatment' principle under the BIT.[46] This is because the reasonable expectations of the investor include a stable legal regime as envisaged under the stabilization clause granted by the host state at the commencement of the investment.[47] As one arbitral tribunal has observed, the "stability of the legal and business environment is directly linked to the investor's justified expectations" and "such expectations are an important element of fair and equitable treatment".[48]

Fair and equitable treatment "involves the obligation to grant and maintain a stable and predictable legal framework necessary to fulfill the justified expectations of the foreign investor".[49] In effect, the principle of fair and equitable treatment imposes an obligation on states "not to alter the legal and business environment in which the investment has been made".[50] It protects foreign investors against subsequent changes to the host state's laws to the extent that "the investor has relied on the laws at the time of the investment".[51] Thus, in *Suez v Argentine Republic*, the tribunal held that failure to implement the legal framework prescribed in a Concession contract amounted to a breach of the fair and equitable treatment principle under the Argentina–France BIT and the Argentina–Spain BIT because it deprived the claimants of their "legitimate expectation that the Argentine authorities would exercise [their] regulatory authority and discretion within the rules of the detailed legal framework that [had been] established for the Concession".[52]

---

45 *LG&E Energy Corp. et al. v Argentine Republic* (ICSID Case No. ARB/02/1), Decision on Liability (3 October 2006) at para. 130.
46 See *LG&E Energy Corp. et al. v Argentine Republic, ibid.* at para. 124 (the "stability of the legal and business framework is an essential element of fair and equitable treatment"). See also *Occidental Exploration and Production Company v Republic of Ecuador* (London Court of Int'l Arb., Case No. UN. 3467) (1 July 2004) at para. 183.
47 *Duke Energy Electroquil Partners & Electroquil S.A. v Republic of Ecuador* (ICSID Case No. ARB/19), Award (18 August 2008) at para. 340.
48 *Ibid.* at para. 340.
49 *LG&E Energy Corp. et al. v Argentine Republic, supra* note 45 at para. 131.
50 *Occidental Exploration and Production Company v Republic of Ecuador, supra* note 46 at para. 191.
51 Rudolf Dolzer, "Fair and Equitable Treatment in International Law" (2006) 100 *American Society of International Law Proceedings* 69 at 72.
52 *Suez v Argentine Republic, supra* note 41 at paras 213–8. See also *LG&E Energy Corp. et al. v Argentine Republic, supra* note 45 at paras 134–8.

2.3.2 Umbrella clause

BITs and MITs often contain 'umbrella clauses' to the effect that the treaty parties shall observe the contractual obligations they have assumed with regard to investments in their territories by investors of the other treaty party. For example, the US–Senegal BIT provides that "[e]ach Party shall observe any engagement it may have entered into with regard to investment of nationals or companies of the other Party",[53] while the USA–DRC BIT states that "[e]ach Party shall observe any obligation it may have entered into with regard to investment of nationals or companies of the other Party".[54] In effect, a treaty-based umbrella clause obliges the treaty parties to comply with the obligations set out in any investment contract signed with investors from the other treaty state.[55] Umbrella clauses "provide assurances to foreign investors with regard to the performance of obligations assumed by the host State under its own law with regard to specific investments", thus helping to "secure the rule of law in relation to investment protection".[56] In the context of mining, one of the obligations commonly assumed by African countries in mining contracts is encompassed in the stabilization clause – that is, an obligation not to change or amend the legal and fiscal regimes governing the mining project.

However, arbitral tribunals differ on the scope of umbrella clauses in BITs. The points of contention are whether umbrella clauses in BITs transform contractual obligations into treaty commitments and whether a breach of contractual obligations amounts to a breach of the relevant treaty. Answers to these questions are significant because they bear directly on an investor's ability to invoke the dispute resolution mechanisms under BITs. Some tribunals have held that umbrella clauses in BITs do not transform contractual obligations into treaty commitments. In *SGS v Pakistan*, for example, the tribunal considered the scope of an umbrella clause in Article 11 of the Switzerland–Pakistan BIT to the effect that "[e]ither Contracting Party shall constantly guarantee the observance of the commitments

---

53 *Treaty between the United States of America and the Republic of Senegal Concerning the Reciprocal Encouragement and Protection of Investment*, art. II(4), (signed December 6, 1983; entered into force October 25, 1990), https://investmentpolicy.unctad.org/international-investment-agreements/treaty-files/2249/download
54 *USA–DRC Treaty, supra* note 39 art. II(4). See also *Agreement on Encouragement and Reciprocal Protections of Investments between the Government of the Republic of Zambia and the Government of the Kingdom of the Netherlands*, art. 3(3), (signed April 30, 2003; entered into force March 1, 2014), https://investmentpolicy.unctad.org/international-investment-agreements/treaty-files/2098/download
55 Jarrod Wong, "Umbrella Clauses in Bilateral Investment Treaties: Of Breaches of Contract, Treaty Violations, and the Divide between Developing and Developed Countries in Foreign Investment Disputes" (2006) 14(1) *George Mason Law Review* 137 at 144.
56 *SGS Société Générale de Surveillance S.A v Philippines* (ICSID Case No. ARB/02/6, Decision on Objections to Jurisdiction) at para. 126.

it has entered into with respect to the investments of the investors of the other Contracting Party".[57] The tribunal held that there is

> [no] convincing basis for accepting the Claimant's contention that Article 11 of the BIT has had the effect of entitling a Contracting Party's investor, like SGS, in the face of a valid forum selection contract clause, to 'elevate' its claims grounded solely in a contract with another Contracting Party, like the PSI Agreement, to claims grounded on the BIT, and accordingly to bring such contract claims to this Tribunal for resolution and decision.[58]

The tribunal opined that a broad interpretation of the umbrella clause portends dire legal consequences, including the danger that umbrella clauses would be susceptible to almost indefinite expansion as well as the danger that state parties would be exposed to endless commitments.[59] Interestingly, the tribunal acknowledged that an umbrella clause in a BIT could elevate contractual obligations to the level of treaty obligations where there is clear and convincing evidence that the parties to the BIT intended to elevate contractual obligations to the level of the BIT.[60] Such would be the case where a BIT expressly provides that breach of contractual commitments by a party shall amount to breach of the umbrella clause in the BIT.

Similarly, in *Joy Mining Machinery Ltd v Egypt*,[61] a case involving the scope of an umbrella clause in the Egypt–United Kingdom BIT, the tribunal declined to ascribe an expansive meaning to the umbrella clause because the language of the clause did not envisage such expansive meaning. The tribunal held that, in the specific context of this BIT,

> it could not be held that an umbrella clause inserted in the Treaty, and not very prominently, could have the effect of transforming all contract disputes into investment disputes under the Treaty, unless of course there would be a clear violation of the Treaty rights and obligations or a violation of contract rights of such a magnitude as to trigger the Treaty protection, which is not the case. The connection between the Contract and the Treaty is the missing link that prevents any such effect. This might be perfectly different in other cases where that link is found to exist, but certainly it is not the case here.[62]

Unlike the decision in *SGS v Pakistan* which interpreted an umbrella clause in a narrow sense, other tribunals have accorded a broad and expansive meaning to

---

57 *SGS Société Générale de Surveillance S.A . Pakistan* (ICSID Case No. ARB/01/13, Decision on Objections to Jurisdiction) at para. 53.
58 *Ibid.* at para. 165.
59 *Ibid.* at para. 166.
60 *Ibid.* at para. 167.
61 *Joy Mining Machinery Ltd. v Egypt* (ICSID Case No. ARB/03/11, Award on Jurisdiction).
62 *Ibid.* at para. 81.

umbrella clauses. This broad interpretation holds that umbrella clauses in BITs not only encompass contractual commitments undertaken by states, but they also confer treaty status on investment contracts. Thus, breach of a contractual commitment by a state party is breach of the relevant treaty. For example, both *Eureko B.V. v Poland*[63] and *Consorzio Groupement L.E.S.I.-DIPENTA v Algeria*[64] acknowledged that umbrella clauses in BITs cover contractual obligations in investor-state contracts. In *Eureko B.V. v Poland*, the tribunal held that the ordinary meaning of "any obligations" in an umbrella clause is that the treaty states shall observe 'any' and 'all' obligations entered into with regard to investments of investors of the other treaty party.[65] In *Consorzio*, the tribunal noted:

> Some treaties contain what is known as an "umbrella clause" that in effect transforms the State's breaches of contract into violations of that provision of the treaty, thereby granting jurisdiction to the Arbitral Tribunal established pursuant to the treaty to consider such violations.[66]

Likewise, in *Sempra Energy International v Argentina*,[67] the tribunal observed that "the specific guarantee of a general umbrella clause" under the BIT in question involves "the obligation to observe contractual commitments concerning the investment".[68] In *SGS v Paraguay*, the arbitral tribunal specifically held that in order to give "purpose and effect" to treaty-based umbrella clauses, they must be interpreted as encompassing "host State commitments of all kinds, including contractual commitments".[69] In *SGS v Philippines*,[70] a tribunal held that, an umbrella clause in the Switzerland–Philippines BIT, which provided that "[e]ach Contracting Party shall observe any obligation it has assumed with regard to specific investments in its territory by investors of the other Contracting Party", encompassed obligations specified in the investor-state contract and that the umbrella clause "makes it a breach of the BIT for the host State to fail to observe binding commitments, including contractual commitments, which it has assumed with regard to specific investments".[71]

Although tribunals differ with regard to the scope of umbrella clauses, overall tribunals ascribe a narrow or broad meaning to umbrella clauses depending on the

---

63 *Eureko B.V. v Poland*, Partial Award (Ad Hoc Arbitration), www.italaw.com/sites/default/files/case-documents/ita0308_0.pdf
64 *Consorzio Groupement L.E.S.I.-DIPENTA v Algeria*, Award (ICSID Case No. ARB/03/08).
65 *Eureko B.V. v Poland*, supra note 63 at para. 246.
66 *Consorzio Groupement L.E.S.I.-DIPENTA v Algeria*, supra note 64 at para. 25(ii).
67 *Sempra Energy International v Argentina* (ICSID Case No. ARB/02/16, Decision on Objections to Jurisdiction, 11 May 2005).
68 *Ibid.* at para. 101.
69 *SGS Société Générale de Surveillance S.A v Paraguay* (ICSID Case No. ARB/07/29, Decision on Jurisdiction) at para. 170.
70 *SGS Société Générale de Surveillance S.A v Philippines* (Case No. ARB/02/6, Decision on Objections to Jurisdiction).
71 *Ibid.* at para. 128.

specific wording of the treaty as well as on other factors that indicate the intention of the treaty parties.[72] Where the language of the treaty is overly broad, tribunals tend to interpret umbrella clauses widely to cover obligations assumed under investor-state contracts. Such broad language is evidence of the parties' intention that the umbrella clause should be given a wide interpretation. Where, however, the wording of the umbrella clause is narrow, tribunals may interpret the clause narrowly as not covering obligations in investor-state contracts. That being said, modern umbrella clauses tend to be couched in broad and inclusive language as they often use phrases such as "any obligation" or "any written undertaking". In such cases, tribunals are likely to interpret these clauses as encompassing obligations assumed by host states in investment contracts.

2.4 Consequences of treaty-based stabilization regimes

Taken together, the 'full protection and security', the 'fair and equitable treatment', and 'umbrella' clauses in BITs serve to stabilize the terms of the investment contracts between host states and foreign investors. However, these BIT clauses have serious legal and financial consequences for developing countries. In the legal sense, because the principles of 'fair and equitable treatment' and 'full protection and security' of investment protect the reasonable expectations of investors arising from the investment contract, including the expectation that the legal regimes prescribed in the contract would not be amended or changed, it could "limit the host state's ability to take legislative action promoting its sustainable development goals".[73] Moreover, breach of a stabilization clause in an investment contract amounts to breach of these BIT clauses, which entitles the investor to sue the host state for damages under the applicable BIT. The amount of damages payable by host states may depend on the losses suffered by the investor as a result of the breach. As held in *Sempra v Argentine Republic*, "the appropriate standard of reparation under international law is compensation for the losses suffered by the affected party".[74] In appropriate cases, the fair market value of the investment may be adopted as the standard for determining the loss suffered by the investor.[75] The host state may be liable to pay damages for the breach of a BIT even where the legal or administrative action constituting the breach is legitimately undertaken by the state, unless, of course, the state is able to prove the defence of 'necessity' under international law.

---

72 Katia Yannaca-Small, "Interpretation of the Umbrella Clause in Investment Agreements" (OECD Working Papers on International Investment, 2006/03) at 22, www.oecd.org/daf/inv/investment-policy/WP-2006_3.pdf
73 Graham Mayeda, "International Investment Agreements between Developed and Developing Countries: Dancing with the Devil? Case Comment on the Vivendi, Sempra and Enron Awards" (2008) *McGill International Journal of Sustainable Development Law & Policy* 189 at 212.
74 *Sempra Energy International v Argentine Republic* (ICSID Case No. ARB/02/16, Award (28 September 2007)) at para. 401.
75 *Ibid.* at para. 404.

That said, investors do not enjoy treaty-based protection where the investment contract was executed in contravention of domestic law or where the contract is void under domestic law. Thus, in the context of mining, a mineral right holder is not protected under a BIT if the mineral right was granted in contravention of the domestic laws of the host country. This assertion finds support in *Cortec Mining Kenya Limited v Republic of Kenya*, where Cortec Mining Kenya Limited alleged that the revocation of its mining licence amounted to expropriation under the United Kingdom–Kenya BIT.[76] Cortec was granted a mining licence over an area comprising the Mrima Hill, which is designated as a forest and nature reserve and a national monument under various statutes in Kenya. While mining could be conducted in the Mrima Hill, Kenyan law require that, prior to the issuance of a mineral right over the Mrima Hill, the applicant shall produce to the licensing authority an environmental impact assessment licence issued to them by the government. In violation of this requirement, Cortec's mining licence was granted without such prior environmental licence or approval. The tribunal held that Cortec's mining licence is *void ab initio*; hence, Cortec could not claim protection under the United Kingdom–Kenya BIT given that "the BIT protects only 'lawful investments'".[77] The tribunal noted further that failure to comply with basic statutory requirements is a serious breach of the investors' obligations.[78] The tribunal elaborated thus:

> [T]he Claimants' failure to comply with the legislature's regulatory regime governing the Mrima Hill forest and nature reserve, and the Claimants' failure to obtain an EIA licence (or approval in any valid form) from NEMA concerning the environmental issues involved in the proposed removal of 130 million tonnes of material from Mrima Hill, constituted violations of Kenyan law that, in terms of international law, warrant the proportionate response of a denial of treaty protection under the BIT and the ICSID Convention.[79]

This arbitral decision (award) is significant not only because it confirms that domestic law governs the validity of an investment contract, but also "for finding that international investment agreements only protect investment made in compliance with domestic law, even in the absence of an explicit legality requirement in the applicable BIT".[80]

---

76 *Cortec Mining Kenya Limited, Cortec (Pty) Limited and Stirling Capital Limited v Republic of Kenya*, ICSID Case No. ARB/15/29, Award (22 October 2018).
77 *Ibid*. at para. 333.
78 *Ibid*. at para. 351.
79 *Ibid*. at para. 365.
80 Lorenzo Cotula & James T. Gathii, "Cortec Mining Kenya Limited, Cortec (Pty) Limited and Stirling Capital Limited v. Republic of Kenya" (2019) 113(3) *American Journal of International Law* 574 at 577.

3 Drivers of stabilization regimes in Africa

Stabilization clauses have become prominent in the African mining industry for a variety of reasons. As noted above, stabilization clauses are often included in mining contracts and in statutory enactments because of the perceived need to protect foreign investors from the economic and political risks in Africa. Investors require stable and predictable legal and fiscal regimes in host countries in order to facilitate a profitable return on investment. Foreign investors worry that the instability of the fiscal regimes in Africa could adversely affect their investment, including financial returns on investment, particularly where the regimes are altered while the project is ongoing.[81] This fear is particularly prevalent in the extractive industries because extractive activities usually require companies to make vast expenditure on geological surveys and exploration for a number of years before the minerals are produced.[82] Besides, the judicial processes in many African countries are often ineffective. Thus, foreign investors may not have confidence in the capacity of African justice systems to resolve disputes arising from investment projects.

To protect against these risks, mining companies usually demand that host governments guarantee the stability of the fiscal regimes for mineral extraction so that the companies are better able to make investment projections.[83] In some instances, foreign investors extract such guarantees from host governments as consideration for their investment. For example, the assurances and guarantees provided by the Nigerian government regarding the LNG project were provided partly "in consideration of the investments which shall have been made [by the foreign investors] in order to prosecute the [LNG] Venture".[84]

The prevalence of stabilization clauses in mining contracts is equally attributable to the power imbalance between foreign investors (usually MNCs) and host African countries. MNCs in the mining industry wield enormous power over African countries not only because of the financial strength of the MNCs but also because of their technological expertise. To the contrary, African countries lack the financial power and the technological expertise to explore and exploit mineral resources. Thus, they rely largely on foreign companies to develop these resources. The power and influence of mining MNCs is particularly effective in those African countries that rely on the mining industry for their economic sustenance. In some instances, MNCs leverage their economic power with the power and influence of their home governments to obtain investment terms that are inimical to the

---

81 See Samuel K.B. Asante, "Stability of Contractual Relations in the Transnational Investment Process" (1979) 28 *International and Comparative Law Quarterly* 401 at 408–9.
82 See Thomas W. Waelde & George Ndi, "Stabilizing International Investment Commitments: International Law versus Contract Interpretation" (1996) 31 *Texas International Law Journal* 215 at 227.
83 Asante, *supra* note 81 at 408–9.
84 *Nigeria LNG (Fiscal Incentives, Guarantees and Assurances) Decree No. 113 of 1993*, 2nd Schedule, Preamble.

interest of African countries, such as stabilization clauses.[85] As discussed above, these investment terms are reinforced by BITs and MITs through provisions that elevate contractual obligations in investor-state contracts into treaty obligations.

The economic liberalization policies forcibly imposed on African countries by the World Bank in the 1980s and 1990s also account for the prevalence of stabilization clauses in Africa. The World Bank believed:

> The future development of the mining sector in Africa will largely depend on attracting new high risk capital from foreign mining companies – large, medium, and small – who have the technical and managerial capabilities to find new deposits and develop new mining operations.[86]

Thus, during this period, the World Bank urged Africa to adopt mining legislation and policies that not only assured security of tenure but also reduced the risk of sudden and unpredictable changes to the laws governing mining projects.[87] More specifically, the Word Bank recommended that African countries should enter into investment contracts that "provide assurances that taxation arrangements will be stable over the life of the mine or for a certain minimum period of mine life (possibly twenty years)".[88]

The intense competition to attract foreign investment often compels African countries to offer generous incentives, such as fiscal stabilization guarantees, to mining companies. Moreover, most African countries lack the institutional capacity to negotiate investment contracts with mining MNCs. Sometimes, African countries rely on lawyers based in Western countries to negotiate and draft mining contracts on their behalf. While foreign lawyers may be well-meaning, these lawyers lack appropriate understanding and knowledge of the peculiarities of the African mining industry and are thus unable to cater to these peculiar features of African mining. The incapacity of African countries to negotiate contracts often leads these countries to agree to contractual terms that are inimical to the economic interest of their citizens, including stabilization clauses that insulate mining companies from new laws.

It is imperative for Africa to optimize resource benefits at the contract negotiating stage, given that "it is difficult to fundamentally renegotiate contracts at a later stage without sending negative signals to investors on the certainty of contracts, with resulting increased negative investment risk perceptions".[89] Thus, Africa must develop the domestic capacity to negotiate mining contracts, taking

---

85 Joseph E. Stiglitz, "Regulating Multinational Corporations: Towards Principles of Cross-Border Legal Frameworks in a Globalized World Balancing Rights with Responsibilities." (2008) 23 *American University International Law Review* 451 at 477.
86 World Bank, *Strategy for African Mining*, supra note 2 at 10.
87 *Ibid.* at 26.
88 *Ibid.* at 26.
89 African Union, *Africa Mining Vision* (February 2009) at 17, www.uneca.org/sites/default/files/PublicationFiles/africa_mining_vision_english.pdf

into account the need for equitable sharing of mineral rents, as well as a "flexible fiscal regime which is sensitive to price movements and stimulates national development".[90] The development of domestic capacity must begin with the teaching of mining law courses at African universities. Capacity-building could equally be "facilitated through ensuring that there is a skills transfer dimension in all contracted consultancies during the lease/license negotiations".[91] Furthermore, African countries can develop institutional capacity through the pooling of resources with neighbouring states and regional economic communities.[92]

4 Deleterious effects of stabilization regimes

As noted above, some stabilization clauses freeze the legal and fiscal regimes applicable to mining projects. Although freezing stabilization clauses are most prevalent in sub-Saharan Africa,[93] in reality freezing stabilization clauses cannot prevent the host country from amending the legal and fiscal regimes governing a project.[94] Even in cases where the host country validly grants a stabilization clause, the legislature of the host country retains the right to alter the legal and fiscal regimes governing the project, subject to the payment of compensation for breach of the stabilization clause.[95] As argued by Wealde and Ndi, "the notion of sovereignty under municipal law means that the legislator can take what he has given" and "nothing would prevent the national legislature from retroactively canceling and revoking rights awarded, possibly subject to constitutional and other legal consequences (e.g., the duty to pay compensation under national law)".[96] Peter Cameron expresses a similar sentiment by observing that "in every country the sovereign retains the power – in spite of any laws or contracts to the contrary – to enact laws that legally will 'trump' previous laws (and contracts)".[97] In effect, stabilization clauses "cannot bind the legislative body from enacting any other law which in any way alters or amends the terms contained therein".[98] This view had earlier been projected by F.A. Mann when he argued that "the express exemption from the effects of future legislation is redundant" because "[s]uch exemption cannot and ought not to preclude the genuine exercise of the state's police power".[99]

---

90 *Ibid.* at 17.
91 *Ibid.* at 18.
92 *Ibid.* at 18.
93 Shemberg, *supra* note 12 at 22.
94 See Waelde & Ndi, *supra* note 82 at 238–40; Yinka Omorogbe, "Law and Investor Protection in the Nigerian Natural Gas Industry" (1996) 14 *Journal of Energy & Natural Resources Law* 179 at 189.
95 Waelde & Ndi, *ibid.*
96 *Ibid.* at 239. See also Cameron, *supra* note 7 at 13 ("in every country the sovereign retains the power – in spite of any laws or contracts to the contrary – to enact laws that legally will 'trump' previous laws (and contracts)").
97 Cameron, *supra* note 7 at 13.
98 Omorogbe, *supra* note 94 at 189.
99 F.A. Mann, "State Contracts and State Responsibility" (1960) 54 *American Journal of International Law* 572 at 587–8.

Thus, notwithstanding a stabilization clause, host states may amend the legal and fiscal regimes prescribed in resource concession contracts where the amendment is required in the public interest. In *BP Exploration Co. (Libya) v The Government of the Libyan Arab Republic*, for example, the arbitral tribunal recognized that although the stabilization clause in that case limited Libya's freedom to alter or amend unilaterally the terms of the concession, Libya nonetheless could do so if the amendment was in the public interest.[100] This position finds support in the fact that a statute enacted by the host state remains valid even if the statute contravenes the stabilization measures prescribed in prior concession contracts signed by the state.

However, the freezing of the legal and fiscal regimes governing mining projects has economic and social implications for host African countries. In the economic sense, the breach of a stabilization clause exposes the host state to financial liability in the form of monetary damages. Thus, while stabilization clauses may not prevent host states from altering the legal and fiscal regimes stipulated in the mining contract, in practice such clauses are functionally useful to foreign investors because their breach by host states creates a cause of action for damages.[101] Stabilization clauses "reinforce the traditional rule that full compensation must be made for property rights taken by the state".[102] The amount of compensation payable by host states for the breach of a stabilization clause depends primarily on two factors: the type of stabilization clause and the nature of the breach. Where the stabilization clause protects against expropriation and where the nature of the breach is such that it amounts to expropriation or 'regulatory taking' of mineral rights, compensation is determined based on "all the circumstances relevant to the particular concrete case".[103] These circumstances include the "loss suffered" by the mineral right holder[104] and "the value of the [expropriated] corporeal property, including all assets, installations, and various expenses incurred" as a result of the breach.[105] More specifically, arbitral tribunals may take into account the value of the mineral rights expropriated by the state,[106] the legitimate expectations of the mineral right holder arising from the mining contract,[107] the anticipated profits from the mineral concession,[108] and the loss of profits resulting from the breach of contract by the host state.[109]

100 (1979) 53 *International Law Reports* 297 at 318–21, 324, & 327.
101 See R. Doak Bishop, Sashe D. Dimitroff, & Graig S. Miles, "Strategic Options Available When Catastrophe Strikes the Major International Energy Project" (2001) 36 *Texas International Law Journal* 635 at 642.
102 Christopher T. Curtis, "The Legal Security of Economic Development Agreements" (1988) 29 *Harvard International Law Journal* 317 at 349.
103 *Kuwait v American Independent Oil Company (Aminoil)*, (1982) 21 I.L.M. 976 at 1033 para. 144 (1982) [*Kuwait v Aminoil*].
104 *Agip Company v Popular Republic of the Congo*, (1982) 21 I.L.M. 726 at 737 para. 98 [*Agip v Congo*].
105 *LIAMCO v Libya*, *supra* note 21 at 67.
106 *Ibid.* at 81.
107 *Kuwait v Aminoil*, *supra* note 103 at 1034 para. 148.
108 *Ibid.* at 1034–5 paras. 152–3.
109 *Agip v Congo*, *supra* note 104 at 737 para. 98.

Compensation is also payable for breach of 'economic equilibrium' clauses. These clauses enjoin host states to restore the economic equilibrium struck in the investor-state contract in the event that the host state breaches the clause. In effect, the host state must restore the legitimate expectations of the investor arising from the investor-state contract. For economic equilibrium clauses, the appropriate compensation is one that restores the affected party to their position prior to the breach of contract by the host state. Thus, where the investor incurs costs because of the breach of the stabilization clause by the host state, the host state must, in order to restore the contractual equilibrium, reimburse the investor the costs incurred by the investor as a result of the breach. Likewise, where an amendment to the host state's laws imposes additional or new regulatory obligations on a mining company that is signatory to a prior stabilization clause, the host state must reimburse the company for costs incurred in complying with the additional or new regulatory obligations.[110]

Economic equilibrium clauses may be economically detrimental to African countries in another sense. Equilibrium clauses may exacerbate the financial obligation of host countries regarding restoration of contractual equilibrium where, as a result of the belligerence or default of a mining company, the cost of compliance with new laws and regulations increases exponentially. A company may incur higher cost of compliance due to its inability or refusal to comply with changes to the law in a timely manner. Given that extant economic equilibrium clauses in the African mining industry do not cater to such situations, the host country would be obliged to pay such cost to the company in order to restore the economic equilibrium, thus rewarding the company for their belligerence or default. To avoid this unpleasant scenario, economic equilibrium clauses in Africa ought to require good-faith compliance with regulatory rules so as not to exacerbate the negative impact of new regulations on the company.

Stabilization clauses could lead to the loss of mining revenues for host states. Zambia is a case in point. In the course of privatizing its mining industry in the 1990s, the Zambian government signed numerous DAs with mining MNCs that not only granted very liberal and generous terms in favour of the MNCs,[111] but also prohibited the government from altering the fiscal regimes prescribed in the agreements. The DAs fixed the royalty rate for copper at 0.6% of the gross revenues of the companies,[112] far below the statutory royalty rate of 3% under the now repealed *Mines and Minerals Act* of 1995.[113] More significantly, the DAs

---

110 See South Africa's *MPRRA*, s. 14.(2).
111 Christopher Adam & Anthony M. Simpasa, "Harnessing Resource Revenues for Prosperity in Zambia" (OxCarre Research Paper 36, Revised Draft) at 25–7, www.economics.ox.ac.uk/materials/working_papers/4930/oxcarrerp201036.pdf
112 See, for example, *Mopani Mines Agreement, supra* note 11, art. 16 read with Schedule 8.
113 John Lungu, "Copper Mining Agreements in Zambia: Renegotiation or Law Reform?" (2008) 117 *Review of African Political Economy* 403 at 407.

froze the fiscal regimes specified under the agreements for 15 to 20 years.[114] Although there was a boom in the price of copper on the international market, the boom did not yield appreciable revenues for Zambia because of the low rate of royalty prescribed under the DAs.[115] In fact, Zambia "incurred losses in tax revenues through the subsidies given to the private mining companies" under the DAs.[116] Zambia realized a royalty of US$20 million from the combined gross proceeds of US$3.4 billion in 2007 because of the low royalty rate of 0.6% stipulated in the DAs.[117]

Stabilization clauses also limit the host government's flexibility to set tax policy, thus constricting the fiscal and tax policy options available to host states.[118] This is particularly so because of the broad scope and long duration of stabilization clauses. For example, some stabilization clauses "extend well beyond the fiscal regime" governing investment projects, including the freezing of exchange controls, foreign investment rules, and regulatory control over projects.[119] In Ghana, a 'stability agreement' can freeze a wide range of issues, including taxes, royalties, fees, exchange control, importation of goods and equipment, transfer of capital, and remittance of dividends.[120] In terms of duration, in Ghana a 'stability agreement' can freeze the law applicable to a mining project for a period of 15 years.[121] Worse still, stabilization agreements in South Africa have effect "for as long the extractor holds the [mineral] right" to which the agreement relates.[122] Similarly, the guarantees and assurances granted under the *Nigeria LNG Act* have effect "so long as the Company, or any successor thereto, is in existence and carrying on the business of liquefying and selling liquefied natural gas and natural gas liquids within and/or outside" Nigeria.[123] In other words, the Nigerian government binds itself not to introduce changes to the legal and fiscal regimes governing the LNG project for as long as the companies involved in the project are implementing the project.

Stabilization clauses enable foreign investors to influence the behaviour of host governments, sometimes to the detriment of their citizens. For example, in order to avoid breaching its commitments under a stabilization clause, the host state may decline to enact new laws and regulations governing a mining project, even though such new laws could be beneficial to its citizens. Where the state enacts

---

114 See *Mopani Mines Agreement*, *supra* note 11, art. 16 read with art. 1, which defines the 'stability period' as "the period commencing as of the date of this Agreement and ending on the fifteenth (15th) anniversary of the date of this Agreement".
115 See Adam & Simpasa, *supra* note 111 at 35–7.
116 Lungu, *supra* note 113 at 409.
117 Adam & Simpasa, *supra* note 111 at 36 (footnote 27).
118 See Thomas Baunsgaard, "A Primer on Mineral Taxation" (IMF Working Paper WP/01/139) at 18, www.imf.org/external/pubs/ft/wp/2001/wp01139.pdf
119 Cameron, *supra* note 7 at 44.
120 *Minerals and Mining Act, 2006* (Ghana), s. 48.
121 *Ibid.* s. 48.
122 *MPRRA*, s. 13.(1).
123 *Nigeria LNG (Fiscal Incentives, Guarantees and Assurances) Act No. 39 of 1990*, 2nd Schedule (as amended by the *Nigeria LNG (Fiscal Incentives, Guarantees and Assurances) (Amendment) Act No. 113 of 1993*, s. 4(a)).

new laws and regulations, it may refuse to enforce the laws for fear that their enforcement could trigger compensation for breach of the stabilization clause. In sum, because host states are desirous of avoiding payment of compensation for breach of stabilization clauses, they may become reluctant to exercise their legislative or regulatory power over investment projects. This is particularly so where a stabilization clause commits the government not to amend or change its laws or regulations in a manner that amounts to expropriation of investment. Given the broad and liberal interpretation of contractual and treaty provisions on 'expropriation', the enforcement of new laws or new regulations against a company that is signatory to a prior contract containing a stabilization clause may be viewed by arbitral tribunals as an indirect expropriation where the new laws or regulations have a material adverse impact on the company's investment. For example, in the context of investment treaties, expropriation has been held to mean:

> not only open, deliberate and acknowledged takings of property, such as outright seizure or formal or obligatory transfer of title in favour of the host State, but also covert or incidental interference with the use of property which has the effect of depriving the owner, in whole or in significant part, of the use or reasonably-to-be-expected economic benefit of property even if not necessarily to the obvious benefit of the host State.[124]

Applying this logic, a new law or regulation that imposes compliance costs on a company that is the beneficiary of a prior stabilization clause may amount to incidental interference with the use of the company's property. This is particularly so where compliance with the new law substantially deprives the company of a reasonably expected economic benefit, such as profits. In effect, a host state that breaches a stabilization clause through the enactment of a law that changes the legal regimes governing a mining project may in fact be guilty of indirect expropriation where the law incidentally interferes with the investor's reasonably expected profits. The host state is liable to pay compensation for such indirect expropriation. Thus, the host state may not want to enact or enforce new laws or regulations in order to avoid paying compensation to the investor.

On the social front, stabilization clauses could freeze the environmental, human rights, and labour laws applicable to a project. The freezing of the legal regimes governing mining projects causes regulatory chill, thus adversely affecting human rights and environmental protection in Africa. In this sense, freezing stabilization clauses restrict the host state's ability to regulate the human rights and environmental practices of mining companies. As noted by some observers, "freezing stabilization clauses may work to reduce the effectiveness of new laws, because they offer [companies] the reasonable opportunity to rely on [the clauses] to avoid

---

[124] *Metalclad Corporation v The United Mexican States* (ICSID Case No. ARB(AF)/97/1), Award (August 30, 2000) at para. 103.

218 Legal stabilization of mining investments

applying new laws".[125] By freezing the legal regimes applicable to mining projects, stabilization clauses hold back host governments from taking steps to improve human rights protection.[126] Furthermore, where stabilization clauses are couched in broad language, host African countries lose "the flexibility to introduce new regulations to promote and protect human rights", labour rights, and the environment, at least in relation to the project.[127] Stabilization clauses can expressly diminish the environmental obligations of mining companies. Moreover, freezing stabilization clauses preclude regulatory agencies from enforcing new laws and regulations against mining companies, thus encouraging social irresponsibility on the part of the companies.

More significantly, freezing stabilization clauses insulate mining companies from complying with future changes to the human rights, labour rights, and environmental protection regimes in African countries. For example, the original version of the Mineral Development Agreement between the Government of Liberia and Mittal Steel contained a stabilization clause that provided that

> any modifications that could be made in the future to the Law as in effect on the Effective Date shall not apply to the CONCESSIONAIRE and its Associates without their prior written consent, but the CONCESSIONAIRE and its Associates may at any time elect to be governed by the legal and regulatory provisions resulting from changes made at any time in the Law as in effect on the Effective Date.[128]

This provision not only shielded Mittal Steel from future changes to Liberian laws, including laws relating to human rights and environmental protection, but also allowed the company "to choose which new laws it will comply with".[129] This clause was particularly detrimental because it served as an incentive for Mittal Steel, a financially powerful MNC, to exert pressure on the Liberian government to enact laws that attenuated the environmental obligations of companies operating in Liberia. This stabilization clause was renegotiated and replaced with a new clause in 2007.[130]

---

125 Shemberg, *supra* note 12 at 35–6.
126 Amnesty International, "Contracting Out of Human Rights: The Chad-Cameroon Pipeline Project" (September 2003) at 11–12, www.amnesty.org/download/Documents/88000/pol340122005en.pdf
127 United Nations Economic and Social Council, *Human Rights, Trade and Investment Report of the High Commission for Human Rights* (E/CN.4/Sub.2/2003/9) at 20, https://documents-dds-ny.un.org/doc/UNDOC/GEN/G03/148/47/PDF/G0314847.pdf?OpenElement
128 Global Witness, "Heavy Mittal? A State within a State: The Inequitable Mineral Development Agreement between the Government of Liberia and Mittal Steel Holdings NV", at 31, https://cdn.globalwitness.org/archive/files/pdfs/mittal_steel_en_oct_2006_low_res.pdf
129 *Ibid.*
130 See *Act Ratifying Mittal MDA, supra* note 10, art. 16(E).

Relatedly, freezing stabilization clauses fetter the exercise of administrative discretion in host countries in the sense that matters covered under these clauses are insulated from the jurisdiction of administrative agencies. In practice, projects covered by stabilization clauses are insulated from administrative scrutiny, thus circumscribing the ability of regulatory agencies to make administrative decisions regarding investment projects. One arbitral tribunal has ruled, for example, that legal stability agreements (LSAs) are subject to private or civil law and not administrative law; hence, the execution, interpretation, and enforcement of the provisions set forth in LSAs are not only subject to the general principles applicable to contracts between private parties but, in addition, "the fundamental rights granted [by the host state] pursuant to an LSA are private contracting rights that are enforceable against the state as if it were a private party".[131]

The deleterious effects of stabilization clauses have prompted some African countries to abolish freezing stabilization clauses, while other countries have terminated BITs which, as argued earlier, reinforce contract-based stabilization clauses. For example, in 2017 Tanzania abolished freezing stabilization clauses in the natural resource industries.[132] To that end, a recent amendment to the *Mining Act* of Tanzania provides that "[i]n any negotiations for the provision of a stabilization regime in the extractive sector, it shall be prohibited to use stabilization arrangements that entail the freezing of laws or contracting away the sovereignty of the United Republic [of Tanzania]".[133] Likewise, South Africa has terminated several BITs with Western countries because these BITs impeded the ability of the government to comply with its human rights obligations under the South African Constitution.[134]

Although stabilization clauses have deleterious effects, in practice stabilization clauses are rarely invoked by investors because "invoking the fiscal stability clause in an agreement is the nuclear option".[135] The invocation of stabilization clauses is certain to damage irretrievably the relationship between the investor and the host state, suggesting, then, "that the real benefit of a fiscal stability clause may be to sow the seed of doubt in the host government that it might be invoked and thereby promote appropriate behavior".[136]

---

131 *Duke Energy International Peru Investments No. 1 Ltd. v Peru*, ICSID Case No. ARB/03/28; IIC 30 (2006), Decision on Jurisdiction, 1 February 2006 at para. 31.
132 *The Written Laws (Miscellaneous Amendments) Act, 2017* (No. 7), s. 25 (incorporating s. 100E into the *Mining Act, 2010*).
133 *Ibid.*
134 See Herbert Smith Freehills LLP, "South Africa Terminates Its Bilateral Investment Treaty with Spain: Second BIT Terminated, as Part of South Africa's Planned Review of Its Investment Treaty", https://hsfnotes.com/arbitration/2013/08/21/south-africa-terminates-its-bilateral-investment-treaty-with-spain-second-bit-terminated-as-part-of-south-africas-planned-review-of-its-investment-treaties
135 Philip Daniel & Emil M. Sunley, "Contractual Assurances of Fiscal Stability" in Philip Daniel *et al.*, eds, *The Taxation of Petroleum and Minerals: Principles, Problems and Practice* (London: Routledge, 2010) 405 at 419.
136 *Ibid.* at 419.

5 Reformation of stabilization regimes in Africa

As noted above, stabilization clauses have adverse economic and social consequences for African countries, including loss of revenues and the hindering of human rights and environmental protection. The rigidity and broadness of stabilization clauses deprive African countries of the opportunity to maximize their mining revenues. Viewed this way, stabilization clauses are detrimental to the achievement of the primary goal of the fiscal regimes for mineral extraction, which is the promotion of economic development through the enhancement of mineral revenues. The rigid fiscal arrangement contemplated by stabilization clauses in many African countries "cannot realistically persist in the face of the dynamic [global] economic changes".[137] This is particularly so because of the increasing global demand for mineral resources and the consequential increase in the price of minerals on the international market. If Africa is to optimize revenues from mineral extraction, the substantive contents of the fiscal stabilization clauses governing mining projects in Africa must be reformed. In the ensuing pages, I offer suggestions on how Africa can ensure that stabilization clauses in the mining industry do not lead to unintended adverse impacts.

5.1 Limiting the scope and duration of stabilization regimes

The starting point is to realign stabilization clauses with reality by ensuring that they take into account the interests of both investors and the host country. While it is no doubt appropriate for foreign investors to seek the protection of their investments through stabilization clauses, the protection of investment ought not to be at the economic expense of African countries. Rather, the interests of investors should be protected in a manner that takes into account the economic and social aspirations of the host countries. In particular, stabilization provisions should be flexible so that they are better able to cater to the interests of mining companies and host countries in an equitable manner.

In this regard, the adverse impacts of stabilization clauses on human rights and environmental protection could be ameliorated by limiting the scope of stabilization clauses. Lorenzo Cotula argues, for example, that a "human rights exception" should be built into stabilization clauses, such that the "host state regulation to promote the full realization of human rights is outside the scope of the stabilization clause".[138] That is, stabilization clauses should not operate to prevent or prohibit the host state from enacting new laws or regulations designed to discharge the state's obligations under international law. Thus, African countries should ensure that stabilization clauses are not open-ended. Rather, stabilization clauses should contain express provisions excepting human rights and environmental protection from the ambit of the stabilization clauses. However, it is

---

137 Asante, *supra* note 81 at 411.
138 Lorenzo Cotula, "Reconciling Regulatory Stability and Evolution of Environmental Standards in Investment Contracts: Towards a Rethink of Stabilization Clauses" (2008) 1(2) *Journal of World Energy Law & Business* 158 at 172.

unlikely that mining companies would agree to such limitation on the scope of stabilization clauses. This is because modern stabilization clauses are designed not only to protect against investment risks, but also to protect the investor's profits. Corporate profits would likely be diminished by the imposition of post-contract regulatory burdens or compliance costs on mining companies. Besides, mining companies may view such express human rights exception as an attempt to impose human rights obligations on companies, a position contrary to extant international law which holds that companies do not bear human rights obligations because they are not subjects of international law.[139]

A complementary strategy is to limit the duration of stabilization clauses to a relatively short period, say five years. For example, a stabilization clause may provide that any changes to the host state's legal regimes are inapplicable to a mining company that is signatory to the contract containing the stabilization clause for a period of five years. Such a provision would afford the company enough time to adapt to the new legal regimes. In that regard, Tanzania now requires stabilization clauses to be time-bound and renders unlawful any stabilization agreement that lasts for the lifetime of a mine.[140] Likewise, the DRC has recently reduced the duration of stabilization clauses from ten years to five years through an amendment to the Mining Code.[141]

In the alternative, flexibility can be achieved by ensuring that stabilization clauses have staggered phases. That is, the exploration and production phases of resource development should be governed by different stabilization clauses, as opposed to the current practice where a single stabilization clause governs both phases. Upon the grant of a concession, a stabilization clause governing the exploration phase of the project may be granted to mining companies. Once minerals are discovered in commercial quantity, a new stabilization clause governing the production phase should then be negotiated between the company and the host state. However, because the production phase lasts for a long period, as discussed below, stabilization clauses governing the production phase should be subject to periodic review and renegotiation to cater to changing circumstances.

Lastly, stabilization clauses should be contingent on compliance with the laws of the host country. Mining statutes and contracts in Africa should contain express provisions to the effect that stabilization clauses are enforceable only if the mineral right holder has complied with the laws of the host country, including the statutory and contractual obligations of the mineral right holder as well as the terms and conditions of the mineral right. Such contingency provision would ensure that a mineral right holder who commits environmental infractions or fails to pay requisite royalties and taxes through tax-avoidance practices is disentitled to rely

---

139 See Evaristus Oshionebo, *Regulating Transnational Corporations in Domestic and International Regimes: An African Case Study* (Toronto: University of Toronto Press, 2009) at 115–6.
140 *The Written Laws (Miscellaneous Amendments) Act, 2017* (No. 7), s. 25 (incorporating s. 100E into the *Mining Act, 2010*).
141 *Law No. 007/2002 of July 11, 2002 Relating to the Mining Code* (as amended by *Law No. 18/001* of 9 March 2018), art. 342.

on a stabilization clause in the contract with the host government. In effect, breach of the domestic laws of the host state should prevent reliance on a stabilization clause. Stabilization clauses are designed to ensure that the host state does not depart from its laws applicable to an investment project; thus, it is logical that investors should be obliged to comply with those same laws in order to sustain the protection afforded under the stabilization clause.

5.2 Periodic renegotiation of stabilization regimes

A further strategy is the renegotiation of stabilization regimes, particularly freezing stabilization clauses. Mining statutes and mining contracts in Africa should contain express provisions regarding renegotiation of stabilization clauses at specified intervals. For example, Tanzania's *Written Laws (Miscellaneous Amendments) Act, 2017* requires all stabilization agreements to "make provision for renegotiation from time to time as may be necessary".[142] Even where a renegotiation clause is not provided for in mining contracts, the contracting parties should renegotiate the stabilization clause in circumstances where a change in economic conditions renders the entire agreement "patently inequitable" to one party[143] and in exceptional circumstances such as national emergencies.

Several countries, including Ghana, Guinea, the DRC, Liberia, Sierra Leone, and Tanzania, have adopted the contract renegotiation strategy. In 2012, Ghana inaugurated a committee to renegotiate stability agreements considered not to be in the best interest of the country.[144] In 2017, Tanzania enacted a new law empowering the government to review and renegotiate natural resource contracts containing unconscionable terms.[145] For this purpose, a contractual term is deemed unconscionable if it restricts the right of the state to exercise full permanent sovereignty over its natural resources, or restricts the right of the state to exercise authority over foreign investment within the country and in accordance with the laws of Tanzania.[146] Unconscionable terms also include terms that are inequitable and onerous to the state; terms that restrict "periodic review of arrangement or agreement which purports to last for life time"; terms securing preferential treatment for an investor; terms that create a separate legal regime for the benefit of a particular investor; terms that restrict the right of the state to regulate the activities of MNCs; and terms that expressly or implicitly undermine the effectiveness of state measures to protect the environment or the use of

---

142 *The Written Laws (Miscellaneous Amendments) Act, 2017* (No. 7), s. 25 (incorporating s. 100E into the *Mining Act, 2010*).
143 Asante, *supra* note 81 at 412.
144 Martin. K. Ayisi, "The Review of Mining Laws and the Renegotiation of Mining Agreements in Africa: Recent Developments from Ghana" (2015) 16 *Journal of World Investment & Trade* 467 at 495.
145 *The Natural Wealth and Resources Contracts (Review and Re-Negotiation of Unconscionable Terms) Act, 2017*.
146 *Ibid.* s. 6.(2).

environment friendly technology.[147] Where the renegotiation does not produce a positive outcome or where a company fails to engage in renegotiation with the government, the unconscionable terms contained in the contract shall cease to have effect and the terms "shall, by operation of this Act, be treated as having been expunged" from the contract.[148] Stabilization clauses (particularly the freezing type) qualify as unconscionable terms in Tanzania because they not only constrain the state's ability to exercise control over its natural resources, but they also secure preferential treatment for mining companies.

In sum, mining contracts in Africa should contain express provisions empowering the parties to mutually review and renegotiate the terms of the contract if certain specified events occur. In particular, stabilization clauses should be mutually renegotiated after a specified period and in situations where the circumstances have changed significantly.[149] Even then, the renegotiated clause should aim at maintaining economic equilibrium rather than freezing the legal regimes governing a mining project. The renegotiation of a stabilization clause after a specified period or after the completion of the exploration stage is beneficial to the contracting parties. Regarding the investor, the economic interest of the investor may be better protected where an investment contract is susceptible to renegotiation. For example, the host state would "be little inclined to abrogate an agreement or nationalize an undertaking if the terms of an investment agreement could be re-opened" and renegotiated at stated intervals.[150] In this sense, the renegotiation of mining contracts ensures the stability of investment projects. As is apparent from the scenario that played out in Tanzania in 2017 and in Zambia in 2008, public opinion could compel African governments to unilaterally repudiate stabilization clauses that lack provisions for renegotiation. As noted previously, Tanzania's legislature recently abolished freezing stabilization clauses and authorized the government to renegotiate or terminate extant stabilization agreements due partly to public pressure by citizens who felt exploited by mining MNCs. Earlier in 2008, the government of Zambia unilaterally terminated several mining DAs, partly because of the refusal of the companies to renegotiate the agreements and partly in order to assuage the citizens of Zambia who were enraged that while the mining companies were making significant profits, Zambia earned low mining revenues as a result of the low royalty rate of 0.6% prescribed under the DAs. Hence, in 2008 Zambia introduced a new fiscal regime for mining "in order to bring about an equitable distribution of the mineral wealth between the Government and the mining companies".[151]

---

147 *Ibid.* s. 6.(2).
148 *Ibid.* s. 7.
149 See Asante, *supra* note 81 at 412–8.
150 *Ibid.* at 411.
151 See "Budget Address by the Honourable Ng'andu P. Magande, MP Minister of Finance and National Planning Delivered to the National Assembly on 25th January 2008", at para. 146, www.parliament.gov.zm/sites/default/files/images/publication_docs/2008%20Budget%20Speech.pdf

The renegotiation of a stabilization clause benefits the host state because, at the time of the renegotiation, the host state would have access to requisite information, such as the density and quality of mineral resources discovered during the exploration stage. Thus, the host state is better able to negotiate the terms of the stabilization clause. Better still, where a significant reserve of mineral resources is discovered within the initial five-year period (suggested above), the host country may choose "not to provide commitments on contract stability at all".[152]

5.3 Independent oversight of stabilization regimes

Aside from reforming the substantive contents of stabilization agreements and clauses, there is a need for oversight of the exercise of the power to grant stabilization agreements and clauses. Oversight is particularly necessary because the statutory discretion granted to the Minister to enter into stabilization agreements with mining companies is susceptible to abuse, given the spectre of administrative inefficiency and corruption in Africa's public service. In Zambia, for example, the power to enter into DAs was indiscriminately exercised by government officials in the 1990s and early 2000s when the government signed several lopsided DAs with foreign mining companies. As noted earlier, these DAs prescribed royalty rates that were lower than the rates specified in the mining statute. In addition, government officials may be influenced by corruption or by their own personal interests to grant generous stabilization terms to mining companies, as was apparently the case in Zambia.[153]

Africa would be better served by vesting the power to negotiate stabilization agreements and clauses in an independent panel of experts, as opposed to the current practice in countries such as Ghana and South Africa, where the power is vested in one person, the Minister of Mines and the Minister of Finance, respectively.[154] Moreover, the process of contract negotiation must be transparent in the sense that it is subject to public scrutiny. In particular, clear guidelines must be enunciated for the exercise of the discretion to grant stabilization agreements and clauses in favour of mining companies. Oversight may take the form of judicial review or approval by an independent body, such as a legislative committee. Better still, as recommended by the ECOWAS, stability agreements should be subject to parliamentary approval and ratification.[155] In this regard, Ghana's *Minerals and Mining Act, 2006* is laudable because it requires some oversight of stability agreements. The Act requires any stability agreement or development agreement signed by the Minister to be ratified by the Parliament.[156] However, the Act does

152 Cameron, *supra* note 7 at 17.
153 Lungu, *supra* note 113 at 408.
154 See *Minerals and Mining Act, 2006* (Ghana), ss. 48 & 49 and South Africa's *MPRRA*, s. 13.(1).
155 ECOWAS, *Directive C/DIR.3/05/09 on the Harmonization of Guiding Principles and Policies in the Mining Sector*, art. 7(4), https://documentation.ecowas.int/wp fb-file/ecowas-directive-and-policies-in-the-minning-sector2-pdf
156 *Minerals and Mining Act, 2006* (Ghana), ss. 48.(2) & 49.(3).

not go far enough because, while it requires parliamentary ratification, it does not expressly authorize the Parliament to vary or alter the terms of the agreement. The Parliament must ratify the agreement as it stands or withhold ratification. Also, the threshold for parliamentary ratification is unclear. The Parliament of Ghana should ensure that its ratification process is transparent and that the public is allowed to participate in the process. In particular, open consultation should be held with the public before the Parliament votes to ratify a stability agreement.

6 Why Africa should desist from incessant grant of stabilization clauses

Overall, African countries should desist from the indiscriminate grant of fiscal stabilization clauses to natural resource extractors. Although fiscal stabilization clauses may have been justifiable in the 1960s and 1970s because of the political and legal uncertainties then prevalent in Africa, there is less justification for such clauses today for a number of reasons. First, African countries are more stable today than they were in the 1960s and 1970s when stabilization clauses were thought to provide a safeguard against the political and legal instability in Africa. In recent years, many African countries have adopted democracy as a governance system, although it must be said that some of the so-called democratic governments in Africa are autocratic and unaccountable. The advent of democracy has positively impacted the legal systems in Africa such that today there is a modest degree of legal predictability and stability in many mineral-rich African countries, including Botswana, Ghana, Namibia, Nigeria, South Africa, Tanzania, and Zambia.

Second, economic and political risks in developing countries can now be insured privately through institutions such as the Multilateral Investment Guarantee Agency (MIGA), the private-sector arm of the World Bank, which grants political risk insurance and guarantees against expropriation; restrictions on currency convertibility and transfer; wars, terrorism, sabotage, and civil disturbances; breach of contract, and non-honouring of financial obligations by host countries.[157] The economic and political risks often associated with natural resource development in developing countries can and are now commonly insured by the MIGA. Since its inception in 1988, the MIGA has issued more than US$30 billion in political risk insurance covering more than 750 projects in a wide variety of industries, including the mining industry.[158] In fact, the value of MIGA's political risk insurance and guarantees has increased steadily in the last few years. For example, in the fiscal year 2018, MIGA issued a record US$5.3 billion in political risk

---

157 See *Convention Establishing the Multilateral Investment Guarantee Agency*, art. 11, www.miga.org/sites/default/files/archive/Documents/MIGA%20Convention%20February%202016.pdf
158 MIGA, "MIGA: Securing Oil and Gas Investments", MIGABRIEF (June 2015), www.miga.org/sites/default/files/2018-06/oil-gas-brief.pdf

insurance and credit guarantees,[159] surpassing the US$4.8 billion worth of insurance and guarantees it issued in 2017.[160]

Third, the demand and competition for access to mineral resources today are considerably higher than they were in past decades. The intensity of the current competition among mineral extraction companies for access to mineral resources plays to the advantage of resource-rich countries. Some companies may be so keen to obtain mineral concessions from host countries that they will outdo the competition by agreeing to fewer fiscal incentives. Also, mineral resources are found in commercial quantities in only a few countries, meaning that mining companies do not have a wide latitude from which to choose the location of their investment. African countries can, in fact, use these factors as the basis for refusing to grant overly generous fiscal incentives.

Fourth, China's increasing demand for, and reliance on, Africa's mineral resources has not only expanded the pool of foreign investors in the mining industry, but, more significantly, also obviates Africa's excessive reliance on Western MNCs for resource exploitation. China is increasingly relying on Africa for the supply of minerals and metals to satisfy its fast-growing economy. China's involvement in Africa's mining industry threatens the dominance of Western MNCs and could diminish the leverage and influence that these MNCs wield over African countries. It is often easy for Western MNCs to obtain generous concessions from African governments because of the fear that the lack of incentives will discourage FDI inflow, coupled with the fear that MNCs will divest from countries that do not provide generous incentives. China's apparently insatiable demand for minerals and metals should allay this fear because it intensifies the competition for Africa's mineral resources. However, Africa must exercise restraint in engaging with Chinese companies, given China's poor record on social issues such as environmental protection and human rights. Mining projects often lead to adverse social consequences, including environmental pollution, involuntary displacement of local communities, and human rights violations. In fact, as noted in Chapter 9, Chinese mining companies have been implicated in human rights and labour rights violations in Africa. Africa should thus engage Chinese mining companies in a transparent and accountable manner. Concessions granted to these companies should contain clear obligations to observe the human rights and environmental rights guaranteed under international law and under African domestic law, including the *African Charter on Human and Peoples' Rights* (the *African Charter*).[161]

Finally, the widespread use of stabilization clauses in Africa is counter to the policy objectives of the *African Charter*, to the effect that African countries "shall recognize the rights, duties and freedoms enshrined in this Charter and shall

---

159 MIGA, *2018 Annual Report*, at 6, www.miga.org/sites/default/files/2018-11/MIGA%202018%20Annual%20Report_0.pdf
160 MIGA, *2017 Annual Report – Insuring Investments, Ensuring Opportunities*, at 9, www.miga.org/sites/default/files/2018-06/Annual-Report-2017e.pdf
161 *African Charter on Human and Peoples' Rights*, O.A.U. Doc. CAB/LEG/67/3/Rev. 5 (adopted in 1981, entered into force in 1986).

undertake to adopt legislative or other measures to give effect to them".[162] The *African Charter* obliges African states to promote and protect human and peoples' rights and freedoms, including the rights to life and the integrity of the human person, liberty and security, and the enjoyment of the best attainable state of physical and mental health.[163] In fact, the *African Charter* imposes an obligation on states to "adopt legislative or other measures to give effect to" the rights enshrined in the Charter.[164] The African Commission on Human and Peoples' Rights has held that the human rights obligations under the *African Charter* oblige African governments "to protect their citizens, not only through appropriate legislation and effective enforcement but also by protecting them from damaging acts that may be perpetrated by private parties".[165] This being so, a stabilization clause that inhibits the ability of a host state to prevent third parties– in this case mining companies – from infringing the human rights of African citizens violates the state's human rights obligations under the *African Charter*. Likewise, a stabilization clause that prevents the host state from taking appropriate legislative or administrative measures towards the full realization of human rights contravenes the human rights obligations of the state under the *African Charter*. Thus, Africa should desist from the indiscriminate grant of stabilization clauses in investment contracts because these clauses are inimical to the *African Charter*.

7 Conclusion

The prevalence of stabilization clauses in mining contracts is unlikely to abate because of the perceived utility of stabilization clauses. They offer functional stability of investment in the sense that their breach attracts monetary compensation or the restoration of the economic equilibrium envisaged under the investment contract. However, stabilization clauses have certain adverse consequences for African countries. As argued in this chapter, stabilization clauses could lead to loss of revenues in host countries. In addition, stabilization clauses adversely impact human rights and environmental protection because they dissuade host countries from regulating the social and environmental practices of mining companies. Given these realities, African countries ought to exercise caution in granting stabilization clauses. As suggested in this chapter, such caution could involve limiting the duration of stabilization clauses to a relatively short period, as opposed to the current practice of granting stabilization clauses for an inordinately long period. Even then, stabilization clauses should be reviewed and renegotiated periodically to cater to the changing economic and social realities in Africa.

---

162 *Ibid.* art. 1.
163 *Ibid.* arts. 4, 6 & 16.
164 *Ibid.* art. 1.
165 *Social and Economic Rights Action Center & Another v Nigeria* (155/96, October 27, 2001) at para. 57, www.achpr.org/sessions/descions?id=134

7 Resource Nationalism in the African Mining Industry

1 Introduction

The phrase 'resource nationalism' refers to policies adopted by resource-producing countries to assert and exert control over the exploitation of natural resources. In the narrow sense, resource nationalism refers to policies that seek to enhance the economic benefits accruing to host countries from the exploitation of natural resources. Resource nationalism encompasses interventionist policies that seek to take "trade and investment in minerals out of international market processes".[1] Viewed purely from an economic angle, resource nationalism involves the transfer of ownership and control of natural resources from foreign entities to domestic entities, including state-owned enterprises. However, resource nationalism goes beyond economic benefits; hence, in the broad sense, resource nationalism encompasses "all policies undertaken by the national government of the producing country that restrict access to resources to a subset of potential players, create separation between the domestic and international market, or directly impose quantitative limitations on production and exports".[2] These policies include those limiting the ability of foreign companies to access minerals; policies protecting against depletion of mineral resources; policies restricting the export of minerals; royalty and taxation policies; and policies regarding the volume and prices of minerals.[3] These policies are "the expression, by states, of their determination to gain the maximum national advantage from the exploitation of natural resources".[4]

---

1 Jeffrey D. Wilson, "Resource Nationalism or Resource Liberalism? Explaining Australia's Approach to Chinese Investment in Its Mineral Sector" (2011) 65(3) *Australian Journal of International Affairs* 283 at 300.
2 Giacomo Luciani, "Global Oil Supplies: The Impact of Resource Nationalism and Political Instability" (CEPS Working Document No. 350, May 2011) at 4, www.ceps.eu/ceps-publications/global-oil-supplies-impact-resource-nationalism-and-political-instability
3 *Ibid.* at 4–15.
4 George Joffé *et al.*, "Expropriation of Oil and Gas Investments: Historical, Legal and Economic Perspectives in a New Age of Resource Nationalism" (2009) 2(1) *Journal of World Energy Law & Business* 3 at 4.

The broad conception of resource nationalism captures the power dynamics between host developing countries and MNCs regarding access to minerals as well as the parameters for conducting mining operations. In this regard, Paul Stevens regards resource nationalism as actions on the part of host countries that not only limit the operations of private international companies, but also assert "a greater national control over natural resource development".[5] Bremer and Johnston define resource nationalism as "efforts by resource-rich nations to shift political and economic control of their energy and mining sectors from foreign and private interests to domestic and state-controlled companies".[6] For Sam Pryke, resource nationalism "concerns the claims and actions of governments over their natural resources *vis-à-vis* another party, usually a foreign company, seeking to extract and benefit from those resources".[7] These claims, which may be economic, political, or social, are often made by governments in exercise of their sovereignty over natural resources within their jurisdictions. Given that most African countries rely on natural resource industries for sustenance, resource nationalism in Africa is akin to economic populism because both resource nationalism and economic populism seek domestic control of the economy, including the means of production.

Irrespective of the perspective from which resource nationalism is viewed, resource nationalism entails the adoption of interventionist laws and policies designed to maximize the economic and social benefits accruing to the host country from resource exploitation. In practice, resource nationalism manifests in government intervention in the natural resource market through the adoption of selective and discretionary policies to attain economic, social, and political benefits that would otherwise be unobtainable in unfettered market conditions.[8] Such interventionist policies include policies seeking greater (or heightened) state ownership and control of mineral resources; greater share of resource rents and profits for host governments; the optimization of resource benefits for citizens in particular and the economy in general; the reduction of foreign ownership and influence; increase in local participation in terms of ownership of mineral rights and use of local goods and services; and the rebalancing of the relationship between host governments and extractive companies.

Resource nationalism in the extractive industries is anchored on state ownership of natural resources, as well as the international doctrine of permanent sovereignty over natural resources. The United Nations (UN) has long adopted a series of resolutions on permanent sovereignty over natural resources. In 1962, the UN General Assembly adopted a resolution which expressly recognized the sovereign right of member countries to exploit natural resources "in the interest of their

---

5 Paul Stevens, "National Oil Companies and International Oil Companies in the Middle East: Under the Shadow of Government and the Resource Nationalism Cycle" (2008) 1 (1) *Journal of World Energy Law & Business* 5.
6 Ian Bremmer & Robert Johnston, "The Rise and Fall of Resource Nationalism" (2009) 51(2) *Survival* 149.
7 Sam Pryke, "Explaining Resource Nationalism" (2017) 8(4) *Global Policy* 474 at 481.
8 Jeffrey D. Wilson, "Understanding Resource Nationalism: Economic Dynamics and Political Institutions" (2015) 21(4) *Contemporary Politics* 399 at 400.

national development and of the well-being of the people of the State concerned".[9] A subsequent resolution adopted in 1966 reiterated that the exploitation of natural resources "should be aimed at securing the highest possible rate of growth of the developing countries" while also acknowledging the right of all countries "to secure and increase their share in the administration of enterprises which are fully or partly operated by foreign capital and to have a greater share in the advantages and profits derived" from natural resources.[10] Developing countries were emboldened further in 1974 when the UN declared that the full exercise of the right to permanent sovereignty over natural resources "requires that action by States aimed at achieving a better utilization and use of those resources must cover all stages, from exploration to marketing".[11]

This chapter assesses resource nationalism in the African mining industry focusing on four key areas – namely, the manifestations of resource nationalism; the factors accounting for the rise of resource nationalism in Africa; the effectiveness of resource nationalism in Africa; and the impact, if any, of resource nationalism on the inflow of FDI to Africa. Resource nationalism is growing in Africa "because of the increased pressure on governments to counter exploitation of natural resources and obtain a greater share of economic rents".[12] Hence, as discussed below, much of the resource nationalism practised in Africa appears to be economic resource nationalism.[13] Quite often, African governments assert greater control over natural resources by implementing policies designed not only to increase resource revenues but also to indigenize the ownership of mineral rights.

2 Manifestations of resource nationalism

The analysis of the various manifestations of resource nationalism in the African mining industry is prefaced with two significant caveats. First, although free market economists often denounce African countries for adopting resource-nationalistic policies, resource nationalism is not peculiar to African countries. Industrialized countries such as Australia, Canada, the United Kingdom, and the United States engage in covert forms of resource nationalism in terms of both regulatory and fiscal frameworks.[14] These countries have adopted foreign investment policies designed to maintain domestic control over strategic industries, including policies

9 *UN General Assembly Resolution 1803 (XVII) of 14 December 1962: Permanent Sovereignty over Natural Resources.*
10 *UN General Assembly Resolution 2158 (XXI): Permanent Sovereignty over Natural Resources,* 22 November 1966.
11 *UN General Assembly Resolution on Permanent Sovereignty over Natural Resources,* A/RES/3171 (XXVIII), 5 February 1974.
12 F.T. Cawood and O.P. Oshokoya, "Resource Nationalism in the South African Mineral Sector: Sanity through Stability" (2013) 113 *The Journal of the Southern African Institute of Mining and Metallurgy* 45 at 46.
13 See Bremmer & Johnston, *supra* note 6 at 150–1 (identifying the four variants of resource nationalism as 'revolutionary resource nationalism', 'economic resource nationalism', 'legacy resource nationalism', and 'soft resource nationalism').
14 *Ibid.* at 152.

that screen FDI inflow and policies imposing entry barriers on foreign companies and entities.[15] Such entry barriers and restrictions are protectionist in nature as they serve to preserve control of strategic industries by domestic enterprises.

Second, many of the practices often associated with resource nationalism in Africa are not undertaken by African countries solely based on nationalistic zeal. Some of these practices are dictated by good governance ethos as well as the desire to ensure equitable utilization of mineral resources. For example, increases in royalty and tax rates are often viewed by mining companies as evidence of resource nationalism, but such increases may be justified on economic grounds, such as high commodity prices. In addition, the termination of mining contracts may be based on any of the grounds provided for in the mining statute, including, for example, the persistent breach of the terms of a mineral right by the holder. Similarly, the renegotiation of mining contracts is not always spurred by sinister nationalistic fervour. Rather, the contracting parties may mutually renegotiate a contract in order to adapt the contract to suit the interests of the parties, as dictated by prevailing circumstances.[16] Furthermore, the local content requirements in Africa are spurred partly by the desire to build the capacity and competency of domestic companies to undertake mining projects.[17] Thus, while the symptoms of resource nationalism are discussed broadly in this section, I caution that not all of these symptoms arise solely from resource nationalism. Rather, the legal and fiscal reforms discussed in the ensuing pages are Africa's modest attempt to rebalance the sharing of mineral benefits between governments and mining companies.

2.1 Fiscal reforms

The most common feature of resource nationalism in Africa over the last decade has been incessant fiscal reforms in the mining industry, including increases in the royalty and tax rates for minerals. The fiscal reforms in the African mining industry are often prompted by economic factors such as high commodity prices buoyed by the growing demand for minerals in China and India and the record profits made by mining companies.[18] Often, these fiscal reforms are driven by a concern that whereas foreign companies take too large a share of resource benefits, local host communities receive little or no benefit from the exploitation of minerals in the land.[19] In countries where this sentiment prevails, governments usually tend to

15 See Wilson, *supra* note 1 at 287–90.
16 Abba Kolo & Thomas W. Walde, "Renegotiation and Contract Adaptation in International Investment Projects: Applicable Legal Principles and Industry Practices" (2000) 1(1) *Journal of World Investment & Trade* 5 at 6.
17 See Damilola S. Olawuyi, "Local Content Requirements in Oil and Gas Contracts: Regional Trends in the Middle East and North Africa" (2019) 37(1) *Journal of Energy & Natural Resources Law* 93 at 98.
18 Lisa E. Sachs, Perrine Toledano, Jacky Mandelbaum, & James Otto, "Impacts of Fiscal Reforms on Country Attractiveness: Learning from the Facts" in Karl P. Sauvant, ed., *Yearbook on International Investment Law & Policy 2011–2012* (New York: Oxford University Press, 2013) 345 at 347.
19 Stevens, *supra* note 5 at 6.

increase the royalty and tax rates for minerals in order to enhance government revenues. In fact, the African Union actively encourages African countries to boost their share of mineral profits by creating "the fiscal environment that enhances mineral development, by optimizing tax packages without discouraging mining investment, as well as through building capacity to negotiate improved fiscal provisions, reduce tax leakages and effectively monitor compliance with taxation laws".[20]

Several African countries, including the DRC, Ghana, Guinea, Kenya, Tanzania, Zambia, and South Africa have reformed their fiscal regimes for minerals by increasing their tax and royalty rates and introducing new taxes altogether. Recently in 2018, the DRC amended its mining code by, among other things, increasing the royalty rates for precious metals from 2.5% to 3.5% and precious stones from 4% to 6%.[21] In addition, the DRC introduced a 10% royalty for "strategic substances".[22] Ghana increased the corporate income tax for mining companies from 25% to 35% in 2012 while also proposing a windfall tax for mining operations.[23] However, at the time of writing, Ghana had yet to formally enact a windfall tax for mining. In 2013, Kenya increased the royalty rate for gold from 2.5% to 5% of gross sales value, while the rate for rare earth minerals such as niobium and titanium was increased from 3% to 10%.[24] In 2010, South Africa introduced a mineral royalty for the express purpose of capturing more resource revenue for the government.[25] In fact, some African countries have increased the royalty rates for minerals multiple times in the last few years. For example, in 2010 Tanzania enacted a new mining statute that increased the royalty rate for metallic minerals (copper, gold, silver, and platinum group minerals) from 3% to 4% of gross value.[26] However, in 2017 Tanzania increased further its royalty rate for metallic minerals from 4% to 6% of gross value.[27] In addition, Tanzania increased the royalty rate for gemstone and diamond from 5% to 6% of gross value in 2017.[28] Similarly, in 2015 Zambia raised its royalty rate for copper from 3% to rates ranging from 6% to 9%.[29] However, in 2016 Zambia reduced the royalty rate

20 *Addis Ababa Declaration on Building a Sustainable Future for Africa's Extractive Industry – From Vision to Action* (16 December 2011), www.africaminingvision.org/amv_resources/AMV/Declaration_Second_Final.pdf [*Addis Ababa Declaration*].
21 *Law No. 007/2002 of July 11, 2002 Relating to the Mining Code*, Art. 241 (as amended by *Law No. 18/001 of 9 March 2018*).
22 *Ibid.*, art. 241.
23 Michael E. Akafia et al., "Ghana" in Michael Bourassa & John Turner, eds, *Getting the Deal Through: Mining 2018* (London: Law Business Research, 2018) 105 at 108.
24 Martin K. Ayisi, "The Review of Mining Laws and the Renegotiation of Mining Agreements in Africa: Recent Developments from Ghana" (2015) 16 *Journal of World Investment & Trade* 467 at 475 ["Renegotiation of Mining Agreements"].
25 *MPRRA*, s. 2.
26 *Mining Act, 2010* (Tanzania), s. 87.(c) (as amended by *The Written Laws (Miscellaneous Amendments) Act, 2017* (No. 7), s. 23).
27 *Ibid.* s. 87.
28 *Ibid.* s. 87.
29 *Mines and Minerals Development Act, 2015*, s. 89 (repealed).

for copper to floating rates ranging between 4% and 6% based on the norm price of copper on the international market.[30]

Mining companies often deride and oppose attempts by African countries to reform the fiscal regimes for mining, especially where such reforms include increases in royalties and taxes. In particular, fiscal reforms often prompt threats of withdrawal of investment and legal actions on the part of mining companies against African governments. Nevertheless, African countries are unfazed by these threats. The increases in royalty and tax rates across Africa are hardly surprising given the widespread dissatisfaction among Africans arising from the inequity in the distribution of the economic benefits flowing from mineral exploitation. While mining companies in Africa often earn large profits, particularly during periods of high commodity prices, the profitability of mining operations has yet to translate into commensurate benefits for mineral-rich African nations and the local communities that host such mining operations.[31]

2.2 Termination of mineral rights

Mineral rights are terminated based on the grounds enumerated in mining statutes and contracts. These grounds include the procurement of a mineral right by fraud or through false statements; the contravention of the terms and conditions of the mineral right; the contravention of the mining statute; non-compliance with the approved plan of mining operations such as the minimum work and minimum expenditure requirements; submission of inaccurate, false, fraudulent, incorrect, or misleading information or data in relation to mining operations; inability to meet financial obligations such as payment of royalties and taxes; and failure to comply with a compliance order to remedy a specific breach.[32] Mineral rights are equally susceptible to termination where the mineral rights holders no longer satisfy the eligibility requirements for holding the mineral rights, such as where they become bankrupt or insolvent,[33] or where the mineral rights were granted in contravention of domestic law. For example, in 2013 Kenya terminated the mining licence held by Cortec Mining Kenya Limited, a subsidiary of Pacific Wildcat Resources Corporation, based on the company's failure to obtain an environmental impact assessment approval prior to the grant of the mineral right.[34] An arbitral tribunal recently held that failure by Cortec to satisfy the statutory condition precedent regarding environmental approval is fatal to the mineral right, and

---

30 *Mines and Minerals Development (Amendment) Act*, No. 14 of 2016, s. 2.
31 UNECA & African Union, *Minerals and Africa's Development: The International Study Group Report on Africa's Mineral Regimes* (Addis Ababa: UNECA, 2011) at 92.
32 *Mines and Minerals Development Act, 2015* (Zambia), s. 72.(1); *Minerals and Mining Act, 2006* (Ghana), ss. 68 & 69; *Mining Act, 2016* (Kenya), s. 147; *MPRDA*, s. 47.
33 *Mines and Minerals Development Act, 2015* (Zambia), s. 72.(1); *Minerals and Mining Act, 2006* (Ghana), s. 68.(1).
34 Martin K. Ayisi, "The Legal Character of Mineral Rights under the New Mining Law of Kenya" (2017) 35(1) *Journal of Energy & Natural Resources Law* 25 at 45.

thus Kenya acted legally in terminating the mineral right.[35] In addition, mineral rights may be suspended or revoked in the public interest.[36] The revocation of a mineral right on any of the above grounds extinguishes the rights of the holder. However, prior to the termination of a mineral right the Minister is obliged to give a written notice to the mineral right holder indicating the reasons for considering termination of the mineral right and affording the holder reasonable opportunity to remedy the breach or show why the mineral right should not be terminated.[37]

The termination of mineral rights may amount to expropriation, which is the lawful dispossession or taking of property from the property owner for public purposes by the host state. Expropriation may be direct – for example, where the host state nationalizes mineral assets – or indirect – such as where the host state undertakes some executive or regulatory actions that render it impossible or impracticable for the mineral right holder to undertake mining operations. Although expropriation is permitted under domestic and international law, expropriation is lawful only when it is undertaken for a public purpose and a fair and reasonable compensation is paid to the mineral right holder.

Explicit forms of expropriation such as nationalization of mining assets are uncommon in today's Africa, but nationalization was rampant in the early years of Africa's independence. This was partly due to the economic atrocities committed by European colonialists, which prompted newly independent African states to decolonize and rebalance their economies through nationalization of mineral assets belonging to foreign governments and enterprises. For example, soon after Ghana gained independence in 1957 it nationalized its mining industry and established state-owned mining companies.[38] Likewise, following Zambia's independence in 1964, the government gradually took control of the mining industry such that, within a few years of Zambia's independence, the government owned a 51% controlling stake in all major copper mining corporations in Zambia.[39] Zimbabwe nationalized its mining industry soon after independence in 1980 by taking direct control of selected mining assets and by creating state-owned companies such as the Zimbabwe Mining Development Corporation and the Zimbabwe Iron and Steel Company.[40]

In the last three decades, however, instances of nationalization of mineral assets have become uncommon in Africa due largely to the adoption of capitalism and

---

35 *Cortec Mining Kenya Limited, Cortec (Pty) Limited and Stirling Capital Limited v Republic of Kenya*, ICSID Case No. ARB/15/29, Award (22 October 2018) at paras 343–65.
36 *Mines and Minerals Development Act, 2015* (Zambia), s. 72.(1)(i).
37 *Ibid*. s. 72.(2); *Minerals and Mining Act, 2006* (Ghana), ss. 68.(2) & 69.(2); *Mining Act, 2016* (Kenya), s. 147.(3); *MPRDA*, s. 47.(2).
38 SAIMM, *The Rise of Resource Nationalism: A Resurgence of State Control in an Era of Free Markets or the Legitimate Search for a New Equilibrium?* at 97, www.saimm.co.za/Conferences/ResourceNationalism/ResourceNationalism-20120601.pdf
39 *Ibid*. at 100–1.
40 *Ibid*. at 115.

market economics across the continent. While nationalization of mineral assets has abated in Africa, the issue of nationalization persists in some African countries. For example, in South Africa there is a raging debate regarding nationalization of mineral assets, a debate spearheaded by the youth wing of the ruling party, the African National Congress (ANC).[41] The debate was temporarily settled in 2017 when the ANC rejected the nationalization of mines on grounds that it would cause massive job losses,[42] but since then prominent opposition politicians in South Africa including Julius Malema have urged the nationalization of mining assets.

Sporadic instances of termination of mineral rights are recurring in the continent. For example, in 2018 the government of Tanzania terminated 11 retention licences including the retention licence for the Kabanga nickel project jointly held by Barrick Gold and Glencore Plc.[43] Similarly, in 2015 Kenya cancelled 65 prospecting and mining licences on grounds of expiry and violations of the terms of the licences and the mining statute.[44] In 2010, Guinea terminated the mineral concession granted to BSG Resources regarding the Simandou iron ore project on grounds that it was procured through bribery.[45] In addition, in 2010 the DRC terminated the mineral rights and assets of First Quantum, a Canadian company, on grounds that the company violated the mining statute and refused to renegotiate its contract with the government.[46] The government sold the assets to a third party who in turn sold the assets to Eurasian Natural Resources Corporation PLC (ENRC). First Quantum filed international arbitral claims against the government and ENRC but, in the end, ENRC reached a settlement agreement with First Quantum and agreed to pay US$1.25 billion for the mineral assets.[47] The Zambian government terminated all existing mineral development agreements in 2008 on grounds that the fiscal regimes prescribed in the agreements, including a paltry royalty rate of 0.6%, were inimical to its economic interest.[48] Earlier, in

---

41 *Ibid.* at 170–208.
42 Reuters, "South Africa's ANC Rejects Calls for Nationalization of Mines", March 12, 2017, https://news.trust.org/item/20170312084924-bovg4/?source=ticker
43 Fumbuka Ng'wanakilala, "Tanzania Cancels License of Barrick, Glencore Nickel Project", Reuters, May 12, 2018, www.reuters.com/article/us-tanzania-mining/tanzania-cancels-license-of-barrick-glencore-nickel-project-idUSKCN1ID0O7
44 Library of Congress, "Kenya: Many Prospecting and Mining Licences Cancelled; New Mining Law Considered", May 22, 2015, www.loc.gov/law/foreign-news/article/kenya-many-prospecting-and-mining-licenses-cancelled-new-mining-law-considered
45 *The Economist*, "Mining and Corruption – Crying foul in Guinea", December 4, 2014, www.economist.com/business/2014/12/04/crying-foul-in-guinea
46 William MacNamara & Christopher Thompson, "Congo Seizes First Quantum Minerals' Assets", *Financial Times*, August 31, 2010, www.ft.com/intl/cms/s/0/27d6c104-b530-11df-9af8-00144feabdc0.html
47 Brenda Bouw, "First Quantum Exits DRC with $1.25-billion Settlement", *The Globe and Mail*, January 5, 2012, www.theglobeandmail.com/globe-investor/first-quantum-exits-drc-with-125-billion-settlement/article4104290
48 Philip Daniel & Emil M. Sunley, "Contractual Assurances of Fiscal Stability" in Philip Daniel *et al.*, eds, *The Taxation of Petroleum and Minerals: Principles, Problems and Practice* (London: Routledge, 2010) 405 at 414.

2007, the DRC terminated 12 mining contracts following the recommendation of the Mining Contract Review Commission.[49]

2.3 Indigenization

As discussed in Chapter 3, the indigenization requirements in the African mining industry are designed to enable African countries to exert greater control over natural resources through increased indigenous participation in the ownership of mineral rights. For example, in Zambia mining rights over areas larger than two cadastre units are granted only to companies with significant domestic ownership, referred to in the mining statute as 'citizen-influenced company', 'citizen-empowered company' and 'citizen-owned company'.[50] In fact, Zambia prohibits foreign entities from engaging in small-scale mining.[51]

Indigenization of the mining industry is more pronounced in the southern part of the continent, particularly South Africa, Namibia, and Zimbabwe. These countries have adopted specific policies designed to promote and empower indigenous participation in the mining industry. South Africa has enacted the *Broad-Based Black Economic Empowerment Act, 2003* with the overarching objective of facilitating black economic empowerment, including the meaningful participation of black people in the economy; achieving a substantial change in the racial composition of ownership and management structures of business enterprises; and increasing the extent to which local communities, workers, and women own and manage business enterprises.[52] Pursuant to this Act, South Africa created the Broad-Based Socio-Economic Empowerment Charter for the Mining and Minerals Industry (Mining Charter), which requires existing mining right holders to achieve a minimum of 26% ownership by persons who qualify as 'historically disadvantaged persons' (HDP).[53] However, a recent amendment to the Mining Charter has increased the indigenous ownership threshold for companies seeking new mining rights. Thus, companies that applied for and acquired their mining rights after the coming into effect of the 2018 amendment to the Mining Charter "must have a minimum of 30% BEE shareholding, which shall include economic interest plus corresponding percentage of voting rights".[54] This minimum 30% BEE (Black Economic Empowerment) shareholding is to be distributed in the following manner: 5% non-transferable carried interest must be allotted to qualifying employees; 5% non-transferable carried interest or a 5% equity equivalent

---

49 TWNA, "Experience of Mining Tax Reforms and Renegotiation of Mining Contracts: The Case of the Democratic Republic of Congo", at 33, www.twnafrica.org/Tax%20Reform%20Congo.pdf.
50 *Mines and Minerals Development Act, 2015* (Zambia), s. 13.(3).
51 *Ibid.* s. 29.(2) & (3).
52 *Broad-Based Black Economic Empowerment Act, 2003*, s. 2.
53 *Broad-Based Socio-Economic Empowerment Charter for the Mining and Minerals Industry, 2018*, art. 2.1.1, www.gov.za/sites/default/files/gcis_document/201809/41934gon1002.pdf [*Mining Charter*].
54 *Ibid.* art. 2.1.3.1.

benefit is assigned to host communities; and 20% interest in the form of shares must be given to a BEE entrepreneur at least 5% of which must preferably be for women entrepreneurs.[55]

Zimbabwe's *Indigenization and Economic Empowerment Act* require that at least 51% of the shares of extractive companies shall be held by Zimbabweans through an "appropriate designated entity".[56] For this purpose, an 'appropriate designated entity' means the Zimbabwe Mining Development Corporation; the Zimbabwe Consolidated Diamond Company; and the National Indigenization and Economic Empowerment Fund.[57] A subsequent amendment to the Act restricted the 51% indigenous shareholding requirement to companies involved in the extraction of diamonds and platinum.[58] However, the government of Zimbabwe indicated recently that it plans to repeal the Act and replace it with a 'business-friendly' economic empowerment statute.[59] Similarly, Namibia has recently adopted the New Equitable Economic Empowerment Framework (NEEEF) embodying the government's resolve to "use all the legitimate market mechanisms at its disposal, in the form of its procurement programmes and licensing regimes, to promote transformation and empowerment" of Namibians in general and 'previously disadvantaged' Namibians in particular.[60] The NEEEF hopes to transform business in Namibia through five empowerment pillars – namely, ownership; management control and employment equity; human resources and skills development; entrepreneurship development; and community investment.[61] These empowerment pillars are to be achieved within the framework of sectoral charters that will prescribe sector-specific requirements.[62] Although the sectoral charters are yet to be developed, the government of Namibia anticipates that upon their adoption these sector-specific charters will complement the NEEEF's objective of indigenizing the Namibian economy, including the mining industry, by prescribing minimum indigenous participation in the ownership of business enterprises in Namibia. When fully operational the NEEEF will mandatorily require companies applying for mining licences and companies tendering for government contracts to have a minimum of 25% of their shares held by 'previously disadvantaged Namibians' – that is, persons who are victims of Apartheid policies.[63] Furthermore, a minimum of 50% of the board of directors and

---

55 *Ibid*. art. 2.1.3.2.
56 *Indigenization and Economic Empowerment Act*, Chapter 14:33, Act 14/2007, s. 3 (as amended by *Finance Act, 2018* (No. 1 of 2018)).
57 *Ibid*. s. 2 (as amended).
58 *Ibid*. s. 3 (as amended).
59 AllAfrica, "Zimbabwe: Govt to Repeal Indigenization Law", https://allafrica.com/stories/201908020500.html
60 Government of the Republic of Namibia, *The New Equitable Economic Empowerment Framework*, at 6, www.ecb.org.na/images/docs/Investor_Portal/NEEEF.pdf. The phrase 'previously disadvantaged' Namibians means victims of Apartheid policies. See NEEEF, *ibid*. at 3.
61 *Ibid*. at 7.
62 *Ibid*. at 7.
63 *Ibid*. at 23.

management staff of these companies shall consist of 'previously disadvantaged Namibians'.[64]

In order to ensure that indigenization policies are not circumvented by foreign investors, African countries impose restrictions on the transfer of the controlling stake in mining companies. These restrictions manifest in the form of discretionary and administrative powers vested in designated government agencies or officers. In most African countries, the transfer of a controlling stake in mining companies cannot be effected without the written consent of the government. In Ghana and Tanzania, for example, the transfer of an interest in a mining company that has the effect of giving the transferee control of the company cannot be validly effected without the written consent of the Minister or the licensing authority.[65] The government may disapprove of the transfer of a controlling interest in a company where it determines that the transfer is not in the public interest. Such would be the case where a proposed transfer of interest in a mining company would reduce indigenous ownership to a level below the prescribed minimum.

2.4 State participation in mining projects

African countries are increasingly playing a direct role in mineral extraction through state-owned mining companies. In fact, in 2011 the African Union called "on member States to explore equity participation in mineral ventures so as to capture a greater share of benefits for the people of Africa".[66] State participation in mineral exploitation in Africa manifests in different forms depending on the policy objectives of the host government. First, host states can engage in full participation through the direct acquisition of mineral rights and the production of minerals through state-owned companies or the acquisition by the state of an equity stake in mining projects.[67] Second, state participation may be implemented through 'carried equity' arrangements in which the mining company bears the costs associated with a project after which compensation is paid to the company out of the project itself.[68] Third, a state may choose to engage in 'free equity participation' encompassing the mandatory grant of an equity stake in a mining project to the government without any financial obligation on the part of the government.[69] Fourth, the host state may engage in a 'production-sharing' arrangement involving the grant of a prescribed percentage of profits or recovered

---

64 *Ibid.* at 23.
65 *Minerals and Mining Act, 2006,* (Ghana), s. 52; *Mines and Mineral Development Act, 2015* (Zambia), s. 67; *Mining Act, 2010* (Tanzania), s. 110.
66 *Addis Ababa Declaration, supra* note 20.
67 See Charles McPherson, "State Participation in the Natural Resource Sectors: Evolution, Issues and Outlook" in Philip Daniel *et al.,* eds, *The Taxation of Petroleum and Minerals: Principles, Problems and Practice* (London: Routledge, 2010) 263 at 266.
68 *Ibid.* at 266–7.
69 *Ibid.* at 267.

minerals to the state after deduction of the cost of production incurred by the company.[70]

These forms of state participation are not mutually exclusive as a state may choose to adopt multiple forms of participation. For example, South Africa participates in mineral exploitation through both the 'full participation' and the 'equity participation' models. In 2008, South Africa established a state-owned mining company, African Exploration Mining and Finance Corporation (Proprietary) Limited (AEMFC) and, as of 2011, this company owned one coal mine, 27 prospecting rights, and one mining right.[71] At the opening of the AEMFC's coal mine on 26 February 2011, Jacob Zuma, the then President of South Africa, reportedly said that the "role of the state cannot merely be confined to that of a regulator" and that the state "must actively participate in the mining industry to ensure that [South Africa's] national interest is protected and advanced".[72] In addition, South Africa undertakes equity participation in mining projects through state-owned companies such as Alexko, the AEMFC, and Industrial Development Corporation.[73] Similarly, Zimbabwe's state-owned mining company, the Zimbabwe Mining Development Corporation, is involved in the production of copper, chromite, gold, platinum, and diamonds.[74]

A more common form of state participation in Africa is the 'equity participation' model. Many African countries, including Botswana, the DRC, Ghana, Guinea, Kenya, Tanzania, and Zambia, actively participate in the production of minerals through acquisition of equity stake in mining projects. The government of Botswana owns a significant equity in several diamond-mining companies, including a 50% stake in Debswana Diamond Company Limited.[75] The DRC's state-owned company, La Générale des Carrières et des Mines (Gécamines), is involved in joint-venture projects with international mining companies such as Katanga Mining Limited.[76] In fact, at some point Gécamines held about 100 mining permits even though the limit on the number of permits that could be held by one entity was 50 permits.[77] In 2009, Namibia established Epangelo Mining Company

---

70 *Ibid.* at 267.
71 Peter Leon, "Whither the South African Mining Industry?" (2012) 30(1) *Journal Energy & Natural Resources Law* 5 at 24.
72 *Ibid.* at 24.
73 World Bank, "Overview of State Ownership in the Global Minerals Industry: Long Term Trends and Future" (2011) at 17, http://documents.worldbank.org/curated/en/339551468340825224/pdf/828480NWP0Extr00Box379875B00PUBLIC0.pdf ["State Ownership"].
74 *Ibid.* at 18.
75 UNCTAD, *World Investment Report 2007: Transnational Corporations, Extractive Industries and Development* (New York & Geneva: United Nations, 2007) at 138.
76 See Katanga Mining Limited, "Katanga Mining Announces Settlement of DRC Legal Dispute with Gecamines and Agreement for the Resolution of KCC Capital Deficiency" (June 12, 2018), www.katangamining.com/media/news-releases/2018/2018-06-12.aspx
77 The Carter Center, "A State Affair: Privatizing Congo's Copper Sector", at 6, www.cartercenter.org/resources/pdfs/news/peace_publications/democracy/congo-report-carter-center-nov-2017.pdf ["State Affair"].

(Pty) Limited, a state-owned mining company, with the goal of becoming "the leading, diversified mining company in Namibia".[78] Namibia also owns a 50% equity in Namdeb, a joint-venture company co-owned by the government of Namibia and De Beers.[79] The government of Ghana holds a 10% free carried interest in mining companies, but the government may increase its participation in any mining operation with the consent of the holder of the mining lease.[80] Similarly, the government of Guinea engages in mining activities "on its own behalf, either directly or through the Public Limited Company responsible for management of the mining patrimony acting alone or in association with third parties in the mining sector".[81] To that end, Guinea's *Mining Code 2011* provides that "the grant by the State of a Mining Operation Title immediately gives the State an ownership interest, at no cost, of up to a maximum of fifteen per cent (15%), in the capital of the company holding the Title".[82] The government of Guinea can in fact acquire additional equity in cash provided that it reaches an agreement with the titleholder and provided further that the total equity (both free carried interest and paid equity) does not exceed 35%.[83]

A recent amendment to the *Mining Act* of Tanzania illustrates the degree to which African governments strive for greater control over mining operations through equity participation. In 2017, Tanzania amended the *Mining Act* by granting the government at least 16% non-dilutable free carried interest in the capital of mining companies.[84] In appropriate cases, the government of Tanzania can acquire up to 50% of the shares of a mining company "commensurate with the total tax expenditures incurred by the government in favour of the mining company".[85] These tax expenditures refer to the revenue costs of the various fiscal allowances and tax exemptions granted to mining companies by the government. Kenya requires holders of mining licences for large-scale mining and 'strategic minerals' to mandatorily cede to the government 10% free carried interest in the share capital of the holder, with a further requirement that the government may acquire additional interest in the mining operations of the holder, provided the terms of such additional participation are agreed with the holder.[86] Finally, it is worth observing that state participation in mineral exploitation in Africa has recently been fuelled by the declaration of certain minerals as 'strategic national assets', an issue discussed next.

---

78 World Bank, "State Ownership", *supra* note 73 at 17.
79 *Ibid.* at 17.
80 *Minerals and Mining Act, 2006* (Ghana), s. 43.
81 *Amended 2011 Mining Code* (Guinea), art. 16.
82 *Ibid.* art. 150-I.
83 *Ibid.* art. 150-I.
84 *Mining Act, 2010,* (Tanzania), s. 10 (as amended by *The Written Laws (Miscellaneous Amendments) Act, 2017* (No. 7), s. 9).
85 *Ibid.*
86 *Mining Act,* 2016 (Kenya), s. 48.

2.5 Declaration of minerals as 'strategic national assets'

A few African countries, including Botswana, the DRC, Kenya, Namibia, Zimbabwe, and South Africa, have recently declared certain minerals to be 'strategic assets' with a view to exerting complete domestic control over such minerals. Minerals may be classified as 'strategic' based on factors such as the economic significance of the minerals and the rarity of the minerals, or based on a combination of economic and social factors, including the desire to exploit strategic minerals "in a manner that benefits the country and protects the environment".[87] In the DRC, "strategic substances" are defined as minerals which, on the basis of the prevailing economic environment, are of special interest to the government given the critical nature of such minerals.[88] Namibia has declared copper, coal, gold, uranium, and zinc to be strategic minerals apparently because of the economic significance of these minerals.[89] In 2018, Zimbabwe amended its mining statute by vesting some discretion in the Minister of Mines to classify any mineral as strategic where the Minister believes that such classification is in the interest of the development of the mining industry.[90]

The declaration of minerals as 'strategic assets' means that the minerals are exploitable based solely on policies and regulations that protect the 'national interest' of the host country. In this context, 'national interest' means the enhancement of economic benefits derivable from such 'strategic mineral assets'. Thus, a higher rate of royalty may be imposed on minerals classified as 'strategic assets' and such minerals may be mined only in collaboration with state-owned mining companies or citizens of the country. For example, the DRC imposes a high royalty rate of 10% on 'strategic substances' in order to derive maximum economic benefits from such substances.[91] In other countries, minerals classified as 'strategic assets' are mined with the active participation of the government. In Kenya, the government acquires a mandatory 10% free carried interest in all mining operations involving strategic minerals.[92] In Namibia, copper, coal, gold, uranium, and zinc are mined only with the equity participation of Epangelo Mining Company (Pty) Limited, the state-owned mining company, because they are strategic minerals.[93] Namibia's declaration of these minerals as 'strategic' means that "exclusive exploration and mining rights to all these 'strategic' minerals will in future be held by Namibia's state-owned mining company [Epangelo]"; hence, "investors will be required to partner with Epangelo should they wish to

---

87 *Mining (Strategic Minerals) Regulations, 2017* (Kenya), reg. 3.(c).
88 *Law No. 007/2002 of July 11, 2002 Relating to the Mining Code*, art. 1 (as amended by *Law No. 18/001 of 9 March 2018*).
89 Leon, *supra* note 71 at 19.
90 Macdonald Dzirutwe, "Zimbabwe Amends Its Mining Bill to Appease Foreign Mining Companies", *BusinessDay*, 1 June 2018, www.businesslive.co.za/bd/world/africa/2018-06-01-zimbabwe-amends-its-mining-bill-to-appease-foreign-mining-companies
91 *Law No. 007/2002 of July 11, 2002 Relating to the Mining Code*, art. 241 (as amended by *Law No. 18/001 of 9 March 2018*).
92 *Mining Act, 2016* (Kenya), s. 48.
93 Leon, *supra* note 71 at 19.

acquire rights to any of the 'strategic' minerals in Namibia".[94] Similarly, Kenya's state-owned company, the National Mining Corporation, is responsible for the exploration of strategic minerals either alone or in association with other persons or companies.[95] More specifically, the National Mining Corporation engages in

(a) ... the reconnaissance, prospecting and mining of a strategic mineral or strategic mineral deposit or any other related mineral activity;
(b) the processing, refining or smelting of a strategic mineral;
(c) the marketing or sale of a strategic mineral;
(d) import and export of a strategic mineral; and
(e) any other functions that the Cabinet Secretary with the approval of the Cabinet may assign to the Corporation in respect of strategic minerals.[96]

However, the National Mining Corporation may seek private-sector participation in strategic minerals where it lacks necessary technical, financial, or other capacity to effectively explore, mine, refine, smelt, process, or market strategic minerals.[97]

2.6 Renegotiation of mining contracts

During the last decade, African countries such as Ghana, the DRC, Guinea, Liberia, Malawi, Mali, Sierra Leone, Tanzania, and Zambia either renegotiated mining contracts or terminated some contracts.[98] The renegotiation of mining contracts in Africa is prompted by a plethora of factors such as the inequitable and lopsided terms of extant contracts; the unequal benefit-sharing inherent in extant contracts; the desire by host countries to increase their mineral revenues; the need to correct historical imbalances in the mining industry; the reversal of state policy from privatization and liberalization to state control of the mining industry; and the long-term nature of mining contracts which exposes the contracting parties to risks and uncertainties such as a significant change in the circumstances or assumptions underlying the original contract.[99] Because of the long duration of mining contracts, the value of minerals is likely to fluctuate during the term of the contract, and if prices rise significantly, host governments tend to seek a larger share of mineral profits through contract renegotiation.[100]

The renegotiation of mining contracts in Africa arises primarily from Africa's desire to obtain a larger share of the proceeds from mineral resources, particularly

---

94 *Ibid.* at 19–20.
95 *Mining (Strategic Minerals) Regulations, 2017* (Kenya), reg. 8.
96 *Ibid.* reg. 8.(2).
97 *Ibid.* reg. 8.(3).
98 See Sarah G. Katz-Lavigne, "The Renegotiation Window: Resource Contract Renegotiations in the Mining Industry in Africa from 2000 to 2013" (2017) 51 *Resources Policy* 22 at 29.
99 See Kolo & Walde, *supra* note 16 at 20–32.
100 Michael Likosky, "Contracting and Regulatory Issues in the Oil and Gas and Metallic Minerals Industries" (2009) 18(1) *Transnational Corporation* 1 at 26.

during times of commodity price boom.[101] Prior to the emergence of the third generation of mining statutes in Africa, many African governments signed mining contracts with MNCs containing lopsided and unfair fiscal provisions that diminished and minimized the resource rent payable to the host countries.[102] Mining contracts signed during periods of military dictatorship, war, and civil unrest in countries such as the DRC, Liberia, and Guinea are universally acknowledged to contain unfair fiscal terms that are inimical to the interest of these countries.[103] In addition, recent instances of renegotiation of mining contracts in Africa are indicative of a shift in structural power from MNCs back towards African countries.[104] In the recent past, the structural power of resource-endowed African countries has been bolstered vis-à-vis mining companies by improvements in domestic resource governance regimes, including the adoption of the principle of transparency which has led to the publication of mining contracts in many African countries. Furthermore, the rise in commodity prices, the increasing demand for mineral resources spurred by industrial revolutions in China and India, the relative stability in the political governance of African countries, and the apparent international support for the renegotiation of inequitable mining contracts in Africa have all coalesced to incentivize Africa to renegotiate mining contracts.[105]

The renegotiation of mining contracts may be 'forced' by the host government or mutually undertaken by the contracting parties.[106] Contract renegotiation is 'forced' where the original contract does not contain a renegotiation clause but the host government unilaterally demands renegotiation of the contract. Investors may succumb to the host government's demand for renegotiation of mining contract because they want to avoid public scrutiny and criticism by citizens and civil society groups. Moreover, at the time of such demand for renegotiation, the investor would already have sunk significant capital into the project and so may want to avoid disruptions in mineral production that would certainly ensue if they do not agree to renegotiate their contract. Furthermore, a refusal to renegotiate a contract could jeopardize the ability of the investor to secure additional mineral rights in the host country. Contract renegotiation may also be 'forced' by statute, as is the case in Tanzania where, as discussed below, the *Natural Wealth and Resources Contracts (Review and Renegotiation of Unconscionable Terms) Act, 2017* compels the renegotiation of natural resource contracts containing unconscionable terms.

---

101 Katz-Lavigne, *supra* note 98 at 23.
102 See Global Witness, "Heavy Mittal? A State within a State: The Inequitable Mineral Development Agreement between the Government of Liberia and Mittal Steel Holdings NV", www.globalwitness.org/documents/17005/mittal_steel_en_oct_2006_low_res.pdf
103 See Ayisi, "Renegotiation of Mining Agreements", *supra* note 24 at 478–9.
104 Katz-Lavigne, *supra* note 98 at 23–4.
105 *Ibid.* at 25–6.
106 Abdullah Al Faruque, "Renegotiation and Adaptation of Petroleum Contracts: The Quest for Equilibrium and Stability" (2008) 9 *Journal of World Investment & Trade* 113 at 119.

Mutual renegotiation of a contract may occur in three distinct scenarios: where the original contract lacks clauses requiring renegotiation but the contracting parties voluntarily and in good faith agree to renegotiate the contract; where the original contract provides expressly for renegotiation of the contract on the occurrence of certain triggering events; and where a statute requires renegotiation of a contract under specific circumstances. For the most part, the renegotiation of mining contracts in Africa falls under 'forced renegotiation', given the absence of renegotiation clauses in mining contracts and given the further fact that contract renegotiation in Africa is often foisted on mining companies by African governments.

Mining companies often resist attempts by African countries to renegotiate mining contracts, partly because many of these contracts do not contain renegotiation clauses and partly because contract renegotiation is contrary to the principle of sanctity of contract – that is, *pacta sunt sevanda*. However, the renegotiation of contracts may be mutually beneficial to both mining companies and host countries because it enables the contracting parties to adapt the contract to prevailing circumstances, thus avoiding disruptions that could arise from a contentious investment relationship.[107] As Sornarajah argues, contract "renegotiation is more sensible as a technique for avoiding disputes and for ensuring that the relationship remains viable in the context of changed circumstances".[108] In practice, contract renegotiation in the mining industry occurs primarily where the fiscal regimes do "not contain instruments that could respond with adequate adaptability and progressivity to changed circumstances".[109] Contract renegotiation is particularly apposite in situations where the contractual equilibrium has been significantly disrupted by external events such that a continuation of performance places an excessively onerous burden or disadvantage on one party.[110] A company can request the renegotiation of a mining contract "if the commerciality of discovery is not viable to continue the project in its original form, or the technical difficulty or geological conditions have made development of the fields unprofitable for the company".[111] Mining agreements are usually negotiated based on speculative projections regarding several variables, including the extent of mineralization in the area, rate of return, and cost of production.[112] If the projections regarding mineralization exceed the parties' expectations upon completion of exploration, the host state may feel short-changed and therefore seek the renegotiation of the contract.[113] Conversely, where the exploration results only in a

---

107 Kolo & Walde, *supra* note 16 at 6.
108 M. Sornarajah, *The International Law on Foreign Investment*, 3rd Edition (Cambridge: Cambridge University Press, 2010) at 244.
109 Daniel & Sunley, *supra* note 48 at 419.
110 Kolo & Walde, *supra* note 16 at 6.
111 Al Faruque, *supra* note 106 at 118.
112 Kapwadi F. Lukanda, "Renegotiating Investment Contracts: The Case of Mining Contracts in Democratic Republic of the Congo" (2014) 5(3) *George Mason Journal of International Commercial Law* 301 at 308.
113 *Ibid*. at 308.

marginal discovery of minerals, an investor may find that the only gateway to escape the devastating effect of such marginal discovery is the renegotiation of the contract.[114] This being the case, it is imperative for African countries "to introduce self-adjusting mechanisms that cater for all phases of resource cycles and attempt to put in renegotiating triggers/milestones within the tenure, to adjust for unforeseen developments".[115]

2.6.1 Renegotiation of mining contracts in Africa: some case studies

In order to address the fiscal imbalance apparent in extant mining contracts, many African countries have, in the last two decades, embarked on contract renegotiation with mining companies. For example, in 2006 Liberia renegotiated its mineral development agreement (MDA) with Mittal Steel with the help of international organizations. The MDA was renegotiated not only because its terms were unfavourable to Liberia, but also because it constrained the ability of Liberia to regulate the activities and operations of Mittal Steel. Among other things, the MDA granted Mittal a five-year tax holiday with an automatic right of renewal; transferred public assets and facilities, including rail infrastructure to Mittal; and froze the laws governing Mittal's operations through a stabilization clause which provided that "any modifications that could be made in the future to the Law as in effect on the effective date shall not apply to [Mittal] and its associates without their prior written consent".[116] The renegotiation resulted in the expungement of many of the offending provisions, including the five-year tax holiday.[117]

Likewise, the DRC inaugurated a Mining Contract Review Commission ("the Commission") in 2007 tasked with the duty of reviewing mining contracts signed by the government and state-owned mining companies with foreign MNCs between 1996 and 2006. The contract review exercise in the DRC was prompted by the lopsided fiscal terms provided for in the contracts, which diminished the mineral rents accruing to the government while mining companies operating in DRC earned large profits. The Commission reviewed 61 contracts, all of which were found to contain unfair and unacceptable terms.[118] The unfair terms in the contracts included the arbitrary allocation of equity interest in joint-venture

---

114 *Ibid.* at 308.
115 African Union, *Africa Mining Vision*, at 17, www.africaminingvision.org/amv_resources/AMV/Africa_Mining_Vision_English.pdf
116 *Mineral Development Agreement between the Government of the Republic of Liberia and Mittal Steel Holdings NV* (August 17, 2005), art. XIX(9).
117 See *An Act Ratifying the Amendment to the Mineral Development Agreement (MDA) Dated August 17, 2005 between the Government of the Republic of Liberia (the Government) and Mittal Steel Holding A.G. and Mittal Steel (Liberia) Holdings Limited (the Concessionaire)*, www.scribd.com/document/152423536/An-Act-Ratifying-the-Amendment-to-the-Mineral-Development-Agreement-MDA-Dated-August-17-2005-between-The-Government-of-the-Republic-of-Liberia-The [*Act Ratifying Mittal MDA*].
118 IPIS, "Democratic Republic of the Congo, Mining Contracts – State of Affairs", www.ipisresearch.be/wp-content/uploads/2015/03/20080325-DRC-miningcontracts.pdf

projects in favour of MNCs without feasibility study and assessment of the parties' contributions; overly generous fiscal and tax incentives; derisory lease premium; and the vesting of power in MNCs to manage partnership and joint-venture projects to the exclusion of state-owned companies.[119] In the end, the Commission recommended the renegotiation of 39 of the 61 contracts, while the remaining 22 contracts were recommended for outright cancellation.[120] One of the contracts recommended for cancellation by the Commission is the contract between the government and Anvil Corporation regarding mineral exploitation in Dikulushi, which granted Anvil and its sub-contractors total exemption from taxes and royalties for 20 years.[121] Equally emblematic of the unfairness of the mining contracts in the DRC is the contract governing the Twangiza mine, which granted the project owners a ten-year tax holiday.[122] In addition, the contract specified a derisory royalty rate of 1% even though at the time of signing the contract the mining code stipulated a royalty rate of 4%.[123] Following the Commission's recommendation, the government renegotiated most of the contracts and terminated at least 12 contracts.[124]

Like the DRC, Tanzania recently commenced the renegotiation of unfair and exploitative mining contracts, a process that intensified with the passage of the *Natural Wealth and Resources Contracts (Review and Renegotiation of Unconscionable Terms) Act, 2017.* The prelude to the enactment of this Act appears to be the report of the Presidential Committee set up to investigate the legal and fiscal regimes governing Acacia Mining Plc's projects. The Committee reportedly found not only that the fiscal terms are unfair to the government, but also that Acacia was engaged in tax-avoidance practices by massively under-reporting the mineral ores it produces in, and exports out of, Tanzania.[125] The report is corroborated by financial disclosures by Acacia Mining Plc, which revealed that in 2016 Acacia exported US$1 billion worth of gold, copper, and silver but only paid 8% in taxes and royalties to the government.[126] Worse still, while Acacia paid out US$444 million in dividends to its shareholders between 2010 and 2015, it did not pay any

---

119 TWNA, *supra* note 49 at 29.
120 IPIS, *supra* note 118.
121 *Ibid.*
122 Ousman Gajigo, Emelly Mutambatsere, & Guirane Ndiaye, "Gold Mining in Africa: Maximizing Economic Returns for Countries" (ADB Working Paper 147, March 2012) at 22, https://pdfs.semanticscholar.org/0002/bfdc1211d8ef7bc3b0e717e5c192a e696db2.pdf?_ga=2.198012972.2030804458.1583516950-168549753.1579644085
123 *Ibid.* at 22.
124 TWNA, *supra* note 49 at 33.
125 Thabit Jacob & Rasmus H. Pedersen, "New Resource Nationalism? Continuity and Change in Tanzania's Extractive Industries" (2018) 5 *The Extractive Industries and Society* 287 at 287–8.
126 Maya Forstater & Alexandra Readhead, "A Brutal Lesson for Multinationals: Golden Tax Deals Can Come Back and Bite You", *The Guardian*, 6 July 2017, www.theguardian.com/global-development-professionals-network/2017/jul/06/a-brutal-lesson-for-multinationals-golden-tax-deals-can-come-back-and-bite-you

corporate income tax in Tanzania during that period.[127] Even the Chief Executive Officer of Acacia Mining Plc has reportedly acknowledged that Acacia's contract with the government is lopsided in favour of Acacia.[128]

The *Natural Wealth and Resources Contracts (Review and Renegotiation of Unconscionable Terms) Act, 2017* authorizes the government of Tanzania to renegotiate natural resource contracts containing unconscionable terms with a view to rectifying the terms and striking a fair deal for the people of Tanzania.[129] Under the Act, unconscionable contractual terms include terms that restrict the right of the state to exercise full permanent sovereignty over its natural resources; terms restricting the right of the state to exercise authority over foreign investment within the country; terms that are inequitable and onerous to the state; terms restricting periodic review of a contract or arrangement which purports to last for life time; terms that secure preferential treatment designed to create a separate legal regime to be applied discriminatorily for the benefit of a particular investor; terms restricting the right of the state to regulate activities of MNCs within the country; terms that deprive the people of Tanzania of the economic benefits derivable from the beneficiation of natural resources in Tanzania; terms which by nature empower MNCs to intervene in the internal affairs of Tanzania; terms that subject the state to the jurisdiction of foreign laws and fora; terms that expressly or impliedly undermine the effectiveness of state measures to protect the environment, including the use of environmentally friendly technology; and terms that are aimed at doing any other act the effect of which undermines or is injurious to the welfare of the people or economic prosperity of Tanzania.[130]

The contract renegotiation process in Tanzania is invoked by the parliament, the National Assembly. Where the National Assembly determines that a natural resource contract contains unconscionable terms, it may pass a resolution advising the government to initiate renegotiation of the contract so as to rectify the unconscionable terms.[131] Within 60 days of the passing of the resolution, the government must serve the other party to the contract a notice of its intention to renegotiate the unconscionable terms.[132] It is anticipated that mining companies would voluntarily agree to renegotiate contracts containing unconscionable terms, but where they fail to do so, the government is authorized to unilaterally expunge the offending terms from the contract. More specifically, where upon the service of a notice of renegotiation a mining company fails to agree to renegotiate the unconscionable terms or where the parties fail to reach an agreement with regard to the unconscionable terms, all unconscionable terms in the contract shall cease to have effect and shall be treated as having been expunged.[133]

127 *Ibid.*
128 *Ibid.*
129 *The Natural Wealth and Resources Contracts (Review and Renegotiation of Unconscionable Terms) Act, 2017*, ss. 4–7.
130 *Ibid.* s. 6.(2).
131 *Ibid.* s. 5.
132 *Ibid.* s. 6.(1).
133 *Ibid.* s. 7.

The contract renegotiation exercises in the DRC and other African countries occurred in the absence of renegotiation clauses and thus could be termed 'forced renegotiations'. Forced renegotiations pose political and economic risks for the contracting parties and are detrimental to the stability of investment contracts.[134] Contract renegotiation ought to be undertaken with the mutual consent of the contracting parties. The consent of the contracting parties could be captured by express renegotiation clauses in mining contracts to the effect that, on the occurrence of the stipulated triggering events, the parties shall mutually and in good faith renegotiate the terms of the contract. In effect, African countries would be well served to insert renegotiation clauses in mining contracts. As some observers have noted, once renegotiation clauses "are prescribed in the contract, they are always consensual in nature and a legally valid means to reach a renegotiation of contract".[135] The events triggering the renegotiation of a contract as well as the procedure governing contract renegotiation should be clearly specified in the contract. In addition, the criteria for contract renegotiation, including good-faith participation in renegotiations, mutual cooperation and trust, and the preservation of the economic equilibrium, should be specified in the renegotiation clause.[136] An alternative arrangement would be to insert clauses requiring the review of mining contracts at specified intervals.

2.7 Beneficiation

Beneficiation refers to the processing and refining of mineral ores within domestic African economies, thus creating value-added products and services within the mining industry, as well as linkages with other sectors of the economy. The domestic beneficiation of minerals in Africa could create forward linkages along and beyond the mining industry value chain, including the development of mineral-processing facilities as well as manufacturing plants.[137] Domestic beneficiation of minerals can provide the feedstock for the manufacturing and allied industries, including the iron, steel, and aluminum industries.[138] In addition, beneficiation of raw minerals in Africa would lead to job creation through the establishment of processing facilities, thus also fostering economies of scale for domestic African economies.[139] Regrettably, most of the mineral ores produced in Africa are exported oversees for processing and refinement. The exportation of raw mineral ores from Africa deprives the continent of the economic value that domestic refinement could obviously add to the local economy.

134 Al Faruque, *supra* note 106 at 119.
135 *Ibid.* at 120.
136 *Ibid.* at 120–2.
137 UNCTAD, *World Investment Report 2007, supra* note 75 at 140.
138 *Africa Mining Vision, supra* note 115 at 13.
139 Cornelius Dube, "The Scope for Beneficiation of Mineral Resources in Zimbabwe: The Case of Chrome and Platinum" at v, www.tips.org.za/research-archive/annual-forum-pap ers/2016/item/3150-the-scope-for-beneficiation-of-mineral-resources-in-zimbabwe-the-case-of-chrome-and-platinum

Recognizing that value addition and linkages with other sectors could be achieved through domestic refining and processing of raw mineral ores, African countries are beginning to adopt beneficiation policies aimed at promoting the establishment of mineral-processing facilities in Africa. For example, Tanzania requires mining companies to establish beneficiation facilities within Tanzania, and hence any arrangement or agreement for the extraction, exploitation, or acquisition and use of natural wealth and resources shall ensure that no raw resources are exported for beneficiation outside Tanzania.[140] In South Africa, the MPRDA requires the Minister to promote the beneficiation of minerals within South Africa on terms and conditions as they deem appropriate, including terms prescribing specific levels for domestic beneficiation.[141] Likewise, Namibia encourages "maximum local beneficiation of mineral products to ensure that as many of the economic benefits as possible are retained in Namibia for the benefit of all its citizens".[142] For purposes of beneficiation, Namibia has established a Mineral Development Fund aimed at broadening the production base of the mining industry into the national economy through diversification, as well as horizontal and vertical integration.[143]

Mineral beneficiation can be promoted through fiscal incentives such as tax credits and fiscal disincentives such as a levy on the export of unprocessed minerals. For example, Namibia imposes a 2% levy on the export of pure unsorted rough diamonds in order to discourage companies from exporting raw diamonds.[144] However, some African countries have attempted to promote beneficiation of minerals through the imposition of restrictions on the export of raw mineral ores. Such restrictions include imposing a limit on the volume of raw mineral ores exportable from the host country or, in extreme cases, an outright ban on the export of raw mineral ores. In this regard, Tanzania requires mining contracts to contain provisions prohibiting the exportation of raw mineral resources for beneficiation outside of Tanzania.[145] Zimbabwe imposed a ban on the export of raw chromite in 2011 in order to promote domestic smelting of chromite, but the ban has since been lifted.[146] Similarly, in 2013 the DRC banned the export of copper and cobalt concentrate, while Zambia banned the export of emeralds.[147]

In other instances, the Minister may be vested with statutory power to compel mining companies to supply or sell raw mineral ores to domestic entities for beneficiation and processing within the country. For example, Namibia's *Diamond Act 13 of 1999* empowers the Minister to issue a written notice to diamond

---

140 *The Natural Wealth and Resources (Permanent Sovereignty) Act, 2017*, s. 9.
141 *MPRDA*, s. 26.
142 *Minerals Policy of Namibia*, at 9, www.chamberofmines.org.na/files/1414/7005/8876/Minerals_Policy_Final.pdf
143 *Minerals Development Fund of Namibia Act, 1996*, s. 3.
144 Namibia Chamber of Mines, "Mining Tax Regimes", www.chamberofmines.org.na/index.php/mining-tax-regime
145 *The Natural Wealth and Resources (Permanent Sovereignty) Act, 2017*, s. 9.
146 Dube, *supra* note 139 at xi–xii.
147 Ayisi, "Renegotiation of Mining Agreements", *supra* note 24 at 478.

producers requiring them to sell to domestic cutters, tool-makers, or researchers, "such quantities, classes, qualities and descriptions of unpolished diamonds as the Minister may by that notice reasonably fix and determine".[148] However, the Minister issues such a notice only where the domestic cutter, tool-maker, or researcher is unable to reach an agreement with the diamond producer "to acquire, on reasonable terms and conditions of sale, a regular supply of unpolished diamonds" as would enable them to carry on without interruption the diamond cutting, tool-making operations, or research and tests authorized by their licence.[149]

Interestingly, both Botswana and Namibia promote domestic beneficiation of minerals through contractual arrangements obliging mining companies to process a percentage of their minerals within the country. For example, in 2006 De Beers and the government of Botswana signed a contract that mandated domestic processing of diamonds, and this contract has reportedly led to an increase in mineral beneficiation within Botswana.[150] A similar contract was signed between De Beers and the government of Namibia in 2016 under which 15% of diamonds produced by Namdeb, a company owned jointly by De Beers and the government, is processed locally in Namibia.[151]

2.8 Local content requirements

Local content requirements in the African mining industry are aimed generally at increasing the use of domestically available factors of production, thus maximizing the social and economic benefits flowing from mineral resource exploitation.[152] Local content policies attempt to foster the development of domestic industries as well as the use of local inputs to create employment opportunities and generate 'spill-over' effects in the domestic economy through consumption linkages.[153] Modern mining statutes in Africa impose local content requirements on mining companies as a prerequisite for the acquisition and maintenance of mineral rights. The employment and training of citizens and the use of local goods and services constitute a significant portion of the statutory requirements for granting and holding mineral rights in Africa. In some African countries, applications for mineral rights must be accompanied by documents describing how the applicant

148 *Diamond Act 13 of 1999*, s. 58.(1).
149 *Ibid.* s. 58.(2).
150 "Diamond Beneficiation in Decline in Namibia and South Africa and Stagnant in Botswana", *Sunday Standard*, 15 December 2014, www.sundaystandard.info/diamond-beneficiation-decline-namibia-and-south-africa-and-stagnant-botswana
151 Ehud A. Laniado, "Beneficiation in the Diamond Industry: Namibia", www.ehudlaniado.com/home/index.php/news/entry/beneficiation-in-the-diamond-industry-namibia
152 Jane Korinek & Isabelle Ramdoo, "Local Content Policies in Mineral-Exporting Countries", *OECD Trade Policy Paper No. 209*, at 10, www.oecd-ilibrary.org/docserver/4b9b2617-en.pdf?expires=1583780718&id=id&accname=guest&checksum=66AE9129609BA25BFAE2DC25FD7F314A
153 *Ibid.* at 12.

intends to employ and train local personnel as well the use of local goods and services. In Africa, mineral rights are granted partly on the condition that the holder complies with the local content requirements.

The local content regimes governing Africa's mining industry require the employment of African citizens; procurement of local goods and services; preferential treatment of local companies in the award of contracts; training of citizens; and the transfer of technology to African citizens and enterprises.[154] The local content regimes in some African countries set specific performance targets, while other countries adopt a relativist position. For example, in Ghana the maximum percentage of expatriate staff allowed for the holder of a mining lease is 10% for the first three years from the commencement of mining operations and 6% after the third year of mining operations.[155] In effect, within the first three years of mining operations, at least 90% of the staff of mining companies must be citizens of Ghana, while the percentage increases to at least 84% after three years of operations. In Tanzania, the minimum local content requirements for mining companies with regard to the procurement of domestic goods and services are 10% at commencement of mining operations, 50% after five years of mining operations, and 60–90% after ten years of mining operations.[156] Similarly, in South Africa, numerous legislative and policy instruments prescribe specific local content targets for mining operations.[157] These include the MPRDA, the Mining Charter, and the Codes of Good Practice for the South African Mining and Minerals Industry. In particular, the Mining Charter redresses historical and social inequities in South Africa's mining industry by expanding opportunities for 'historically disadvantaged persons' to participate in mining operations, thus benefiting from the exploitation of South Africa's vast mineral endowment. Originally adopted in 2004 but revised in 2010 and 2018, the Mining Charter requires mining companies to create employment equity, including workplace diversity and equitable representation at all levels.[158] To this end, mining companies must appoint to their boards of directors a minimum of 50% 'historically disadvantaged persons' (HDP), 60% HDP in senior management positions, 60% HDP in middle management positions, and 70% HDP in junior management positions.[159] In addition, the Mining Charter enjoins mining companies to procure a minimum of 70% of goods from South African manufacturers, while a minimum of 80% of

---

154 See *Minerals and Mining (General) Regulations*, 2012, LI 2173 (Ghana), regs 1–2; *Amended 2011 Mining Code* (Guinea), arts 107–9; *Mining Act*, 2016 (Kenya), s. 47; *Mining (Local Content) Regulations, 2018* (Tanzania), regs 13–14, 20–4; *Mines and Minerals Development Act, 2015* (Zambia), s. 20.
155 *Minerals and Mining (General) Regulations*, 2012, LI 2173 (Ghana), reg. 1.(9).
156 *Mining (Local Content) Regulations, 2018* (Tanzania), reg. 13 read with the First Schedule.
157 These include the *MPRDA*; the *Broad-Based Black Empowerment Act, 2003*; the *Mining Charter*, and the Codes of Good Practice for the South African Mining and Minerals Industry.
158 *Mining Charter*, *supra* note 53, s. 2.4.
159 *Ibid.* s. 2.4.

expenditure on services, excluding non-discretionary expenditure, must be spent on services provided by companies based in South Africa.[160]

Unlike Ghana, Tanzania, South Africa, and Zambia which prescribe mandatory local content requirements for mining companies, some African countries adopt discretionary or relativist regimes that enjoin compliance based on prevailing circumstances. In Botswana, mining companies are urged to employ citizens and procure goods and services "to the maximum extent possible", meaning that, depending on the prevailing circumstances, a mining company may not employ or procure local personnel and goods.[161] In that country, holders of mineral rights are required to give "preference, to the maximum extent possible consistent with safety, efficiency and economy, to (a) materials and products made in Botswana; and (b) service agencies located in Botswana and owned by Botswana citizens or bodies corporate established under the Companies Act".[162] The Kenyan regime is similarly relativist in terms of procurement of local goods and services. Holders of mineral rights in Kenya are obliged to give preference, to the maximum extent possible, to materials and products manufactured in Kenya; services offered by members of the host communities and Kenyan citizens; and companies or businesses owned by Kenyan citizens.[163] The relativist regime is also apparent in a mining contract in Liberia which provides that:

> The CONCESSIONAIRE and its Associates shall, when purchasing goods and services required with respect to the Operations, give first preference, at comparable quality, delivery schedule and price, to goods produced in Liberia and services provided by Liberian citizens or businesses, subject to technical acceptability and availability of the relevant goods and services in Liberia. Subject to the foregoing, the CONCESSIONAIRE may freely contract with such Persons as it desires.[164]

3 Drivers of resource nationalism in Africa

The dominance of foreign enterprises in Africa's mining industry, coupled with the realization that foreign entities often place their economic interests above those of the host African countries, fuels resource nationalism in Africa. Africa's resource nationalism is a reaction to the incessant imposition of unfavourable and undesirable investment terms on African countries by MNCs in the mining industry. For example, as discussed in Chapter 6, MNCs often leverage their financial clout to prevail on poverty-stricken African countries to grant fiscal stabilization clauses in mining contracts. This economic reality fuels the desire of African countries to

---

160 *Ibid*. s. 2.2.
161 *Mines and Minerals Act, 1999* (Botswana), s. 12.(2) & (3).
162 *Ibid*. s. 12.(1).
163 *Mining Act*, 2016 (Kenya), s. 50.
164 *Act Ratifying Mittal MDA*, *supra* note 117, art. 11.

enact new laws to remedy the perceived imbalance between mineral rents accruing to African countries and the enormous profits made by mining companies. The nationalistic impulses apparent in many resource-rich developing countries explains why Paul Stevens characterizes resource nationalism as a "battle between national interests and foreign influences", a battle that is fiercely fought by MNCs (supported by the governments of their home countries) on the one hand, and the governments of host developing countries on the other hand.[165] This section analyzes the factors accounting for the growing rate of resource nationalism in Africa and lays a foundation for the subsequent discussion of the impacts, if any, that resource nationalism could potentially have on foreign direct investment in the continent.

3.1 Desire to exert greater control over mineral resources

The primary reason for the rise of resource nationalism in Africa is the desire to wrest control of mineral resources from foreign MNCs. Reminiscent of what Amy Chua describes as "nationalism directed against the external foreigner",[166] African countries often engage in resource nationalism in an attempt to prevent or at least diminish foreign domination of the natural resource industry. Through resource nationalism, African countries hope to checkmate the economic and social excesses of foreign companies which, in an unbridled bid to enhance their profits, often externalize the adverse impacts of their operations to local host communities. The desire to assert greater control over mineral resources is more apparent in countries where mineral resources are regarded as gifts of nature that ought to be exploited solely for the benefit of citizens. As non-renewable resources, mineral resources generate emotional attachments in African countries which perceive the exploitation of the mineral resources by foreign entities as a one-off extraction and siphoning of their wealth.[167]

Thus, resource nationalism is a matter of state policy in mineral-rich African countries, and this position is traceable directly to the vesting of ownership and custodianship of mineral resources in governments in trust for African citizens. Even in South Africa, where private ownership of mineral resources was permitted prior to the enactment of the MPRDA, the shift from private ownership to state custodianship under the MPRDA signalled the advent of resource nationalism as state policy in that country. As a matter of state policy, resource nationalism in mineral-rich African countries is often reflected in legal and policy instruments such as statutes, regulations, contracts, and ministerial declarations. Mining

---

165 Stevens, *supra* note 5 at 8.
166 Amy L. Chua, "The Privatization-Nationalisation Cycle: The Link between Markets and Ethnicity in Developing Countries" (1995) 95(2) *Columbia Law Review* 223 at 263.
167 A. Butler, "Resource Nationalism and the African National Congress" (2013) 113 *The Journal of the Southern African Institute of Mining and Metallurgy* 11 at 13.

statutes across Africa require that mineral resources be exploited for the economic and social wellbeing of citizens. In Zambia, for example, mineral resources are required to be explored and developed in a manner that promotes the socio-economic development of Zambians, including citizen access to, and benefit from, mineral resources.[168] Similarly, the foundation for South Africa's MPRDA is the attainment of economic and social advancement, including the promotion of local and rural development and the social upliftment of communities affected by mining.[169] Tanzania has expressly proclaimed resource nationalism through a recent statute that makes it "unlawful to make any arrangement or agreement for the extraction, exploitation or acquisition and use of natural wealth and resources except where the interests of the People and the United Republic [of Tanzania] are fully secured".[170] In fact, in Tanzania mining contracts must guarantee "returns into the Tanzanian economy from the earnings accrued or derived from" mining operations.[171]

3.2 The developmental state

The phrase 'developmental state' refers to states that actively intervene in their domestic economies in order to guide the direction and pace of economic growth.[172] African countries increasingly view themselves as developmental states whose role is to facilitate economic development. As is evident in the AMV, African countries believe that the harnessing of resource endowment is key to the economic emancipation of the continent.[173] Hence, they declare that a "resource-based African industrialization and development strategy must be rooted in the utilization of Africa's significant resource assets to catalyze diversified industrial development".[174] The prevailing view in Africa is that natural resources can be leveraged as catalysts for economic development through the optimization of resource rents as well as the creation of linkages with other sectors of the economy.[175] Thus, African countries seek to attain greater control of their natural resources with the expectation that "once the state is in (greater) control or outright ownership it can more easily direct revenues from resource extraction, of which it is now able to retain a much greater share, to national development projects".[176] The recent reforms of mining laws in Africa must thus be situated within

---

168 *Mines and Minerals Development Act, 2015* (Zambia), s. 4.
169 *MPRDA*, Preamble.
170 *The Natural Wealth and Resources (Permanent Sovereignty) Act, 2017*, s. 6.(1).
171 *Ibid.* s. 7.
172 Esteban P. Caldentey, "The Concept and Evolution of the Developmental State" (2008) 37 *International Journal of Political Economy* 27 at 28.
173 See *Africa Mining Vision, supra* note 115 at 2–6.
174 *Ibid.* at 3.
175 Chris W.J. Roberts, "The Other Resource Curse: Extractives as Development Panacea" (2015) 28(2) *Cambridge Review of International Affairs* 283 at 284–92.
176 Stefan Andreasson, "Varieties of Resource Nationalism in Sub-Saharan Africa's Energy and Mineral Markets" (2015) 2 *The Extractive Industries and Society* 310 at 313.

this context of the developmental state.[177] These reforms not only reflect greater state involvement in the mining industry but they also demonstrate Africa's willingness to engage in "activist industrialization policy-making" aimed at challenging liberal market policies championed by the World Bank and Western powers.[178]

3.3 Exploitative relationship between MNCs and African governments

Some of the policy initiatives termed by critics as resource nationalism ought properly to be viewed in the context of domestic circumstances in individual African countries, including the lopsidedness of the mining regimes in these countries. Mining regimes and concession contracts in Africa contain terms that disproportionately favour mining companies to the detriment of African economies. As discussed in Chapters 5 and 6, such unfavourable terms include tax holidays for MNCs, low royalty rates, and fiscal stabilization. For example, under South Africa's sliding-scale royalty scheme, which ranges from 0.5% to 7%, the average rate of royalty paid by mining companies is only 2%.[179] Thus, while in 2011 South Africa produced 12% of all rough diamonds in the world, it received a meagre US$11 million in mining royalties.[180] A similar scenario played out earlier in Zambia because in the 1990s Zambia indiscriminately signed DAs with foreign MNCs that prescribed generous terms, including a paltry royalty rate of 0.6% of gross revenue. Thus, while mining MNCs earned gross revenues amounting to US3.4 billion in 2007, only US$20 million was paid to Zambia as royalty in 2007.[181] This prompted the Zambian government to increase the royalty rate for minerals and corporate tax rate in 2008.[182] In addition, the government introduced a windfall tax to cater to the phenomenal profits made by mining companies resulting from the exponential rise in commodity prices on the international market.[183] At the time the DAs were signed, the average price of copper was US $1,814.00 per tonne but the price increased significantly during the term of the DAs and peaked at US$7,126 per tonne in 2007.[184] Mining companies fought against the windfall tax, threatening to scale down their mining operations and lay

---

177 See Sara Ghebremusse, "New Directions in African Developmentalism: The Emerging Developmental State in Resource-Rich Africa" (2016) 7(1) *Afe Babalola University Journal of Sustainable Development Law & Policy* 1.
178 Roberts, *supra* note 175 at 284.
179 Andreasson, *supra* note 176 at 315.
180 *Ibid*. at 315.
181 Christopher Adam & Anthony M. Simpasa, "Harnessing Resource Revenues for Prosperity in Zambia" (OxCarre Research Paper 36), at 36 (footnote 27), www.economics.ox.ac.uk/materials/working_papers/4930/oxcarrerp201036.pdf
182 *Ibid*. at 40.
183 *Ibid*. at 41.
184 Sangwani Ng'ambi, "Stabilization Clauses and the Zambian Windfall Tax" (2011) 1 (1) *Zambia Social Science Journal* 107 at 113.

off many of their employees. As a result, the government was compelled to abandon the windfall tax.[185]

The lopsidedness of mining regimes, particularly the fiscal terms provided for in mining contracts, accounts partly for the widespread dissatisfaction with mining operations in many African countries. The lopsidedness of mining contracts in Africa is hardly surprising given that some of these contracts were executed by corrupt, autocratic, and unaccountable governments, while others were executed by governments operating a one-party dictatorship with little or no accountability. Worse yet, some of these contracts were signed by warlords masquerading as presidents during times of war and civil unrest, as is the case in Liberia, the DRC, and Sierra Leone. In addition, these lopsided contracts are often the products of other factors, such as the power imbalance between the financially and technologically buoyant MNCs and impoverished African states;[186] Africa's lack of capacity and expertise to negotiate mining contracts; and the information asymmetry between MNCs and African governments regarding mineralization in concession areas.[187] This exploitative relationship, coupled with the mismanagement of mineral revenues, has ensured that Africa's vast mineral resources benefit mining companies and the ruling African elites, while African citizens are confined to misery and penury. Faced with these unfavourable contractual terms Africa governments are often compelled to address the dissatisfaction of the local population either through renegotiation of contracts or through resource-nationalistic policies that seek to capture more mineral rents.

3.4 Redressing historical inequalities

Most African countries suffered economic and social exploitation at the hands of their European colonizers. The exploitation of Africa has never been explicitly acknowledged by European governments, nor has Africa been compensated for being exploited. The exploitation of Africa was effected partly through MNCs owned by European governments and these companies were deeply entrenched in Africa during and after European colonial rule. Today, most of the companies operating in the African mining industry are based in Western countries. Given this antecedent, the operations of Western mining companies in Africa evoke memories of the economic exploitation and plundering of Africa during the colonial era.[188] This long history of exploitation breeds resentment against Western mining companies which are viewed by Africans as the modern plunderers of

---

185 Sangwani P. Ng'ambi, "Mineral Taxation and Resource Nationalism in Zambia" (2015) 2(1) *Southern African Journal of Policy and Development* 6 at 12.
186 Evaristus Oshionebo, "Corporations and Nations: Power Imbalance in the Extractive Sector" (2018) 77(2) *American Journal of Economics and Sociology* 419 at 421–30.
187 Gajigo et al., *supra* note 122 at 14.
188 David Humphreys, "Transatlantic Mining Corporations in the Age of Resource Nationalism", *Transatlantic Academy Paper Series*, May 2012, at 13, www.yumpu.com/en/document/read/21474563/transatlantic-mining-corporations-in-the-age-of-resource-nationalism

Africa's mineral resources. Stefan Andreasson captures this sentiment when he asserts:

> [T]he divisive historical experience of colonialism coupled with more recent fears of neo-colonial exploitation by former colonial powers (and now also the emerging powers), play an important role in explaining why countries may push to extend control over their natural resources.[189]

It is hardly surprising, then, that some mining regimes in Africa were consciously enacted to redress certain historical inequalities embedded in Africa's mining industry. The desire to redress historical wrongs and inequalities often propel African countries to adopt policies that manifest in the form of resource nationalism. A case in point is South Africa where the MPRDA was enacted with the fundamental objective of promoting "opportunities for historically disadvantaged persons, including women and communities, to enter into and actively participate in the mineral and petroleum industries and to benefit from the exploitation of the nation's mineral and petroleum resources".[190] As noted by some observers, South Africa's resource nationalism is based on "the history of the mining industry in which whites dominated and blacks were exploited as unskilled labourers".[191] In fact, the MPRDA was itself enacted as part of a broader effort to discharge the government's "obligation under the Constitution to take legislative and other measures to redress the results of past racial discrimination".[192] Hence, the MPRDA embodies nationalistic provisions empowering the government to redress past racial discrimination within the mining industry by promoting indigenous ownership of mineral rights. Similarly, Zimbabwe's indigenization policy for the mining industry is based on the desire to redress historical disadvantages suffered by indigenous Zimbabweans on grounds of their race during colonial rule.[193]

In the context of mining, historical wrongs are not confined to economic exploitation but include the adverse social and environmental impacts of mining operations across much of Africa. For the most part, mining companies do not bear the cost of the adverse environmental impacts of their operations; rather, these companies often externalize such cost to the host governments and the local communities in which they operate. Some countries have attempted to address the adverse social and environmental impacts of mining on local communities by adopting policies requiring mining companies to share the economic benefits of resource exploitation directly with the communities. As discussed in Chapter 8, several African countries, including Guinea, Kenya, Mozambique, Nigeria, Sierra

---

189 Andreasson, *supra* note 176 at 314.
190 *MPRDA*, s. 2.(d).
191 Stephen Burgess & Janet Beilstein, "This Means War? China's Scramble for Minerals and Resource Nationalism in Southern Africa" (2013) 34(1) *Contemporary Security Policy* 120 at 133.
192 *MPRDA*, preamble.
193 See *Indigenization and Economic Empowerment Act*, Chapter 14:33, Act 14/2007 (as amended).

Leone, and South Sudan, have enacted statutory provisions compelling mining companies to provide infrastructural facilities for their local host communities through CDAs.

3.5 High commodity prices

There appears to be a correlation between commodity prices and incidents of resource nationalism in Africa. In fact, the intensity of resource nationalism in Africa's mineral resource industry often correlates with the price of commodities on the international market. As is evident from the commodity booms of the 1970s and mid-2000s, incidents of resource nationalism tend to spike or trend upwards during times of high commodity prices. Governments tend to want to maximize resource rents when commodity prices are high, while they tend to retreat from rent maximization when prices are low.[194] In essence, the financial incentives for resource nationalism are accentuated when the prices of commodities are high.

Commodity prices equally affect the power dynamics between the governments of mineral-rich African countries and the MNCs engaged in mineral exploitation in these countries. The pendulum of power shifts back and forth between host governments and MNCs depending on the price of commodities on the international market. Host governments have more leverage during times of high commodity prices because, during these times, MNCs are more likely to tolerate nationalistic policies since they are still able to maintain high profit margins. In essence, commodity prices not only motivate governments regarding whether or not to engage in resource nationalism, but also determine the tolerance of mining companies for nationalistic policies.[195] As Sam Pryke puts it: "If, when market prices are high, a government moves to increase royalties or impose a wind fall tax to take greater benefit from rents, the [MNCs] may reluctantly accept the imposition as [their] profits will remain strong despite the hit."[196] Thus, host states often attempt to capture a larger share of resource rent during commodity price boom by increasing the royalties and taxes payable by mining companies.

Conversely, the ability of governments to engage in resource nationalism weakens considerably during periods of low commodity prices, particularly because of the fear that nationalistic policies could dissuade investment in the resource industry during such periods. When there is a downturn in commodity prices, mining companies are reluctant to invest in new projects or expand production at existing facilities and so, in an attempt to incentivize investment in the mining industry, host states generally shy away from actions that could increase the financial burden borne by mining companies. Moreover, during periods of low

---

194 Luciani, *supra* note 2 at 10. See also Sergei Guriev, Anton Kolotilin, & Konstantin Sonin, "Determinants of Nationalization in the Oil Sector: A Theory and Evidence from Panel Data" (2011) 27(2) *Journal of Law, Economics & Organization* 301 at 302.
195 Pryke, *supra* note 7 at 479.
196 *Ibid*. at 479.

commodity prices MNCs are less amenable to nationalistic policies that could impact negatively on their profits. Host governments are less likely to increase royalties and taxes "when prices are comparatively low as the investor may not tolerate the impact on profits".[197] Also, MNCs are able to exert greater influence on host governments during periods of low commodity prices because these governments want to retain MNC investment in order to ensure a steady source of resource revenues. In effect, a sustained period of low commodity prices shifts power "back to international companies and away from host governments, as international companies can now afford to be more selective about the fiscal terms and regulatory conditions they are willing to accept from host governments".[198] Hence, there are fewer incidents of resource nationalism when commodity prices are low.

3.6 Domestic politics

Domestic politics and, in particular, the political ideology of the ruling class play a significant role in resource nationalism.[199] Depending on the political ideology of the ruling party, a government may adopt either neo-liberal and pro-market policies to encourage the inflow of FDI to the mining industry or interventionist policies designed to reduce foreign dominance of the mining industry. In Africa, governments that are the offshoot of socialist-oriented liberation movements tend to engage in resource nationalism in an attempt to fulfil their populist manifestos.[200] For example, Zimbabwe's indigenization law is informed largely by the political ideology of the ruling Zimbabwe African National Union–Patriotic Front, a liberation movement during the colonial era, which aims to redistribute ownership of mineral resources. Similarly, South Africa's governing party, the ANC, has adopted resource-nationalistic policies such as the Mining Charter in part because of its desire to fulfil its populist electoral promises.

Ruling elites in Africa engage in resource nationalism because it very often boosts the popularity and legitimacy of incumbent governments. As Stefan Andreasson notes, resource nationalism could "enhance the legitimacy of post-colonial governments that are thus seen as doing something concrete to rectify exploitative relationships between their own societies and the formal colonial powers in which many of the international resource companies are headquartered".[201] In this sense, resource nationalism could help an incumbent government to consolidate its power base among the electorate.[202] The correlation between politics and resource nationalism is particularly apparent in semi-democratic countries with weak institutions. By semi-democratic countries, I mean developing countries that practice multi-party democracy and regularly conduct

197 *Ibid.* at 479.
198 Bremmer & Johnston, *supra* note 6 at 149–50.
199 Andreasson, *supra* note 176 at 313–4.
200 Burgess & Beilstein, *supra* note 191 at 121.
201 Andreasson, *supra* note 176 at 313.
202 Jacob & Pedersen, *supra* note 125 at 290.

elections, but lack the requisite institutional framework to check the excesses of the government. Thus, these countries are often governed by irresponsive governments who, in order to boost their popularity and enhance their re-election prospects, adopt resource-nationalistic policies, particularly during election periods.

Moreover, opposition parties in Africa often campaign against incumbent governments using the perceived generosity of mining contracts in favour of foreign companies as evidence that the government is not maximizing resource revenues. In turn, incumbent governments attempt to blunt such attacks by engaging in resource nationalism particularly close to election time. This appears to be the case in Tanzania where the increasingly competitive presidential election is one of the primary drivers of resource nationalism.[203] In 2017, Tanzania enacted several statutes that not only assert permanent sovereignty over natural resources but also authorize the renegotiation of resource contracts.[204] These statutes, which were enacted under 'a certificate of urgency' (an expedited legislative procedure), are "aimed at maximizing the benefits Tanzania can draw from the sector and thus address the perception, initially promoted by the opposition parties, that the country was not benefiting enough from its resources".[205] Where an opposition party successfully wrests power from the government through an election, it may be tempted to engage in resource nationalism in an attempt to demonstrate that it is keeping its election promises to the public. For example, the government may renegotiate mining contracts made with foreign investors by previous governments, particularly where the contracts were procured through corruption or where the legitimacy of the previous government is doubtful, as is the case in non-democratic or war-torn countries.[206]

3.7 Competition for access to mineral resources

The intense competition for access to mineral ores, spurred largely by China's insatiable appetite for minerals, serves to boost African countries in their quest for higher resource revenues. Chinese companies are "prepared to make long-term engagements in producer states, especially in Africa in the hope to ensuring long-term access" to mineral resources.[207] In some instances, Chinese mining companies readily engage in 'off-take agreements' involving the provision of infrastructure in host African countries in exchange for access to minerals.[208] The willingness of Chinese companies to enter into 'off-take agreements' with African

---

203 Ibid. at 288.
204 See *The Natural Wealth and Resources Contracts (Review and Renegotiation of Unconscionable Terms) Act, 2017*; *The Natural Wealth and Resources (Permanent Sovereignty) Act, 2017*; and *The Written Laws (Miscellaneous Amendments) Act, 2017* (No. 7).
205 Jacob & Pedersen, *supra* note 125 at 289.
206 Sornarajah, *supra* note 108 at 75.
207 Joffé *et al.*, *supra* note 4 at 7.
208 Burgess & Beilstein, *supra* note 191 at 120.

countries weakens the ability of Western mining companies to exert economic pressure on African countries to obtain overly generous financial incentives, as was the case in decades past. The recent influx of Chinese companies into the African resource industry broadens the pool of foreign investors, thus emboldening African governments in their quest for higher revenues from mineral resource exploitation.

4 Effectiveness of resource nationalism in Africa

As is apparent from the analysis above, resource nationalism is deeply embedded in the legal and fiscal regimes governing mining in Africa. However, it is unclear whether resource nationalism in Africa's mining industry is achieving the economic and social objectives informing the adoption of resource-nationalistic policies in the continent. This prompts several questions. To what degree have African countries successfully exerted control over mineral resources? Have African countries wrested control of the mining industry from foreign MNCs? Have African countries optimized their share of resource rents? Has the increase in resource rents transformed the economic and social wellbeing of African citizens? How effective are the policies on indigenization, beneficiation, and local content across the continent? These are some of the questions animating the analysis in this section of the book.

The assertion of greater control over natural resources by African governments occurs primarily through mandatory state participation in resource exploitation. Unfortunately, in many instances, the participation of state-owned mining companies in resource exploitation in Africa has failed to enhance resource revenues due primarily to corruption and lack of managerial expertise.[209] While state-owned mining companies in Africa are often provided with competitive advantages such as easy access to mineral rights and 'free carried interest' in mining operations, these companies are rarely profitable and they create very little trickle-down effects for African economies.[210] Moreover, some state-owned mining companies in Africa are undercapitalized and are sometimes unable to pay for their equity stake in mining projects. Even South Africa's state-owned mining company, the AEMFC, has incurred significant financial losses, particularly in 2012 when it posted a net loss of R48.4 million, while its total liabilities exceeded its assets by about R96.7 million.[211]

Regarding resource rents, as argued in Chapter 5, there is evidence that mining revenues have increased in Africa over the last few years due to increases in royalty and tax rates. In South Africa, for example, mining revenues have increased exponentially due to the enactment of the MPRRA which, for the first time, introduced royalty for mineral production in that country. More specifically,

---

209 See The Carter Center, "State Affair", *supra* note 77.
210 Burgess & Beilstein, *supra* note 191 at 131.
211 Andre Janse van Vurren, "AEMFC Readies for Second Mine as Losses Mount", October 18, 2012, www.miningmx.com/news/energy/19222-aemfc-readies-for-second-mine-as-losses-mount

mining revenues in South Africa "increased from R10 billion in 2009 (the year before the Royalty Act was introduced) to R25.7 billion in 2011 (the year after the Royalty Act was introduced)".[212] In fact, in South Africa mining companies pay more taxes than companies in other sectors in part because of the nationalistic provisions in the MPRRA.[213] Similarly, mining revenues in Zambia increased significantly in 2008 when the royalty and tax regimes were revised upwards. In that year, Zambia adjusted its fiscal regimes for minerals by, among other measures, increasing corporate income tax from 25% to 30%; increasing the royalty rate for base metals; imposing a withholding tax of 15% on all mining companies; and reducing capital allowances for mining operations from 100% to 25% of expenses.[214] These fiscal measures led to a significant increase in mining revenues from a paltry US$20 million in 2007 to an estimated US$400 million in 2008.[215] In addition, due largely to nationalistic policies, mining revenues now constitute a significant percentage of government revenues in other African countries. For example, between 2000 and 2013, mining revenues constituted 45%, 23%, and 14% of government revenues in Botswana, Guinea, and Zambia, respectively.[216]

In many African countries, however, mineral rents are not being optimized because of weak and ineffective governance, including the inefficiency and corruption of public officials and institutions. Weak governance

> impacts on the state's share of the resource rents to the extent that Africa states with weak governance generally fail to impose resources tax regimes that ensure an equitable share of the rents, particularly windfall rents, due either to a lack of state capacity or the subversion of that capacity to produce overly investor friendly outcomes.[217]

Across Africa, royalty and tax laws are not effectively enforced by tax administrators and, in some cases, mining companies engage in tax-avoidance schemes such as transfer pricing. Worse yet, even where resource nationalism leads to an increase in resource rents, the benefits of resource rents hardly flow to African citizens due to mismanagement of revenues by inept and corrupt government officials. The inefficiency and ineffectiveness of tax administrators defeat the policy objective of African countries to leverage mining revenues for economic development. In addition, the optimization of mining revenues is adversely impacted by the 'grandfathering' provisions encompassed in fiscal reforms in some African countries. These provisions insulate mining companies from fiscal regimes enacted after the commencement of a mining project. For example, when Tanzania

212 Cawood & Oshokoya, *supra* note 12 at 51.
213 *Ibid.* at 51.
214 Adam & Simpasa, *supra* note 181 at 40.
215 *Ibid.* at 41.
216 ICMM, *Role of Mining in National Economies*, 3rd Edition (2016) at 35, www.icmm.com/website/publications/pdfs/social-and-economic-development/161026_icmm_romine_3rd-edition.pdf
217 *Africa Mining Vision*, *supra* note 115 at 14.

adopted a new *Income Tax Act* in 2004, "there was a general 'granfathering' rule for companies that [had] binding agreements with the government", a rule that prevented the Act from applying to these companies.[218]

The indigenization policies adopted by African governments are designed to enhance the participation of African citizens in mineral exploitation. The hope is that an increase in domestic ownership of mineral rights would in turn benefit the masses through job creation and other spill-over effects. Regrettably, these indigenization policies have yet to bear the desired outcomes.[219] For example, South Africa's attempt to redress historical discrimination within the mining industry through the MPRDA and the Mining Charter has only been modestly successful due largely to administrative inefficiency and political intervention.[220] A recent review of the Mining Charter reveals that, more than a decade after its adoption, "the mining industry is still unacceptably far from being fully transformed" in regards to the historical and socio-economic inequalities in the industry.[221] On the contrary, the Mining Charter appears to be promoting crony capitalism in South Africa, given that its implementation thus far has enriched "the well-connected few, as opposed to the economic empowerment of workers, as well as poor and marginalized black communities, who should be the principal beneficiaries of" the Charter.[222]

Regarding beneficiation, attempts by African countries to promote the beneficiation of minerals have yet to bear appreciable dividends. Rather, mineral ores produced in Africa are still being shipped to foreign countries such as India for refinement. For a variety of reasons, including the non-availability of critical inputs necessary for competitive beneficiation such as electricity, mining MNCs operating in Africa "often prefer to send crude resources to a central beneficiation facility in another country".[223] Legislative and policy instruments requiring domestic beneficiation may not be enough to spur the value addition desired by Africa. In addition to legislation, African countries should adopt contractual provisions that impose minimum levels of domestic beneficiation on mining companies.[224] Such contractual provisions would compel mining companies to domestically process and refine a portion of their mineral ores within Africa, thus creating jobs and spill-over effects for other sectors of the economy. However, the effectiveness of beneficiation policies in Africa depends on prevailing domestic circumstances such as the availability of capital to build and operate mineral-processing facilities; adequate supply of electricity and energy to power these facilities; the training of

---

218 Daniel & Sunley, *supra* note 48 at 415.
219 Burgess & Beilstein, *supra* note 191 at 131.
220 Cawood & Oshokoya, *supra* note 12 at 48.
221 Department of Mineral Resources, *Draft Broad-Based Socio-Economic Empowerment Charter for the Mining and Minerals Industry, 2018* (Government Gazette No. 41714 of 15 June 2018) at v, www.gov.za/sites/default/files/gcis_document/201806/41714gon611.pdf
222 Leon, *supra* note 71 at 17–18.
223 *Africa Mining Vision*, *supra* note 115 at 14.
224 *Ibid*. at 14.

citizens regarding cutting and polishing of minerals; and the technical, scientific, and managerial capacity of African citizens.

As discussed earlier, the local content provisions in Africa's mining industry are designed to create upstream and downstream linkages with other sectors of the economy, particularly with regard to the domestic supply of capital goods, consumables, and services. However, the local content provisions in the mining industry are at best modestly successful in some African countries and shambolic in other countries. While mining companies source some goods and non-tradeable services from domestic suppliers, almost all hardware, equipment, and machinery in the African mining industry are imported.[225] Mining MNCs tend to procure goods and services from their subsidiaries and affiliated companies based in Western countries, sometimes at inflated prices in order to reduce their tax liability to African governments. Local content laws and policies have also failed to improve the employment of African citizens in the mining industry. For example, the local content provisions in Tanzania have neither significantly increased the employment of Tanzanian citizens nor led to any appreciable improvement in the procurement of local goods and services.[226] A similar situation obtains in Ghana where, between 2000 and 2007, the mining industry employed a negligible 0.2% of the non-agricultural labour force, even though mining accounted for 5.5% of Ghana's GDP and 40% of its exports.[227]

The ineffectiveness of local content laws in Africa can be traced to a myriad of factors, including lack of enforcement arising from the incapacity of monitoring and enforcement agencies;[228] lack of good-faith compliance on the part of MNCs who, for the most part, prefer to hire foreign personnel and goods for their operations; and the incapacity of local suppliers to meet the product and service quality demanded by MNCs.[229] However, beyond such institutional incapacity, mining MNCs in Africa often treat local content laws with levity. Although mining companies readily agree to comply with local content requirements during mineral right acquisition, once mineral rights are granted to these companies, their desire to track local content performance declines.[230] Moreover, local suppliers in Africa are not competitive on quality and pricing of products and services. In fact, local manufacturers lack the human resources and technological expertise required to produce the quality and quantity of products demanded by MNCs.[231] In Tanzania, for example, some local suppliers are unable to do business with MNCs not

225 Gajigo et al., supra note 122 at 13.
226 Siri Lange & Abel Kinyondo, "Resource Nationalism and Local Content in Tanzania: Experiences from Mining and Consequences for the Petroleum Sector" (2016) 3 *The Extractive Industries and Society* 1095 at 1098–9.
227 Gajigo et al.,, supra note 122 at 13.
228 Berryl C. Asiago, "Fact or Fiction: Harmonizing and Unifying Legal Principles of Local Content Requirements" (2016) 34(3) *Journal of Energy & Natural Resources Law* 337 at 344.
229 Lange & Kinyondo, supra note 226 at 1100.
230 Simon White, "Regulating for Local Content: Limitations of the Legal and Regulatory Instruments in Promoting Small Scale Suppliers in Extractive Industries in Developing Economies" (2017) 4 *The Extractive Industries and Society* 260 at 264.
231 *Africa Mining Vision*, supra note 115 at 14.

only because of their inability to satisfy the high safety and technical standards required by MNCs but also because they lack the capacity and expertise to meet the timelines for delivery of goods and services.[232] Similarly, in Ghana local companies lack the financial and technical expertise to undertake resource extraction projects which are complex in nature; hence, they are rarely engaged by MNCs as suppliers and sub-contractors.[233] In some instances, the economy of scale weighs heavily against locally manufactured goods because these goods are more expensive than imported goods, particularly goods imported from China.

The lack of provisions regarding performance targets in some African countries such as Botswana makes implementation and monitoring of local content laws difficult. It is impracticable to determine compliance with local content laws in the absence of hard targets set by legislation or policy instruments. African countries that introduced local content laws in recent years face an additional obstacle in implementing these laws. Stabilization clauses in mining contracts signed before the enactment of such local content statutes preclude, at least in theory, host African governments from enforcing the statutes against MNCs that are signatories to such contracts. African countries may be fearful that should they enforce local content laws against these MNCs, they could be sued by the MNCs for breach of the stabilization clauses in the mining contracts.

Investment treaties constrain the ability of African countries to implement and enforce local content laws and regulations. These treaties usually contain provisions that are in direct conflict with local content statutes, including provisions regarding non-discrimination such as 'national treatment' clauses and 'pre-establishment' protection clauses. Local content laws requiring foreign mining companies to source a percentage of their goods and services from the domestic market could be interpreted as discriminatory against foreign mining companies, particularly in countries where similar local procurement requirements are not imposed on companies in other sectors of the economy. In addition, 'pre-establishment' protection clauses in BITs and MITs enjoin state parties not to impose investment conditions on foreign investors that are not imposed on domestic investors. Local content laws in the African mining industry are directed primarily at foreign companies, given that most of the mining companies operating in Africa are foreign companies. However, 'pre-establishment' protection clauses in BITs and MITs protect these foreign companies from discrimination by host governments. For example, the Canada–South Africa BIT prohibits the state parties from enforcing a law or regulation requiring nationals of the state parties to achieve a given level or percentage of domestic content; or purchase, use, or accord a preference to goods produced or services provided in its territory; or purchase goods or services from persons in its territory.[234] In the face of this BIT, the local

---

232 Lange & Kinyondo, *supra* note 226 at 1100.
233 White, *supra* note 230 at 262.
234 *Agreement between the Government of the Republic of South Africa and the Government of Canada for the Promotion and Protection of Investments*, art. V(2), (signed on November 27, 1995, not in force).

content provisions in South Africa's Mining Charter requiring Canadian mining companies operating in South Africa to purchase domestic goods and services could be said to violate the above-stated clause. The negative impact of BITs on the implementation of South Africa's Mining Charter partly informs South Africa's decision to review and in some cases terminate its BITs. As of August 2020, South Africa had 50 BITs, of which 12 are in force, 27 are signed but not in force, and 11 are terminated.[235] Interestingly, most of the BITs terminated by South Africa involve developed countries that are home to many of the foreign companies operating in South Africa's mining industry.[236]

On a positive note, there is some evidence that the renegotiation of mining contracts produces better contractual terms for African countries. In fact, the terms of the renegotiated contracts are better protective of the rights of mineral-producing African countries than the terms of the original contracts, which are often lopsided against African countries. For example, through contract renegotiation Liberia succeeded in almost completely rewriting the MDA with Mittal Steel, thus extracting more favourable terms from Mittal Steel.[237] Among other things, the renegotiated MDA abolished the five-year tax holiday previously granted to Mittal Steel; set the royalty rate for iron ore based on the prevailing international market price; restricted the scope of the stabilization clause to fiscal matters; and changed the governing law from the laws of the United Kingdom to the laws of Liberia.[238]

Relatedly, the renegotiation of mining contracts in Africa appears to have yielded dividends with regard to mineral rents. For example, contract renegotiation in the DRC raised for the government US$307,283,040 as lease premium and US$5,206,000 representing duties on plot area.[239] In addition, the renegotiated contracts increased the rate of royalty payable to the government; increased the government's equity interest in partnerships and joint-venture projects; enabled the participation of state-owned companies in the management of partnerships and joint-venture projects; and expressly committed MNCs to contribute to the socio-economic development of local host communities.[240] However, the renegotiation of contracts in Africa is plagued by the incapacity of governments to negotiate favourable terms. Even in the DRC where there appears to be some modest successful outcomes, the negotiators that acted on behalf of the DRC

---

235 UNCTAD, "South Africa – Bilateral Investment Treaties (BITs)", https://investmentpolicy.unctad.org/international-investment-agreements/countries/195/south-africa
236 As of March 2020, South Africa had terminated its BITs with Argentina, Austria, BLEU (Belgium-Luxembourg Economic Union), Denmark, France, Germany, Italy, the Netherlands, Spain, Switzerland, and the United Kingdom. See UNCTAD, *ibid*.
237 Global Witness, "Update on the Renegotiation of the Mineral Development Agreement between Mittal Steel and the Government of Liberia" (August 2007), www.globalwitness.org/documents/17638/mittal_steel_update_en_aug_07.pdf
238 See *Act Ratifying Mittal MDA*, *supra* note 117, arts. 16, 19, 20, & 35.
239 TWNA, *supra* note 49 at 33–4.
240 *Ibid*. at 34.

were politicians who lacked requisite skills to handle the complex legal, economic, and social issues involved in mining contract renegotiation.[241] Hence, these negotiators settled for short-term benefits such as a one-time increase in royalties and taxes at the expense of long-term improvements in the mining industry.[242]

5 Resource nationalism and foreign direct investment

Good governance, political stability, and the quality and predictability of the legal regimes in a country determine the degree to which that country attracts FDI. It is generally accepted that political and legal instability in the host country pose significant risks for foreign investors. As Baek and Qian have argued, "a host economy with high political risk tends to discourage FDI flows into its market, since political volatility hurts the profitability of foreign investment".[243] In this sense, resource nationalism, particularly the extreme forms of resource nationalism such as nationalization and expropriation of assets, portends significant risks for foreign investors.[244]

Market proponents have long argued that resource nationalism threatens the continuity of investment projects and thus undermines the future viability of business entities.[245] Viewed this way, resource nationalism can discourage the inflow of FDI to developing countries because, by impeding the operations of foreign investors, resource nationalism hurts the profitability of investment projects, thus alienating prospective investors. In addition, radical forms of resource nationalism cause capital and skills flight from the host country, resulting subsequently in capital shortages and skills crises in the host country.[246] Resource nationalism deprives a country of the foreign technology and expertise needed for the growth of the mineral resource industry, thus stagnating the mineral output and revenue streams they need for long-term survival.[247] For example, the enactment of a new mining code by Guinea in 2011 containing resource-nationalistic provisions resulted in BHP Billiton's withdrawal from the country,[248] thus depriving Guinea of BHP Billiton's capital and technological expertise. Capital shortages and skills crises in the host country could eventually cause a decline in the mineral production capacity of the country. When Zambia nationalized

---

241 Lukanda, *supra* note 112 at 357.
242 The Carter Center, "The Mining Review in the Democratic Republic of the Congo: Missed Opportunities, Failed Expectations, Hopes for the Future" (April 2, 2009), www.cartercenter.org/news/pr/drc_040309.html
243 Kyeonghi Baek & Xingwan Qian, "An Analysis on Political Risks and the Flow of Foreign Direct Investment in Developing and Industrialized Economies", at 6, http://faculty.buffalostate.edu/qianx/index_files/PoliRiskFDI_Baek&Qian.pdf
244 Sornarajah, *supra* note 108 at 71–2.
245 Vlado Vivoda, "Resource Nationalism, Bargaining and International Oil Companies: Challenges and Change in the New Millennium" (2009) 14(4) *New Political Economy* 517 at 532.
246 Butler, *supra* note 167 at 13.
247 Bremmer & Johnston, *supra* note 6 at 152.
248 Katz-Lavigne, *supra* note 98 at 23.

foreign-owned mines in the early 1970s, its mining industry suffered a significant decline due largely to the inability of its citizens to manage the mines successfully. The decline led the government to change course in 1992 when, following the enactment of the *Privatization Act, 1992*, the government privatized and sold all previously nationalized mines to foreign investors.[249]

Resource nationalism impacts negatively on investor confidence, thus threatening the future viability of mining projects. For the most part, foreign investors regard resource nationalism as a "market risk which could involve the enforced transfer of value or ownership without fair compensation, thereby undermining the economics of their projects".[250] In addition, radical forms of resource nationalism such as nationalization and expropriation are inimical to the sanctity of mineral concession contracts, and thus represent a constant systemic risk to investors in the mining industry.[251] Furthermore, resource nationalism could cause market imbalance and disruptions because it is in essence an attempt by states to interfere arbitrarily with the free market. The Fraser Institute's survey of mining companies demonstrates that the risks arising from resource nationalism negatively affect investor confidence and account partly for the low ranking of mineral-producing African countries as destinations for foreign investment.[252] According the survey, Africa is "the worst performing region in terms of policy environment for mining activities" and "[i]n terms of overall investment attractiveness, as a region, Africa ranks as the least attractive jurisdiction for investment".[253] To be sure, a number of other factors, including the instability of legal regimes, corruption, and institutional incapacity account for the poor ranking of African countries on the Fraser Institute survey, but it is demonstrably the case that resource-nationalistic policies also contribute to the low ranking of these countries as investment destinations.

Although resource nationalism is often viewed by developing countries as a means to assert greater control over natural resources, resource nationalism could be counter-productive during times of low commodity prices. Incidents of resource nationalism such as increases in royalties, rents, and taxes could dissuade the production of minerals during periods of low commodity prices as companies seek to minimize their financial liability to the government. In addition, resource nationalism could delay the commencement of mining projects as foreign investors are more likely to be hesitant to execute such projects.[254] Depending on the natural resource in question, resource nationalism in a given country could dissuade foreign investors from investing in that country and instead shift their investment

---

249 See John Craig, "Putting Privatisation into Practice: The Case of Zambia Consolidated Copper Mines Limited" (2001) 39(3) *Journal of Modern African Studies* 389.
250 Cawood & Oshokoya, *supra* note 12 at 46.
251 Joffé *et al.*, *supra* note 4 at 3.
252 See *Fraser Institute Annual Survey of Mining Companies 2019*, at 31–3, www.fraserinstitute.org/sites/default/files/annual-survey-of-mining-companies-2019.pdf
253 *Ibid.* at 31.
254 Bremmer & Johnston, *supra* note 6 at 155.

to other countries with less nationalistic policies. As Bremmer and Johnston have observed, "the resource nationalism of the OPEC countries in the 1970s triggered a massive shift in exploration and production to areas such as the North Sea and Alaska".[255] Thus, the risks stemming from resource nationalism, coupled with the political risks associated with sub-Saharan Africa, could compel companies that have significant operations in the region to diversify their operations by shifting some of their investment to less risky regions of the world.[256]

The preceding observations should, however, be tempered with two significant caveats. First, resource nationalism alone does not account for a company's decision to defer or delay extractive projects. Nor does resource nationalism solely account for the decline in FDI inflow to a country. Investment decisions regarding deferment or delay of projects are usually informed by a multitude of factors, including the prevailing commodity prices; the potential economic value of mineral deposits; the legal and political uncertainties and risks in the host country; the degree of corruption and insecurity in the host country; and the social and human rights conditions in the host country.[257] As John Bray argues, "the overriding concern for would-be extractive industry investors remains security of tenure and the extent to which this might be threatened by political unrest, legal uncertainties, changes in government policy, or expropriation".[258] Likewise, a decrease in FDI may arise from factors such as lacklustre economic performance; low commodity prices; and high costs of production in the domestic economy.[259] Second, the resource-nationalistic policies adopted by African countries are similar to those adopted by other developing countries, and thus it is unlikely that companies would divest from Africa and move to other regions based on these policies alone. In fact, the fiscal regimes for mining in Africa are comparable to the regimes in other developing countries, and despite recent increases in royalty and tax rates in some African countries, Africa's fiscal regimes remain generous to investors given the allowances, deductions, and exemptions afforded investors in the mining industry.[260] Thus, only a negligible number of companies would seriously consider divesting from the African mining industry based solely on the fiscal regimes.[261]

While it is undeniable that resource nationalism has some negative impact on FDI inflow to African countries, it remains unclear whether resource nationalism in a given country has long-lasting negative impact on the inflow of FDI to the country. There appears to be evidence that resource nationalism diminishes the

---

255 *Ibid.* at 155.
256 Andreasson, *supra* note 176 at 312.
257 See John Bray, "Attracting Reputable Companies to Risky Environments: Petroleum and Mining Companies", in Ian Bannon & Paul Collier, eds, *Natural Resources and Violent Conflict: Options and Actions* (Washington, DC: World Bank, 2003) 287 at 289–96.
258 *Ibid.* at 292.
259 UNCTAD, *World Investment Report 2016 – Investor Nationality: Policy Challenges* (Geneva: United Nations Publication, 2016) at 40–1.
260 Gajigo *et al.*, *supra* note 122 at 28.
261 *Ibid.* at 28.

volume of FDI inflow to host countries in the short term. Consequently, countries engaging in incessant resource nationalism may experience temporary stagnation or decline in their mining industry. For example, during the global commodity boom of 2001 to 2008 the South African mining industry declined by an average of 1% a year while the mining industry in other top mining-export countries grew by about 5% per year.[262] The decline in the South African mining industry during this period is due partly to resource nationalism, particularly "the increasingly arbitrary role of the government and the rising levels of political uncertainty, caused by a number of resource nationalist actions by the government".[263] The adverse impact of resource nationalism on FDI inflow is exacerbated during periods of low commodity prices. This appears to be the case currently in mineral-dependent African countries. The drastic fall in commodity prices in the last few years has caused a consequential drop in the volume of FDI inflow to Africa.[264] The UNCTAD reports that in 2015, FDI inflow to Africa was US$54.1 billion, a decrease of 7.2%,[265] while in 2016, FDI inflow to Africa was US$59.4 billion, representing a 3.5% decline.[266] Likewise, in 2017, FDI inflow to Africa stood at US$41.8 billion, a decline of 21.5%.[267] Figures on mining-related FDI inflow to Africa are unavailable but, as the UNCTAD reports, low commodity prices not only shrunk investor interest in sub-Saharan Africa,[268] but the "lingering macroeconomic effects from the commodity bust" continue to weigh negatively on FDI inflow to sub-Saharan Africa.[269]

Although FDI inflow to Africa has decreased in the last few years, some regions and countries within the continent have seen sporadic uptick in FDI inflow. For example, in 2016, East Africa attracted US$7.1 billion in FDI, representing a 13% increase, while FDI inflow to West Africa increased by 12% to US$11.4 billion.[270] Within the West African sub-region, FDI inflow to Ghana increased by 9% to US$3.5 billion in 2016, even though Ghana has engaged extensively in resource nationalism.[271] That FDI inflow to a country could increase in the face of resource nationalism should come as no surprise. This is because although investors consider resource nationalism in making investment decisions, numerous other factors such as the profitability of a project, the political stability of a country, and the

---

262 Burgess & Beilstein, *supra* note 191 at 131.
263 *Ibid.* at 132.
264 See UNCTAD, *World Investment Report 2017: Investment and the Digital Economy* (Geneva: United Nations Publication, 2017) at 44–7; UNCTAD, *World Investment Report 2018: Investment and New Industrial Policies* (Geneva: United Nations Publication, 2018) at 38–42.
265 UNCTAD, *World Investment Report 2016, supra* note 259 at 38.
266 UNCTAD, *World Investment Report 2017, supra* note 264 at 44.
267 UNCTAD, *World Investment Report 2018*, supra note 264 at 38.
268 UNCTAD, *World Investment Report 2017, supra* note 264 at 46.
269 UNCTAD, *World Investment Report 2018, supra* note 264 at 40.
270 UNCTAD, *World Investment Report 2017, supra* note 264 at 46–7.
271 *Ibid.* at 47

stability and predictability of the legal frameworks could neutralize the negative impact of resource nationalism on investment decisions.

This assertion is in tandem with an earlier study conducted by Lisa Sachs *et al.* which shows that the attractiveness of a country to foreign investors is not necessarily tarnished by the fiscal reforms undertaken by the country.[272] While FDI inflow may be adversely impacted by resource nationalism in the short term, resource nationalism does not seem to dissuade FDI inflow into a country in the medium and long terms. On the contrary, FDI inflow to a country may increase in the medium and long terms despite resource nationalism in the country.[273] As observed by Sachs *et al.*, "there is generally a decrease in the perceived attractiveness of each country's taxation regime during and immediately after the reform but one year later, the attractiveness returns to pre-reform levels".[274]

It is also worth observing that the negative impact of resource nationalism on FDI is attenuated by certain unique features of the mining industry. First, because mineral resources are non-renewable, the ability of foreign investors to shift investment from one country to another is severely compromised. Second, the capital-intensive nature of mineral exploration and production makes it unwise for foreign investors who have already sunk large capital into mining projects to divest from the host country even in the face of resource nationalism in that country. As presciently observed by John Bray:

> [C]ompanies from these [extractive] industries are obliged to make major capital investments before they can expect to reap any profit. These investments amount to a kind of "hostage". Once the companies have paid for multi-million-dollar fixed assets, they cannot lightly withdraw from the host country.[275]

One final observation ought to be made regarding the impact of resource nationalism on FDI inflow to host countries. This observation is that although resource nationalism is often opposed by industry proponents, aspects of resource nationalism could be good for business. For example, nationalistic policies requiring mining companies to provide social amenities and infrastructure for local host communities could help companies to establish and maintain a cordial relationship with the communities, thus providing a conducive environment for mining projects to thrive. As discussed in Chapter 8, the financial and social assistance provided by mining companies through community development agreements could form the basis for host communities to buy-in on mining projects. The provision of such infrastructural facilities could be the springboard for mining companies to obtain the notional 'social licence to operate' from the host communities.

---

272 Sachs *et al.*, *supra* note 18 at 365–7.
273 *Ibid*. at 366.
274 *Ibid*. at 365.
275 Bray, *supra* note 257 at 292.

6 Conclusion

Although resource nationalism poses risks for investors, foreign investors in the African mining industry should be pragmatic enough to realize that resource nationalism will not dissipate in the near future. The drivers of resource nationalism in Africa, including the exploitative tendencies of MNCs which has denied mineral-rich African countries of the benefit of mineral resources, are too deeply engrained in the psyche of African leaders and peoples to be overthrown by concerns over FDI inflow to these countries. MNCs and other companies involved in mineral exploitation in Africa can stem the growing trend of resource nationalism in Africa by nurturing a non-exploitative and mutually beneficial relationship with host governments. Resource nationalism, as some authors have observed, does not necessarily spell woe for investors; rather, investors ought to "forge more sustainable bonds with these mineral-rich countries by providing them with more benefits, thereby working in favour of their longer-term interests by improving the stability of their projects".[276] Until such a time as mining contracts in Africa strike a fair balance between the interest of mining companies and host African countries, resource nationalism in the African mining industry is unlikely to abate.

276 Cawood & Oshokoya, *supra* note 12 at 46.

8 Management and utilization of mining revenues

1 Introduction

The exploitation of Africa's vast mineral resources ought to yield enough rents for the economic transformation of mineral-rich African countries. However, the reality is that, with the notable exceptions of Botswana and South Africa, there is little evidence that resource revenues have been, or are being, utilized for the benefit of ordinary citizens. Mineral resource endowment has in reality not translated into economic development in Africa. Rather, most resource-rich countries in Africa are plagued with poverty and economic hardships primarily due to the poor governance and mismanagement of resource revenues by corrupt and inept public officials. The negative correlation between natural resource endowment and economic growth in Africa has led some authors to conclude that Africa is suffering a 'resource curse'.[1]

This chapter analyzes the management of mining revenues in Africa, focusing on the legal and institutional mechanisms for revenue management, including public agencies and semi-autonomous bodies such as sovereign wealth funds and mineral development funds. The chapter also analyzes community-based revenue-sharing schemes such as community development agreements, foundations, and trusts. The chapter notes that these legal and institutional mechanisms are ineffective in promoting the prudent management of mining revenues in Africa. For the most part, institutions responsible for managing mining revenues in Africa are corrupt, ineffective, and unaccountable for their actions. Moreover, much of Africa's mineral wealth is deliberately hidden from public view, thus facilitating theft from the public purse, misallocation, and waste of revenues.[2]

---

1 See Terry L. Karl, *The Paradox of Plenty: Oil Booms and Petro-States* (Berkeley, CA: University of California Press, 1997); Richard M. Auty, *Sustaining Development in Mineral Economies: The Resource Curse Thesis* (London & New York: Routledge, 1993).
2 Africa Progress Panel, *Equity in Extractives: Stewarding Africa's Natural Resources for All*, at 71, https://static1.squarespace.com/static/5728c7b18259b5e0087689a6/t/57ab29519de4bb90f53f9fff/1470835029000/2013_African+Progress+Panel+APR_Equity_in_Extractives_25062013_ENG_HR.pdf

The mismanagement of mining revenues in Africa is unlikely to abate unless Africa adopts policies that act as disincentives for corruption and mismanagement of revenues. Hence, this chapter recommends the adoption of transparency and accountability mechanisms requiring mining companies and African governments to disclose payments made to the governments, as well as how mining revenues are utilized by governments. If nothing else, the disclosure of information will enable public scrutiny of mining revenues and perhaps propel African citizens to hold their governments to account. In addition, the chapter recommends that African countries should allow host communities to participate directly in the utilization of mining revenues. Such participation could be achieved by establishing independent trust funds to manage and utilize mining revenues on behalf of host communities and by allowing host communities to own equity interest in mining operations.

2 Revenue management agencies and institutions

As owners of mineral resources, the governments of African countries possess constitutional powers to manage and utilize mining revenues. Thus, across the continent, the power to manage mining revenues is vested primarily in public agencies such as the Ministry of Finance and the Central Bank. In some instances, a designated proportion of mining revenues is managed by semi-autonomous bodies such as sovereign wealth funds and trust funds usually referred to as mineral development funds. In other instances, state-owned mining companies, such as the DRC's Gécamines, participate in revenue management, particularly where they act as representatives of the government in joint-venture projects. Often, these state-owned companies receive royalties from mining companies on behalf of the government.

2.1 Public agencies and state-owned mining companies

Mining revenues in Africa are collected and managed by government agencies and bureaucrats. In practice, the management of mining revenues in most African countries is undertaken by the government through the Ministry of Finance, the Central Bank, revenue collection agencies, and state-owned mining companies. The Ministry of Finance has responsibility for receiving and collecting mining revenues as well as the setting of national economic priorities and the parameters for budgetary allocations. The Ministry of Finance collects mining revenues through subordinate agencies such as the Ghana Revenue Authority, the Zambia Revenue Authority, and the Federal Inland Revenue Service of Nigeria. In addition, state-owned mining companies occupy a pivotal position in mining revenue management as they exercise control over concessions; are involved in joint ventures, partnerships, and production-sharing agreements; and act "as a conduit for foreign investment, export earnings and domestic market activities".[3] State-owned

3 *Ibid.* at 55.

mining companies are equally able to dispose of mining assets and receive revenues under joint ventures, partnerships, and production-sharing agreements.

While African governments are responsible for managing mining revenues, the power of management is exercisable in trust for African citizens. The trusteeship relationship between governments and citizens enjoins African governments not only to create an environment conducive for the prudent management of mining revenues, but also to promote accountability in managing these revenues. However, a common feature across Africa is the poor management of mining revenues. A large proportion of mining revenues in Africa is lost due primarily to mismanagement and corruption on the part of government officials. In the DRC, for example, about US$750 million, representing 20% of the DRC's mining revenues, was reportedly lost between 2013 and 2015 due to mismanagement.[4] The mismanagement of mining revenues occurs in various forms, including diversion and non-remittance of revenues to the treasury; non-collection of revenues; under-payment of royalties and taxes by mining companies with the connivance of revenue collection agencies; under-reporting of revenues; misappropriation and misapplication of revenues; and outright embezzlement of revenues.

The diversion of mining revenues is endemic across mineral-rich African countries.[5] Corrupt public officers strike secret deals with mining companies for the private benefit of such officials. Foreign mining companies are known to engage in opaque practices such as the extensive use of offshore companies to hide and facilitate the illicit diversion of mining revenues into the private bank accounts of African political leaders and public officials.[6] The diversion and non-remittance of mining revenues is particularly prevalent in countries with weak governance systems. For example, the DRC's state-owned mining company, Gécamines, is reported to have remitted to the DRC treasury less than 5% of the US$1.5 billion it earned from its partnership ventures with other mining companies between 2009 and 2014.[7]

Aside from the diversion of mining revenues, mineral-rich African countries often lose revenues through the undervaluation of mineral assets. Mineral assets in Africa are often undervalued and sold to foreign investors at prices far below the prevailing market prices. In some cases, the identity of the foreign investors is unknown because the companies purchasing the mining assets are incorporated in foreign offshore jurisdictions, such as British Virgin Island, with strict secrecy laws. For example, mining concessions in the DRC are systematically undervalued and

---

4 Global Witness, "Regime Cash Machine – How the Democratic Republic of Congo's Booming Mining Exports Are Failing to Benefit Its People", at 6, www.globalwitness.org/en-gb/campaigns/democratic-republic-congo/regime-cash-machine. See also BBC News, "Congo's Mining Revenue 'Missing' – Global Witness", July 21, 2017, www.bbc.com/news/world-africa-40680795
5 Africa Progress Panel, *supra* note 2 at 55.
6 *Ibid.* at 55.
7 The Carter Center, "A State Affair: Privatizing Congo's Copper Sector" at 7, www.cartercenter.org/resources/pdfs/news/peace_publications/democracy/congo-report-carter-center-nov-2017.pdf

"have been sold on terms that appear to generate large profits for foreign investors, most of them registered in offshore centres, with commensurate losses for public finance".[8] Between 2010 and 2012, the DRC lost at least US$1.36 billion from five mining transactions due to the underpricing of mining assets.[9] The magnitude of the financial loss sustained by the DRC from these transactions is captured succinctly thus:

> Across the five deals, assets were sold on average at one-sixth of their estimated commercial market value. Assets valued in total at US$1.63 billion were sold to offshore companies for US$275 million. The beneficial ownership structure of the companies concerned is unknown. Offshore companies were able to secure very high profits from the onward sale of concession rights. The average rate of return across the five deals examined was 512 per cent, rising to 980 per cent in one deal.[10]

The gross undervaluation of mining assets is equally noticeable in Guinea where the government granted a concession to mine iron ore in Simandou to Beny Steinmetz Group Resources (BSGR), but two years later BSGR sold a 51% stake in the concession to Vale for US$2.5 billion, thus earning a 3,000% profit.[11] The profit earned by BSGR from this transaction was two times higher than the entire national budget of Guinea in 2011.[12]

The above observation begs the question: why are mining revenues poorly managed in Africa? The answer to this question is multifaceted, but, in a nutshell, the mismanagement of mining revenues in Africa is due primarily to the institutional incapacity of, and corruption within, the revenue management agencies and, in particular, corruption within the government. Furthermore, revenue management agencies in Africa are plagued with significant problems such as lack of independence, weak oversight, and lack of expertise, funds, and equipment to monitor production level.

2.1.1 Institutional incapacity and lack of independence

Most of the institutions responsible for managing mining revenues in Africa are weak in terms of managerial expertise.[13] For example, as argued in Chapter 5, revenue collection agencies in Africa are often unable to detect transfer pricing and other illicit financial practices adopted by mining MNCs due to lack of expertise. State-owned mining companies are unable to manage joint-venture projects,

---

8   Africa Progress Panel, *supra* note 2 at 56.
9   *Ibid.* at 56.
10 *Ibid.* at 56.
11 *Ibid.* at 60.
12 *Ibid.* at 60.
13 Kobena T. Hanson, "Managing Africa's Natural Resource Endowments: New Dispensations and Good-Fit Approaches" (2017) *Journal of Sustainable Development Law & Policy* 121 at 127–9.

including the revenues accruing from these projects, due to lack of capacity and expertise.[14] The institutional weakness of state-owned mining companies fuels corruption, revenue losses, and inefficiency within these companies.[15] As the United Nations Economic Commission for Africa (UNECA) has observed, the institutional weakness of mineral-rich African countries "provides ample scope for inefficient policies, discretionary behaviour, and outright corruption, all of which could contribute to poor growth performance and eventual dissipation of national [resource] wealth".[16] Institutional weakness also makes it easy for the political and economic elites to "capture rents and use them for unpopular or illegal objectives, including self-enrichment".[17] Moreover, the secrecy surrounding mining revenues in Africa compounds the institutional weakness of public agencies. As argued below, many African countries do not publicly disclose details regarding the payments they receive from mining companies, thus making it difficult for public scrutiny of revenue management agencies.

Perhaps more significantly, revenue management agencies lack operational independence and are, in fact, controlled by the executive arm of government. Revenue managers lack security of tenure and could be relieved of their employment at the whim of the President. For example, the directors of state-owned mining companies in Africa lack operational independence given that many of these directors are political operatives loyal to the incumbent government. Likewise, the managers and staff of revenue management agencies are appointed by, and serve at the pleasure of, the executive arm, in particular the President of the country. Staff of public agencies in Africa are often appointed not on the basis of qualifications and competence but based on ethnic and political affiliation. Moreover, these institutions are often undercapitalized and starved of personnel to discharge their statutory duties. The government determines the funds available to the agencies, and thus the President could intentionally starve the agencies of the funds required to effectively discharge their management responsibilities. The insecurity of tenure and the fear of starvation of funds ensure that there is very little resistance to executive orders regarding the utilization of mining revenues. Moreover, because the government is itself controlled by corrupt political elites, there is always the problem of political interference in the management of mining revenues in Africa.

Relatedly, revenue management agencies are sometimes 'captured' by mining companies due to the power and influence of these companies over African governments. In this context, 'capture' means the exertion of influence over public agencies in a manner that compels the agencies to adopt administrative choices

---

14   Africa Progress Panel, *supra* note 2 at 55.
15   *Ibid*. at 55.
16   IMF, *Guide on Resource Revenue Transparency (2007)*, at 5, www.imf.org/external/np/pp/2007/eng/101907g.pdf
17   UNECA, *Natural Resource Governance and Domestic Revenue Mobilization for Structural Transformation*, at 34, www.uneca.org/sites/default/files/PublicationFiles/agr-v_en.pdf [*Resource Governance*].

that suit the interests of mining companies rather than the interest of citizens.[18] Revenue collection agencies may decline to impose penalties on tax defaulters due to pressure exerted on them by mining companies. In Uganda, for example, the Directorate of Geological Survey and Mines is said to be "controlled by a shadow system which benefits predatory investors and politically powerful Ugandans" – hence its inability to detect and punish "the underpayment or complete failure to pay taxes, mineral rents and royalties by many of [the companies] operating in the [mining] sector".[19]

2.1.2 Weak oversight and audit mechanisms

In ideal settings, oversight over natural resource revenues should include assessment and monitoring of compliance with laws and regulations; monitoring of the performance of government agencies responsible for managing resource revenues; scrutinizing revenue projections, allocations, and utilization; and querying and investigating public spending of resource revenues.[20] In most African countries, the legislature and the public audit agency are responsible for supervising the management of public funds, including resource revenues. Legislative oversight is particularly significant because, "through the 'power of the purse', legislators can shape the allocation of revenues in ways that promote fiscal discipline and limit funding for high-profile projects that have little impact on citizens' quality of life".[21] In addition, the legislature has general power to conduct investigations and hold public hearings on the management of resource revenues. In fact, some parliaments in Africa, including Nigeria's National Assembly, have established resource-specific committees to oversee the management and utilization of resource revenues.[22] In appropriate cases, the legislature may recommend the imposition of sanctions and other penalties on erring managers of resource revenues. The legislature may also enact laws to enhance the proper management of resource revenues, including, as discussed below, laws requiring accountability and transparency in revenue management.

The problem, however, is that legislative bodies in Africa are weak, corrupt, and often susceptible to the influence of the executive arm of government. The reality is that "in the vast majority of natural resource-rich countries [in Africa], the

---

18 See Jean-Jacques Laffont & Jean Tirole, "The Politics of Government Decision-Making: A Theory of Regulatory Capture" (1991) 106(4) *The Quarterly Journal of Economics* 1089–127.
19 Global Witness, "Uganda: Undermined – How Corruption, Mismanagement and Political Influence is Undermining Investment in Uganda's Mining Sector and Threatening People and Environment" (June 5, 2017), www.globalwitness.org/en/campaigns/oil-gas-and-mining/uganda-undermined
20 UNECA, *Resource Governance*, supra note 17 at 42.
21 Shari Bryan & Barrie Hofmann, eds, *Transparency and Accountability in Africa's Extractive Industries: The Role of the Legislature* (Washington, DC: National Democratic Institute for International Affairs, 2007) at 30, www.ndi.org/sites/default/files/2191_extractive_080807.pdf
22 UNECA, *Resource Governance*, supra note 17 at 41.

executive wields considerable control over the legislature, in law and practice".[23] The control of the parliament by the executive arm stems primarily from two factors. First, members of the ruling parties constitute the majority in the parliaments of mineral-rich African countries.[24] Second, some mineral-rich African countries are one-party states wherein the parliament consists predominantly of members of the ruling party. Given the depth of party loyalty in Africa, it is rare for members of parliament to vote against the executive arm of government. In fact, African parliamentarians often vote along party lines. As some observers have noted, in Africa "party loyalty often takes precedence over basic legislative functions, especially in countries" operating proportional representation systems wherein members of parliament are tied "more closely to their parties than to their constituents".[25]

A more worrisome observation is that in African countries "where the ruling party holds a large majority, executive initiatives are rarely questioned or debated in legislative bodies, as opposition legislators often lack the ability or motivation to raise divergent points of view".[26] Thus, in most instances, African governments are able to obtain parliamentary approval for revenue expenditure even where there is no legal or fiscal justification for the expenditure.[27] Even in countries where there is parliamentary oversight of the public agencies responsible for managing resource revenues, the members of parliament may be corrupt and thus susceptible to influence by the executive arm of government. The susceptibility of the legislature to executive influence effectively renders parliamentary checks and balances ineffective in Africa.[28]

With regard to auditing, while most African countries rely on the office of the Auditor-General to audit the financial dealings of the government, a few countries have created specialized audit agencies for resource revenues. For example, Tanzania has established the Tanzania Minerals Audit Agency to monitor not only the quality and quantity of minerals produced in, and exported from, Tanzania but also the profits made by mining companies. These specialized audit agencies are intended to ensure that the government obtains appropriate data regarding the quantity and value of minerals produced in the country. In turn, such data enables the government to maximize resource rents by ensuring that mining companies pay their fair share of rents to the government.

Although audit agencies have power to audit the financial dealings of governments, audit agencies have yet to make any appreciable impact regarding the prudent management of resource revenues in Africa. This is hardly surprising given that audit agencies in Africa lack most of the basic prerequisites for an effective public audit institution. The International Organization of Supreme Audit Institutions has identified eight core principles as essential requirements for an effective

23 *Ibid.* at 42.
24 *Ibid.* at 42.
25 Bryan & Hofmann, *supra* note 21 at 27.
26 *Ibid.* at 27.
27 UNECA, *Resource Governance*, *supra* note 17 at 42.
28 Hanson, *supra* note 13 at 129.

public audit agency. These are: the existence of an appropriate and effective constitutional or statutory framework; the independence of the audit agency, its head, and staff, including security of tenure and legal immunity in the normal discharge of their duties; a sufficiently broad mandate and full discretion in the discharge of their functions; unrestricted access to information; the right and obligation to report on their work; the freedom to decide the content and timing of audit reports, including the publication and dissemination of the reports; the existence of effective follow-up mechanisms on the recommendations of the audit agency; and financial and administrative autonomy, including the availability of appropriate human, material, and monetary resources.[29]

Public audit agencies in Africa are weak and ineffective, and lack independence from the executive arm of government. Staff of audit agencies, including the Auditor-General, are public officials who serve at the mercy of the President. In Ghana, for example, the President not only appoints the Auditor-General[30] but also determines the salary, allowances, and other emoluments of the Auditor-General.[31] Similarly, the President of Nigeria has power to appoint and dismiss the heads of ministerial and extra-ministerial departments, including the Auditor-General of Nigeria.[32] The President's power to appoint and dismiss the Auditor-General and the lack of financial autonomy compromise the independence of audit agencies in Africa. Thus, audit agencies in Africa often operate with fear and in favour of corrupt African governments. For example, despite the widespread corruption in natural resource governance in Africa, public auditors seldom uncover evidence of financial mismanagement or impropriety by African governments. Even where they uncover financial impropriety, public auditors are unlikely to report such impropriety to the public for fear of stepping on the toes of powerful political elites. If audit agencies in Africa are to be effective, they must be granted functional and financial independence, including security of tenure for their staff.[33] Such independence would insulate and protect these agencies against outside influence.[34]

2.1.3 Executive lawlessness and corruption

The executive arm of the governments of African countries is largely responsible for the mismanagement of resource revenues. African governments are for the most part fiscally irresponsible and corrupt – hence the plundering of Africa's

---

29 International Organization of Supreme Audit Institutions, *INTOSAI-P 10: Mexico Declaration on SAI Independence*, www.intosai.org/fileadmin/downloads/documents/open_access/INT_P_1_u_P_10/INTOSAI-P-10_en.pdf
30 *The Constitution of the Republic of Ghana, 1992*, s. 70.(1).
31 *Ibid*. s. 71.(1).
32 *The Constitution of the Federal Republic of Nigeria 1999*, s. 171.
33 International Organization of Supreme Audit Institutions, *The Lima Declaration*, s. 5.(2), www.intosai.org/fileadmin/downloads/documents/open_access/INT_P_1_u_P_10/issai_1_en.pdf
34 *Ibid*. s. 5.(1).

mineral wealth. Resource revenues are hardly utilized for projects and programmes that could enhance the standard of living of African peoples. Rather, many African governments spend resource revenues on projects that are neither fiscally justifiable nor viable. Moreover, African governments are unable to formulate policies that could create linkages between the resource sector and other sectors of the economy, and therefore Africa has yet to experience positive spill-over effects from the resource sector.

The mismanagement of resource revenues in Africa is compounded by the fact that, many resource-rich African countries have been governed consistently by dictatorial, authoritarian, and irresponsible governments.[35] While in the recent past there has been some movement towards democracy, the reality is that even the so-called democratic governments in Africa are, for the most part, autocratic and unaccountable to their citizens. African governments seldom take the interest of citizens into consideration in managing resource revenues. The autocratic and unaccountable nature of political governance in Africa breeds corruption, self-enrichment, and waste of public funds on extravagant and unnecessary projects.[36]

The deficiencies inherent in the management of resource revenues in Africa can be traced partly to the centralized and unitary governance structure prevalent in Africa, which itself is a legacy of colonialism in Africa. The primary aim of colonial rule in Africa was the extraction of resources to meet the needs of European countries. European colonialists not only created a unitary form of government in African countries by concentrating power on the executive, but they also neglected to provide mechanisms for checking and balancing the discretionary power vested in the executive arm of government.[37] Upon independence, African countries inherited a public administration system that had "little or no tradition of legislative or judicial checks and balances" and in which the central government "continued to be powerful, isolated, repressive, and locally alienating".[38] Such unchecked system of governance suited the autocratic desires of many post-colonial governments and dictators in Africa – hence its retention by African countries.

The desire of Africa's ruling elites to exploit the centralized governance structure for their personal benefit partly explains why, upon attaining political independence, African countries vested ownership and control of natural resources in the central government rather than local and regional governments as is the case in developed countries such as Canada and the United States. The centralization of governance and the "lack of countervailing legislative and judicial checks and balances" have enabled ruling African elites to capture the state and its resources

---

35 See generally Arthur A. Goldsmith, "Donors, Dictators and Democrats in Africa" (2001) 39(3) *Journal of Modern African Studies* 411.
36 See Xavier Sala-i-Martin & Arvind Subramanian, "Addressing the Natural Resource Curse: An Illustration from Nigeria" (National Bureau of Economic Research, Working Paper No. 9804, June 2003), www.nber.org/papers/w9804.pdf
37 Mamadou Dia, *Africa's Management in the 1990s and Beyond: Reconciling Indigenous and Transplanted Institutions* (Washington, DC: World Bank, 1996) at 42–3.
38 *Ibid.* at 43.

for the private benefit of the elites.[39] The unchecked power of the executive arm of government culminates ultimately in the irresponsible and unresponsive governments across Africa – hence the plundering and mismanagement of resource revenues in Africa. The ill-effects of the centralized governance system in Africa is particularly acute with regard to the utilization of resource revenues. Because governance in Africa lacks checks and balances, resource revenues are spent at the discretion of the Presidents of African countries. African leaders rarely set development priorities, and even where they do, the priorities are often neglected in favour of grandiose and unproductive projects. Surely, African countries have specific criteria for allocating and utilizing revenues while budgets are approved by African parliaments annually. The problem, however, is that because the executive arm of government is the supreme authority with unchecked power, budgetary appropriations in Africa often reflect the perverse preferences of the ruling elites rather than the economic priorities of the citizens of Africa.[40] As rightly observed by the UNECA, "the budgetary process in some African countries is marked by irregularities, abuses and political pressure in the form of unjustifiable extra-budgetary expenditure and disregard for budgetary rules".[41]

2.1.4 Lack of transparency and accountability

Transparency and accountability are universally recognized as essential pillars for the prudent management of resource revenues.[42] Transparency refers primarily to the timely publication and disclosure of information regarding the management and utilization of resource revenues. The more information is available to the public, the more the public is able to hold governments and extractive companies accountable, thus ensuring that resource revenues are utilized for the benefit of citizens of resource-rich countries.[43] In essence, transparency is a powerful disincentive for corruption, abuse of power, and mismanagement of resource revenues.[44] Information disclosure heightens the prospect of public scrutiny of public agencies, which deters wasteful expenditure and embezzlement of public funds.[45]

The concept of accountability relates to the duty of managers of resource revenues to explain or justify their actions and give reasons for taking any particular decision or course of action. Accountability encompasses "the duty to give

39 *Ibid*. at 44.
40 *Ibid*. at 47.
41 UNECA, *Resource Governance*, supra note 17 at 43.
42 See Joseph C. Bell & Teresa M. Faria, "Critical Issues for a Revenue Management Law", in Macartan Humphreys *et al.*, eds, *Escaping the Resource Curse* (New York: Columbia University Press, 2007) 286 at 305–8; Abdullah Al Faruque, "Transparency in Extractive Revenues in Developing Countries and Economies in Transition: A Review of Emerging Best Practices" (2006) 24 *Journal of Energy & Natural Resources Law* 66.
43 Peter D. Cameron & Michael C. Stanley, *Oil, Gas, and Mining: A Sourcebook for Understanding the Extractive Industries* (Washington, DC: World Bank, 2017) at 221.
44 *Ibid*. at 221.
45 *Ibid*. at 223.

account for one's actions to some other person or body".[46] In the legal sense, the concept of accountability holds public officers and agencies "to the democratic will" and promotes "fairness and rationality in administrative decision making".[47] Thus, accountability presupposes the existence of a superior and independent body with power to hold revenue managers to account.

Regrettably, managers of resource revenues in Africa are neither transparent nor accountable for their actions. Information regarding mining revenues is rarely disclosed, and even in the few instances where such information is disclosed, the disclosures are sporadic, incomplete, and untimely. The non-disclosure of mining-related information permeates the entire mining cycle. In most African countries, mining contracts are undisclosed, thus inhibiting the ability of citizens to determine the financial obligations owed by mining companies to the government. African governments seldom disclose payments made by companies, while mining companies actively hide financial information from the public. Moreover, the operations of state-owned mining companies in Africa "are hidden behind opaque financial management systems, with limited legislative oversight, restricted auditing procedures and, in the worst cases, a comprehensive disregard for transparency and accountability".[48] Some of the foreign mining companies operating in Africa intentionally hide behind the secrecy surrounding offshore jurisdictions and use their incorporation in these jurisdictions to avoid disclosure of information and facilitate the transfer of illicit funds.[49]

In some instances, disclosure of information is deliberately undermined by contractual provisions requiring confidentiality of mining-related information. Confidentiality clauses in mining contracts not only hinder the ability of revenue managers to disclose information but also constrain the ability of oversight bodies to perform their oversight duties. For example, confidentiality clauses in joint-venture agreements prevent state-owned mining companies from disclosing the royalties and other payments they receive from mining companies, thus hindering the ability of the parliament to perform its oversight duties. Worse yet, some African countries criminalize the disclosure of mining-related information, thus creating a perverse incentive for public agencies and officials to withhold information from the public.[50]

With regard to accountability, Africa lacks the superior and independent institutions that are required to hold African governments accountable for their actions. As noted earlier, African parliaments are weak and susceptible to the influence of the executive arm of government. Likewise, the judiciary lacks independence given that judges in many African countries serve at the mercy of the President. Thus, government officials responsible for managing resource revenues in Africa are hardly held to account. Even in cases where it is clear that revenues

46 Colin Scott, "Accountability in the Regulatory State" (2000) 27(1) *Journal of Law and Society* 38 at 40.
47 Ibid. at 39.
48 Africa Progress Panel, *supra* note 2 at 55.
49 Ibid. at 60.
50 Ibid. at 75.

are being mismanaged, erring public officials are seldom prosecuted. Besides, because the public is denied access to mining-related information, public scrutiny of mining revenues is lacking in much of Africa. The effect, then, is that African governments are unaccountable for mismanaging mining revenues.

2.2 Semi-autonomous bodies

2.2.1 Sovereign wealth funds

Sovereign wealth funds (SWFs) are special purpose investment funds created by the government for macroeconomic purposes, and charged with responsibility to "hold, manage, or administer assets to achieve financial objectives, and employ a set of investment strategies which include investing in foreign financial assets".[51] Although SWFs are established primarily to promote efficient management and utilization of resource revenues, some SWFs provide stabilization support for the domestic economy in times of economic stress as well as preserving and saving resource revenues for future generations.[52] Several African countries, including Algeria, Angola, Botswana, Chad, Equatorial Guinea, Libya, Gabon, Ghana, Mauritania, Nigeria, Tanzania, Sao Tome and Principe, South Sudan, and Zimbabwe, have established resource-based SWFs, while other countries such as Kenya are contemplating establishing their SWFs.[53] However, the vast majority of SWFs in Africa manage oil and gas revenues. In fact, of the SWFs in Africa only the Pula Fund of Botswana, the Minerals Income Investment Fund of Ghana and the Sovereign Wealth Fund of Zimbabwe manage revenues accruing from the mining of hard-rock minerals. When established, Kenya's SWF will be unique in Africa because it will manage both oil and gas revenues and mining revenues.[54]

The governance structure of SWFs in Africa falls into two categories.[55] The first category comprises of SWFs that are vested with a separate legal personality. SWFs in this category are either created by statute as independent entities or incorporated as state-owned companies. The Nigeria Sovereign Investment Authority, the Sovereign Fund of Angola, the Minerals Income Investment Fund of Ghana, and the Sovereign Wealth Fund of Zimbabwe are examples of such SWFs. Common features of this variant of SWFs include perpetual succession; power to acquire and own property; power to sue and be sued in their names; and a governing board of directors.[56]

---

51 International Working Group of Sovereign Wealth Funds, *Sovereign Wealth Funds: Generally Accepted Principles and Practices* at 27 [*Santiago Principles*], www.ifswf.org/sites/default/files/santiagoprinciples_0_0.pdf
52 See *Nigeria Sovereign Investment Authority (Establishment, Etc.) Act,* 2011, s. 3; *Sovereign Wealth Fund of Zimbabwe Act, 2014,* s. 4.
53 See *Draft Kenya Sovereign Wealth Fund Bill, 2019.*
54 See *Draft Kenya Sovereign Wealth Fund Bill, 2019,* s. 6.
55 Evaristus Oshionebo, "Managing Resource Revenues: Sovereign Wealth Funds in Developing Countries" (2015) 15 *Asper Review of International Business and Trade Law* 217 at 226–35.
56 See *Sovereign Wealth Fund of Zimbabwe Act, 2014,* ss. 5, 6–8; *Minerals Income Investment Fund Act, 2018* (Ghana), s. 1.(2).

The second category of SWFs, usually referred to as the "manager model",[57] consists of SWFs that are managed by the Ministry of Finance or its delegate, the Central Bank, but lack a separate legal personality. Unlike SWFs that are vested with a separate legal personality, this model of SWFs lacks operational independence from the government as the SWFs are managed directly or indirectly by the Ministry of Finance. The Ministry of Finance may establish a management unit within the ministry to manage the SWF or, in the alternative, delegate the power to manage the SWF to the Central Bank under an Operational Management Agreement. In some instances, the Ministry of Finance may appoint external and independent fund managers to manage the SWF.[58] Botswana's Pula Fund and the Ghana Petroleum Funds are examples of the 'manager model' of SWFs. In Ghana, the Minister of Finance is vested with the power to manage the Ghana Petroleum Funds, although in practice the Minister delegates the operational management of the SWF to the Bank of Ghana. Likewise, Botswana's Pula Fund established pursuant to the *Bank of Botswana Act* is managed by the Bank of Botswana.[59]

SWFs in the first category are preferable because the vesting of a separate legal personality in SWFs promotes operational independence on the part of the board of directors and managers of the SWFs. However, the independence of the board of directors of SWFs in Africa may well be superficial for two reasons. First, in some African countries the Minister of Finance has statutory power to interfere with the performance of the duties of the board of the SWF. In Zimbabwe, the Minister may, in consultation with the President, "give the board such directions in writing of a general character relating to the exercise by [the board] of its functions as appear to the Minister to be requisite in the national interest".[60] Thus, the Minister can order the board to undertake or desist from undertaking an action if they think that it is in the national interest. Such interference undermines the independence of the board of the SWF. Second, members of the boards of SWFs lack security of tenure given that they are appointed and dismissed at the whim of the executive arm of government, in particular the President. The fear of being dismissed if a director disagrees with the government's position regarding the affairs of an SWF has a chilling effect on the independence of the board of directors. Furthermore, even in instances where SWFs are statutorily declared to be independent of the government, the directors, officers, and employees of the SWFs are often deemed to be public servants as defined under domestic law. For example, the Chief Executive Officer and other members of staff of the Sovereign Wealth Fund of Zimbabwe are public officers.[61] Likewise, the directors, officers, and employees of the Nigeria Sovereign Investment Authority are public officers

---

57 See Abdullah Al-Hassan *et al.*, "Sovereign Wealth Funds: Aspects of Governance Structures and Investment Management" (2013) IMF Working Paper WP/13/231 at 10, www.imf.org/en/Publications/WP/Issues/2016/12/31/Sovereign-Wealth-Funds-Aspects-of-Governance-Structures-and-Investment-Management-41046
58 *Ibid.* at 10.
59 See *Bank of Botswana Act*, Chapter 55:01.
60 *Sovereign Wealth Fund of Zimbabwe Act, 2014*, s. 11.(1).
61 *Ibid.* s. 8.(2).

within the meaning of the *Public Officers Protection Act*.[62] The implication, then, is that such directors are subject to the rules governing public officers, including rules relating to termination of employment.

Irrespective of the model of SWFs, the effectiveness of SWFs depends not only on the expertise of their managers but also on the soundness of their institutional frameworks, including the objectives of the SWFs and their general approach to withdrawal and spending of funds.[63] In this regard, the institutional frameworks and objectives of the two prominent mining-related SWFs in Africa are a cause for concern. As discussed above, the Sovereign Wealth Fund of Zimbabwe lacks operational independence from the government. While the Pula Fund has thus far been creditably managed by the Bank of Botswana, the objectives of the Pula Fund appear to be in flux, particularly because its enabling statute, the *Bank of Botswana Act*, does not prescribe specific objectives for the Fund.[64] The Act does not stipulate whether the Pula Fund is a fiscal stabilization fund, a savings fund, or a development fund. The lack of clarity regarding the objectives of the Pula Fund has enabled the government to make incessant and unnecessary withdrawals from the Fund.[65]

Moreover, the Pula Fund and other mining-related SWFs in Africa lack transparency particularly in regard to the investment activities of the SWFs. While countries such as Ghana and Nigeria have enacted clear rules regarding the transparency and accountability of oil- and gas-based SWFs, in reality the few solid-minerals-based SWFs in Africa are neither transparent nor accountable. For example, there is a culture of secrecy surrounding Botswana's Pula Fund. In fact, the Pula Fund "is managed by a secretive committee" consisting of unnamed officers of the Bank of Botswana.[66] In addition, the Pula Fund's reporting and disclosure requirements are unclear; rather, its activities are disclosed within the conscripted rules of the Bank of Botswana. Hence, there is limited public information on the assets of the Pula Fund.[67] Likewise, the statute establishing the Sovereign Wealth Fund of Zimbabwe prohibits and criminalizes the disclosure of information regarding the SWF. The statute provides that

> no member of the Board or employee or agent of the Board shall disclose to any person any information relating to the affairs of the Board or Fund or any person which he or she has acquired in the performance of his or her duties or the exercise of his or her functions.[68]

62 *Nigeria Sovereign Investment Authority (Establishment, Etc.) Act, 2011 (Act No. 15)*, s. 53.
63 *Santiago Principles*, supra note 51 at GAPP 4.
64 Columbia Center on Sustainable Investment, "Natural Resource Funds: Botswana Pula Fund", at 3, http://ccsi.columbia.edu/files/2014/09/NRF_Botswana_July2013.pdf
65 Joel Konopo et al., "Botswana Repeatedly Raids Preservation Fund", https://mg.co.za/article/2016-02-04-botswana-repeatedly-raids-preservation-fund
66 *Ibid*.
67 Columbia Center on Sustainable Investment, *supra* note 64 at 2.
68 *Sovereign Wealth Fund of Zimbabwe Act, 2014*, s. 28(1).

Although there are exceptions to this prohibition, such as disclosure of information where it is necessary for the performance or exercise of an officer's duties and functions or pursuant to the order of a court,[69] this prohibition denies the public the requisite information to hold members of the board accountable for their actions.

A corollary issue is that SWFs in Africa lack appropriate oversight and may thus be prone to mismanagement. While the statutes establishing some SWFs in Africa require auditing and submission of annual reports to the parliament, parliamentary oversight is lacking in Africa because, as argued earlier, African legislators often succumb to the influence of the executive arm of government.

2.2.2 Government-funded mineral development funds

The fiscal regimes for mineral extraction in Africa are, for the most part, silent on the question of equitable utilization and distribution of resource revenues. However, a few African countries have established statutory mechanisms for the internal distribution of resource revenues. Some countries, such as Nigeria and South Africa, have enacted constitutional provisions on the internal distribution of resource revenues.[70] The Nigerian Constitution empowers the national legislature to establish a revenue-sharing formula for the country, "provided that the principle of derivation shall be constantly reflected in any approved formula as being not less than thirteen per cent of the revenue accruing to the Federation Account directly from any natural resources".[71] Thus, under the derivation principle currently applicable in Nigeria, no less than 13% of the revenue accruing to the government from mining is allocated to the states in which the minerals are produced.[72] The *Constitution of the Republic of South Africa, 1996* does not stipulate a specific formula for the distribution of revenues; rather, it confers broad powers on the parliament to set the parameters for the equitable division and utilization of revenues among the national, provincial, and local governments.[73] Pursuant to this power, the parliament enacts legislation each year not only to "provide for the equitable division of revenue raised nationally among the three spheres of government", but also to promote transparency, equity, and accountability in the allocation and utilization of revenues.[74]

Unlike Nigeria and South Africa, some African countries, such as Burkina Faso, Cameroon, Ghana, the DRC, Kenya, Sierra Leone, and Uganda, directly allocate resource revenues to local host communities. In both Burkina Faso and Ghana,

---

69 *Ibid.* s. 28(1).
70 See *The Constitution of the Federal Republic of Nigeria 1999*, s. 162.(2); *The Constitution of the Republic of South Africa, 1996*, s. 214.(1).
71 *The Constitution of the Federal Republic of Nigeria 1999*, s. 162.(2).
72 *Ibid*, s. 162.(2).
73 *The Constitution of the Republic of South Africa, 1996*, s. 214.(1).
74 *Division of Revenue Act, 2009*, s. 2.

20% of all mining royalties is allocated to mining-affected communities.[75] In Cameroon, 75% of mineral royalties is allocated to the central government, while 25% is allocated to local councils and communities.[76] Madagascar allocates 60% of mining revenues to host communities, 30% to the regional governments, and 10% to the autonomous provinces.[77] In Kenya, the central government retains 70% of mining royalties while county governments and mining communities receive 20% and 10%, respectively.[78] Similarly, in Uganda 80% of mineral royalties belongs to the central government, while the local governments and the owner or lawful occupier of the land (from which the minerals are produced) receive 17% and 3%, respectively.[79]

The direct allocation of mining revenues to mining-affected communities often necessitates the creation of community-based revenue management bodies such as mineral development funds (MDFs). MDFs have been established in several African countries, albeit with varying and disparate functions. Some MDFs are policy-based in the sense that they are responsible for coordinating and promoting the mining industry in general, while other MDFs are management-based because their primary role is to manage mining revenues on behalf of host communities. An example of a policy-based MDF is Namibia's MDF, which is responsible for safeguarding the production and earning capacity of the mining industry as well as the broadening and integration of the production base of the mining industry into the national economy.[80] Unlike Namibia's MDF, Ghana's MDF is management-based as it is charged with managing mining revenues for the benefit of mineral-producing communities.[81] In some instances, MDFs may be tasked with responsibility to manage the monetary compensation paid to host communities by mining companies for the adverse social and environmental impacts arising from mining operations.

Government-funded MDFs are specialized semi-autonomous agencies consisting of representatives of the government, host communities, and independent third parties. MDFs may be vested with responsibility to manage all mining revenues accruing to the government or a specified percentage of mining revenues, as is the case in Burkina Faso, Ghana, and Sierra Leone. Burkina Faso's *Mining Code 2015* creates a Community Development Fund (CDF) which is funded partly through mining royalties and partly through a special levy imposed on mining companies.[82] In fact, 20% of mining royalties is transferred by the government of Burkina Faso to the CDF, while mining companies are obliged to contribute 1% of their monthly pre-tax earnings or 1% of the value of the minerals produced

---

75 See Burkina Faso's *Mining Code 2015, Law No. 036-2015/CNT*, art. 26; and Ghana's *Minerals Development Fund Act, 2016*, s. 3.
76 UNECA, *Resource Governance, supra* note 17 at 39 (Table 2.1).
77 *Ibid.* at 47.
78 *Mining Act, 2016* (Kenya), s. 183.(5).
79 *Mining Act, 2003* (Uganda), s. 98.(2) & 2nd Schedule.
80 *Minerals Development Fund of Namibia Act 19 of 1996*, s. 3.
81 *Minerals Development Fund Act, 2016* (Ghana), ss. 2 & 7.
82 *Mining Code 2015, Law No. 036-2015/CNT* (Burkina Faso), art. 26.

monthly to the CDF.[83] In Sierra Leone, 25% of the export tax on diamonds produced by artisanal miners is paid into the Diamond Area Community Development Fund (DACDF), and funds in the DACDF are used exclusively for the needs of mining communities.[84] The DACDF funds are distributed at prescribed intervals to district councils, city councils and chiefdom councils. District and city councils receive 20% of the DACDF funds, while chiefdom councils receive 80% of the funds. The DACDF funds are spent on projects approved by a multitude of bodies, including the Chiefdom Project Committee, the Chiefdom Committee, and the Local Review Committee.[85]

Ghana established a MDF in 1992 but it was redesigned recently with the passage of the *Minerals Development Fund Act, 2016*, which vests in the MDF the status of a body corporate with power to acquire property and enter into contracts.[86] Ghana's MDF derives its capital from a variety of sources, including 20% of mineral royalty received by the government; moneys approved by the Parliament; grants, donations, gifts, and other voluntary contributions; moneys that accrue to the MDF from investments made by its board of directors; and other money that may become lawfully payable to the MDF.[87]

The objectives of Ghana's MDF include the provision of financial resources for the direct benefit of mining communities, holders of interest in land within mining communities, and traditional and local government authority within mining communities.[88] Thus, a significant portion of the money received by the MDF from mineral royalty is disbursed to the mineral-producing communities. More specifically, 50% of the money accruing to the MDF from mineral royalty is disbursed to mining communities through the Office of the Administrator of Stool Lands, while 20% is allocated to the Mining Community Development Scheme (MCDS).[89] A MCDS has been established for each mining community in Ghana in order to "facilitate the socio-economic development of communities in which mining activities are undertaken and that are affected by mining operations".[90] Although Ghana's MDF acts primarily through the Administrator of Stool Lands and the MCDS, it can participate directly in redressing the harmful effects of mining on communities and, in appropriate cases, initiate local economic development projects in mining communities.[91] In addition, the board of the MDF can

---

83 *Ibid.* art. 26.
84 UNECA & African Union, *Minerals and Africa's Development: The International Study Group Report on Africa's Mineral Regimes* (Addis Ababa: UNECA, 2011) at 98.
85 See "A Simplified Handbook on the Government of Sierra Leone's New Operational Procedures and Guidelines for the Diamond Area Community Development Fund (DACDF)", http://documents.worldbank.org/curated/en/699581468167370138/pdf/523990BRI0P1131oor0Simplified0DACDF.pdf
86 *Minerals Development Fund Act, 2016* (Ghana), s. 1.
87 *Ibid.* s. 3.
88 *Ibid.* s. 2.
89 *Ibid.* s. 21.
90 *Ibid.* ss. 16 & 17.
91 *Ibid.* s. 5.

"invest some of the moneys of the Fund in safe securities that the Board considers financially beneficial to the Fund".[92]

The management structure of Ghana's MDF is two-tiered. The first tier comprises the board of directors consisting of government officials, representatives of mining communities, and representatives of the Ghana Chamber of Mines.[93] The second tier is the MCDS which is managed by a Local Management Committee (LMC) comprising of the chief executive of the District Assembly of the mining area, traditional rulers of the mining community, one representative of the District Office of the Minerals Commission of Ghana, one representative of each mining company operating within the district, and one representative each of women and youth in the community.[94]

Prior to its reformation in 2016, Ghana's MDF was beset with significant administrative problems, including lack of transparency, accountability, and oversight. The *Minerals Development Fund Act, 2016* appears to have addressed some of these problems by establishing a hierarchical order for the operations of the MDF. For example, the board of the MDF is the superior decision-making body with responsibility for ensuring "accountability of the moneys of the Fund by defining appropriate procedures for accessing and monitoring the Fund".[95] The MCDS and the LMCs are not only subject to the authority of the board, but the operations and decisions of the LMCs must be approved by the board prior to implementation. Moreover, by diversifying the membership of the LMCs, the Act promotes robust debate within the LMCs regarding their activities, including project execution.

With regard to transparency, the Act requires the board of the MDF to "prepare and publish in a daily newspaper of national circulation the criteria for the disbursement and utilization of the moneys from the Fund".[96] The board is also empowered to "receive and examine reports from designated persons or institutions in respect of financial assistance granted those persons".[97] This presupposes that persons and institutions, including the LMCs, to which the MDF grants financial assistance are obliged to disclose to the board how moneys allocated to them are spent. Oversight over the MDF is provided by the Auditor-General of Ghana and the Parliament. The MDF is audited annually by the Auditor-General of Ghana, and after each audit the board of the MDF is obliged to submit an annual report to the Minister.[98] In turn, the Minister submits the board's annual report to the Parliament in order to facilitate exercise of parliamentary oversight.[99]

---

92 Ibid. s. 7.(g).
93 Ibid. s. 6.
94 Ibid. s. 19.
95 Ibid. s. 7.(d).
96 Ibid. s. 7.(e).
97 Ibid. s. 7.(i).
98 Ibid. ss. 23 & 25.
99 Ibid. s. 25.(3).

3 An appraisal of mineral development funds in Africa

The ability of MDFs to effectively manage mining revenues for the benefit of host communities depends on a plethora of factors, including the expertise and competence of the managers of the MDFs. As well, MDFs are effective only where they are infused with transparency and accountability mechanisms. In that regard, some of the MDFs in Africa fall short of the ideal because the enabling statutes do not facilitate transparency within the MDFs. Many of the MDFs in Africa do not have in-built transparency, accountability, and oversight mechanisms to ensure the judicious use of mining revenues. Thus, there is little oversight over the utilization of the funds allocated to local governments and mineral-producing communities. For example, despite the recent reform of Ghana's MDF, neither the Office of the Administrator of Stool Lands nor the communities to whom it disburses funds is obliged to disclose how the funds are spent.

The lack of transparency and accountability mechanisms has in the past led to the misuse and embezzlement of MDF funds in Africa.[100] In fact, MDFs in Cameroon, Ghana, Liberia, Sierra Leone, and Zimbabwe have engaged in corrupt practices such as bribery, misuse of funds, and favouritism in contracting.[101] For example, prior to the recent reformation of Ghana's MDF, local chiefs in Ghana engaged in wasteful spending of the funds allocated to their communities under the MDF. These traditional leaders used "the MDF money for conspicuous consumption rather than for community development or disbursing it to those most adversely affected by mining".[102] Similarly, in Sierra Leone many chiefs were reportedly unable to explain how allocations to their chiefdoms under the DACDF were spent.[103] In Zimbabwe, some chiefs were found to have paid themselves inflated sitting-fees, thus depleting the funds of the MDF.[104]

Although MDFs in Africa are managed by a consortium of participants, including members of the host community, in reality these managers lack independence from the executive arm of government. Thus, the managers are susceptible to interference from government officials. For example, the board of Ghana's

---

100 Kendra Dupuy, "Corruption and Elite Capture of Mining Community Development Funds in Ghana and Sierra Leone", in Aled Williams & Philippe Le Billon, *Corruption, Natural Resources and Development: From Resource Curse to Political Ecology* (Cheltenham and Northampton, MA: Edward Elgar Publishing, 2017) 69–79.
101 Michael Nest, "Preventing Corruption in Community Mineral Beneficiation Schemes" (U4 Issue No. 3, February 2017) at 15–18, www.u4.no/publications/preventing-corruption-in-community-mineral-beneficiation-schemes.pdf
102 World Bank, "Project Performance Assessment Report: Ghana Mining Sector Rehabilitation Project (Credit 1921-GH) Mining Sector Development and Environment Project (Credit 2743-GH)" (July 1, 2003) at 21, http://documents.worldbank.org/curated/en/120891468749711502/pdf/multi0page.pdf ["Assessment Report: Ghana"].
103 Roy Maconachie, "The Diamond Area Community Development Fund: Micropolitics and Community-led Development in post-war Sierra Leone" in Päivi Lujala & Siri Aas Rustad, eds, *High-Value Natural Resources and Peacebuilding* (London: Earthscan, 2012) 261 at 269.
104 Nest, *supra* note 101 at 17.

292 *Management of mining revenues*

MDF is subject to the authority of the Minister of Mines and the board must comply with any directive issued by the Minister on matters of policy.[105] While funds are statutorily allocated to MDFs, the remittance of funds to MDFs depends on the will of government officials. In some instances, government officials could withhold funds from MDFs, thus thwarting performance of their statutory duties. For example, in the past Ghana's Ministry of Finance retained the MDF's share of mineral royalties for its own general budgetary use rather than disbursing the royalties to the MDF.[106]

The direct allocation and transfer of resource revenues to local host communities could "significantly change the level of public infrastructure, the quality of service provision and improve people's lives" in Africa, especially where the revenues are managed prudently.[107] Thus, Africa needs to build the capacity of local communities to manage mining revenues, including the capacity to invest in post-mining activities.[108] As observed by the African Union, investment in post-mining activities is essential because mineral deposits are finite and non-renewable. Thus, "the economy of any local community, which depends substantially on mining, could in time grind to a halt if the use and management of the community's share of revenues is not planned properly".[109] Regrettably, revenue-sharing arrangements in Africa not only fail to address investment in post-mining activities but also lack specifics regarding how the revenues allocated to mining communities are to be utilized. Moreover, in much of Africa, mineral-producing communities lack the right to determine how the revenues allocated to them are spent. Rather, funds allocated to mineral-producing communities are sometimes utilized on behalf of these communities by local and municipal governments and, in some cases, by a committee chosen by the government.

4 Towards the prudent management of mining revenues

Transparency and accountability are the cornerstones of good governance of resource revenues. This underscores the recent emergence of international and domestic initiatives to promote transparency in resource revenue management. For example, the Extractive Industries Transparency Initiative (EITI)[110] and the Publish What You Pay (PWYP) campaign[111] require governments and companies to publish all payments and receipts derived from resource exploitation, while the International Monetary Fund (IMF) urges that "[r]eports on government receipts

---

105 *Minerals Development Fund Act, 2016* (Ghana), s. 13.
106 World Bank, "Assessment Report: Ghana", *supra* note 102 at 21.
107 International Finance Corporation, "The Art and Science of Benefit Sharing in the Natural Resource Sector" (Discussion Paper) at 38, https://commdev.org/wp-content/uploads/2015/07/IFC-Art-and-Science-of-Benefits-Sharing-Final.pdf
108 African Union, *Africa Mining Vision*, at 23, www.africaminingvision.org/amv_resources/AMV/Africa_Mining_Vision_English.pdf
109 *Ibid*. at 23.
110 See https://eiti.org
111 See www.pwyp.org

of company resource revenue payments should be made publicly available as part of the government budget and accounting process".[112] Likewise, the *Santiago Principles* urge SWFs to publicly disclose their "policies, rules, procedures, or arrangements in relation to the SWF's general approach to funding, withdrawal, and spending operations".[113] The question is, how can Africa ensure the management of mining revenues in a transparent and accountable manner?

4.1 Mandatory transparency, accountability, and oversight regimes

Africa must enact legislation requiring the disclosure of information regarding resource revenues as well as providing accountability and oversight frameworks for revenue management. Information disclosure has the potential to promote transparency and accountability in the management and utilization of resource revenues by enhancing public debate. Accountability can be achieved through a multi-layered approach involving the publication of information and the deployment of auditing and oversight mechanisms.[114] These accountability mechanisms could promote checks and balances in the management of resource revenues. As the IMF has rightly noted, statutory enactments should not only require full disclosure of all resource-related revenues, but they should also ensure oversight of revenue flows by establishing a national audit office or other independent organization which should report regularly to parliament on the revenue flows between extractive companies and the government and on any discrepancies between different sets of data on these flows.[115]

Checks and balances within revenue management institutions can be promoted through oversight mechanisms consisting of both internal and external structures. Internal oversight could take the form of an oversight committee constituted by members of the management board, an audit unit established within the institution, or an advisory body or audit body separate and distinct from the management board of the institution. External oversight can be provided by the parliament, or a specialized parliamentary committee, or by independent auditors and monitors.

A few African countries including Liberia, Nigeria, and Tanzania have enacted statutory provisions aimed at promoting transparency and accountability in resource revenue management. These statutes, some of which create autonomous oversight and audit bodies, were enacted in response to pressures exerted on African countries by international organizations such as the EITI. Among other things, the statutes compel the disclosure of information regarding resource revenues and attempt to promote accountability in the utilization of revenues. In that regard, the *Nigeria Extractive Industries Transparency Initiative Act, 2007*

---

112 IMF, *Guide on Resource Revenue Transparency, supra* note 16 at 9.
113 *Santiago Principles, supra* note 51 at GAPP 4.
114 See Bell & Faria, *supra* note 42 at 300–4 (discussing various mechanisms for overseeing and controlling the activities of SWFs).
115 See IMF, *Guide on Resource Revenue Transparency, supra* note 16 at 10.

outlines its primary objectives as ensuring due process, transparency, and accountability in the payments made by all extractive industry companies to the government; eliminating corrupt practices in the determination, payments, receipts, and posting of resource revenues; and ensuring transparency and accountability in the application of resource revenues.[116] Similarly, the overarching objective of the transparency statute in Liberia is to ensure that

> all benefits due to the government and people of Liberia on account of the exploitation and/or extraction of the country's minerals and other resources are (1) verifiably paid or provided; (2) duly accounted for; and (3) prudently utilized for the benefits of all Liberians and on the basis of equity and sustainability.[117]

While the transparency statute in Nigeria is governed and implemented by a body corporate with perpetual succession and capable of suing and being sued in its name, Liberia and Tanzania have taken a slightly different approach as the governing bodies in those countries are autonomous agencies of the government.[118] However, the governing agency for the Liberia Extractive Industries Transparency Initiative (LEITI) can sue and be sued in its name, enter into contracts, and acquire and hold property.[119] These governing bodies and agencies are vested with broad powers to promote transparency and accountability in resource revenue management, including the power to compel production of documents by governments and companies; monitor, investigate, and audit all resource revenues paid to the government; ensure disbursement of statutory allocations to local and municipal governments; and publish all information concerning resource revenues, including taxes, royalties, bonuses, dividends, and other fees paid to the government.[120] Apparently to address the problem of tax avoidance by extractive companies, the statutes empower the governing bodies to obtain from extractive companies records relating to the cost of production, capital expenditures, volume of production, and export and sales data.[121] These governing bodies must be transparent; hence, they are required to disclose their activities to the

---

116 *Nigeria Extractive Industries Transparency Initiative (NEITI) Act, 2007*, s. 2.
117 *An Act Establishing the Liberia Extractive Industries Transparency Initiative, 2009*, s. 3.1.
118 *An Act Establishing the Liberia Extractive Industries Transparency Initiative, 2009*, s. 2.1; *The Tanzania Extractive Industries (Transparency and Accountability) Act, 2015*, s. 4.
119 *An Act Establishing the Liberia Extractive Industries Transparency Initiative, 2009*, s. 2.3.
120 See *Nigeria Extractive Industries Transparency Initiative (NEITI) Act, 2007*, s. 3; *An Act Establishing the Liberia Extractive Industries Transparency Initiative, 2009*, s. 4; *The Tanzania Extractive Industries (Transparency and Accountability) Act, 2015*, s. 10.
121 *Nigeria Extractive Industries Transparency Initiative (NEITI) Act, 2007*, s. 3.(d); *The Tanzania Extractive Industries (Transparency and Accountability) Act, 2015*, s. 10.(2c).

public, including the submission of periodic reports to the President and the parliament.[122]

Given the primacy of transparency and accountability in resource management, it is surprising that very few African countries have enacted transparency statutes. These statutes lay a foundation upon which Africa can realistically begin to tackle the crisis of revenue management. The power to investigate and compel production of documents is particularly significant because it not only enables the tracking of resource revenue streams, but also aids the detection of corruption and sharp practices such as non-payment of taxes and royalties and non-remittance of revenues by government agencies. For example, through its audit functions the Nigeria Extractive Industries Transparency Initiative (NEITI) recently discovered that, between 2009 and 2011, there was a discrepancy of about US$8 billion between what oil and gas companies paid to the Nigerian government and what the government reported receiving from these companies.[123] But for the NEITI, this staggering amount would have been lost permanently, thus depriving Nigerians of the economic and social benefits that could arise from the utilization of the funds.

Revenue management laws ought to be enacted and enforced across Africa. In order to promote the dissemination of information regarding mining revenues, the proposed laws should outlaw confidentiality clauses in mining contracts, as is the case in Sao Tome and Principe where confidentiality clauses are statutorily declared null and void and contrary to public policy.[124] Revenue management laws must provide clear frameworks for oversight, auditing, and reporting of revenue management by independent third parties. In countries where parliamentary oversight is weak, revenue management laws should establish special oversight bodies to oversee the utilization of resource revenues. Such oversight bodies must be independent in terms of staffing and funding, and must be vested with investigative power as well as quasi-judicial power to compel the production of documents relating to resource revenues. African countries may want to emulate Sao Tome and Principe where a special commission, the Petroleum Oversight Commission, consisting of representatives of the government and civil society groups has been created to provide oversight for oil revenues.[125] This Commission is a legal entity with administrative and financial autonomy, and it has broad power to investigate and monitor the management of oil revenues.[126] The Commission can initiate investigations on its own volition or based on complaints by third parties; conduct searches and inspections; and compel the production of relevant

---

122 *Nigeria Extractive Industries Transparency Initiative (NEITI) Act*, 2007, s. 4.(4); *An Act Establishing the Liberia Extractive Industries Transparency Initiative*, 2009, s. 4.1(i).
123 Cameron & Stanley, *supra* note 43 at 222.
124 *National Assembly, Law No. 8/2004, Oil Revenue Law* (Sao Tome and Principe), art. 20.(1).
125 *Ibid*. arts. 23 & 24.
126 *Ibid*. arts. 23 & 24.

documents and information.[127] In countries where mining revenues are allocated directly to host communities, the proposed independent oversight body should have broad powers to investigate and monitor the utilization of revenues allocated to host communities.

4.2 Capacity-building

The mismanagement of resource revenues in Africa will not be eradicated through transparency statutes alone. In fact, mismanagement of resource revenues persists even in Nigeria, Liberia, and Tanzania where transparency statutes have been enacted. Thus, much more is required if African countries are to manage resource revenues prudently. Aside from the statutory guarantees regarding transparency and accountability, there is a need to train personnel on the prudent management and utilization of resource revenues in Africa. Currently, African states do not have the institutional capacity and expertise to manage resource revenues effectively for the good of their citizens. Institutional incapacity in Africa is exacerbated by rampant corruption within administrative agencies across the continent. If African countries are to ameliorate the 'resource curse', they must begin to build long-term human and institutional capacities to design and effectively implement revenue management policies.[128] Management capacity can be enhanced by the training of personnel on fiscal management; the granting of autonomy and independence to management agencies; and by instituting checks and balances within the management system.

In addition, the capacity of African parliaments to provide oversight for the management of resource revenues must be enhanced through training on budgetary planning, provision of accurate information relating to the utilization of resource revenues, and the funding of parliamentary committees. Some legislators in Africa are poorly educated, and therefore they are unable to understand the complex technical and financial issues arising from resource revenues. Oversight training programs could help such legislators to enhance their knowledge of the natural resource industry. In turn, such knowledge would enable the legislators to provide appropriate oversight for resource revenues. Parliamentary committees should be adequately funded through the national budget, including funds for hiring research staff as well as independent experts on resource revenue management.[129]

5 Benefit-sharing schemes

The participation of local communities in the sharing of mineral revenues may occur through the direct allocation of revenues to the communities by central governments (discussed above), or indirectly through benefit-sharing schemes designed to improve the socio-economic conditions in mineral-producing

---

127 *Ibid.* art. 24.
128 UNECA, *Resource Governance, supra* note 17 at 15.
129 Bryan & Hofmann, *supra* note 21 at 28.

communities. Such benefit-sharing schemes include the payment of compensation to host communities and the provision of infrastructural projects through independent vehicles such as community development agreements (CDAs), foundations, and trusts.[130] Benefit-sharing may also be achieved through equity participation in mining operations by local communities. Benefit-sharing schemes may be mandated by statutes or created voluntarily under private contracts between mining companies and local communities. Mining companies are inclined to engage in benefit-sharing with local host communities in order to obtain and maintain a social licence to operate. As discussed below, CDAs are tools for sharing the benefits of mining with host communities, and in that regard CDAs aid companies in acquiring a 'social licence' from host communities, thus ensuring a conducive atmosphere for mining projects to thrive.

5.1 Community development agreements

CDAs are contractual agreements between mining companies and host communities that expressly provide that mining companies shall provide development assistance to host communities.[131] CDAs are frequently adopted by mining companies as mechanisms for benefit-sharing with host communities, particularly in countries with a large population of indigenous peoples.[132] CDAs in Africa's mining industry may be mandated by statute, or arise from a legal obligation to provide development assistance to host communities, or be voluntarily negotiated by mining companies and local host communities. CDAs are mandated by mining statutes in Nigeria, Guinea, Kenya, Mozambique, Sierra Leone, and South Sudan.[133] Some of these countries mandatorily require CDAs with regard to all mining companies. For example, holders of mining concessions in Guinea are required to enter into Local Development Agreements with communities in the immediate vicinity of their mining operations.[134] Local Development Agreements in Guinea must establish

> conditions that are conducive to the efficient and transparent management of the contribution to local development paid by the holder of the Mining

---

130 Elizabeth Wall & Remi Pelon, *Sharing Mining Benefits in Developing Countries: The Experience with Foundations, Trusts, and Funds* (World Bank, 2011) at 6–18, http://documents.worldbank.org/curated/en/359961468337254127/pdf/624980NWP0P1160ns00trusts0and0funds.pdf
131 See Stefan Matiation, "Impact Benefits Agreements Between Mining Companies and Aboriginal Communities in Canada: A Model for Natural Resource Developments Affecting Indigenous Groups in Latin America?" (2002) 7 *Great Plains Natural Resources Journal* 204.
132 *Ibid.* at 205–12.
133 See *Nigerian Minerals and Mining Act, 2007*, s. 116; *Amended 2011 Mining Code* (Guinea), arts 37.(II) & 130; *Mining Act, 2016* (Kenya), s. 109; *Mining Law No. 20/2014*, dated 18th August 2014 (Mozambique), art. 8; *Mines and Minerals Act, 2009* (Sierra Leone), s. 139; *Mining Act, 2012* (South Sudan), s. 68.
134 *Amended 2011 Mining Code* (Guinea), arts 37.(II) & 130.

Operation Permit, and to strengthen the capacities of the local community in the planning and implementation of the community development program.[135]

Unlike Guinea where all concession holders must sign Local Development Agreements, other African countries require CDAs based on specified circumstances, such as the type of mineral right; production capacity of mining companies; seriousness of the adverse impacts of a mining project; the size of the workforce of mining companies; and the magnitude of a mining project. For example, in South Sudan, only large-scale mining licence holders are required to enter into CDAs with host communities.[136] In this regard, South Sudan's *Mining Act, 2012*, provides that the holder of a large-scale mining licence shall

> (a) assist in the development of communities near to or affected by its operations to promote the general welfare and enhance the quality of life of the inhabitants living there; and (b) enter into Community Development Agreements with such communities in cooperation with relevant government authorities.[137]

Likewise, Nigeria requires CDAs based on the type of mineral right, as only holders of mining leases are obliged to have CDAs with host communities. Thus, the holder of a mining lease, small-scale mining lease, or quarry lease in Nigeria shall,

> prior to the commencement of any development activity within the lease area, conclude with the host community where the operations are to be conducted an agreement referred to as a Community Development Agreement or other such agreement that will ensure the transfer of social and economic benefits to the community.[138]

In Sierra Leone, CDAs are required based on several factors, including the magnitude of a mining project, the production capacity of the project, and the size of the workforce of mining companies. Mining licence holders in Sierra Leone are required to have CDAs with their primary host community only if the approved mining operation will or does exceed the prescribed threshold. Thus, in the case of extraction of mineral from primary alluvial deposits, a CDA is required where annual throughput is more than 1 million cubic metres per year; while companies engaged in underground mining operations are required to have CDAs where their annual combined run-of-mine ore and waste production is more than 100,000 tonnes per year.[139] Likewise, companies engaged in opencast mining

---

135 *Ibid.* art. 130.
136 *Mining Act, 2012* (South Sudan), s. 68.(1).
137 *Ibid.* s. 68.(1).
138 *Nigerian Minerals and Mining Act, 2007*, s. 116.(1).
139 *Mines and Minerals Act, 2009* (Sierra Leone), s. 139.(1).

operations relating primarily to non-alluvial deposits must have CDAs with the host community if the annual combined run-of-mine ore, rock, waste, and overburden production is more than 250,000 tonnes per year.[140] Lastly, mining companies in Sierra Leone are required to have CDAs if they employ or contract more than 100 employees or workers at the mine site on a typical working day.[141]

Aside from CDAs that are mandated by statutes, CDAs commonly arise in African countries where the statutes governing mineral exploitation impose obligations on mining companies to assist host communities with infrastructural projects without specifying the means by which such obligations shall be fulfilled.[142] As well, CDAs could arise where the government grants a mineral right on the condition that the mineral right holder shall provide development assistance to the host communities. Mining companies often utilize CDAs as tools for fulfilling their legal and contractual obligations to provide development assistance to host communities.[143] For example, Central African Republic, the DRC, Ethiopia, Equatorial Guinea, Ghana, Niger, South Africa, and Zimbabwe impose a statutory obligation on mining companies to provide community development assistance to host communities.[144] Similarly, the Mining Act of Rwanda requires holders of mining or quarry licences to prepare and submit to the licensing authority a plan for the development and social welfare of host communities.[145] The Mining Law of Equatorial Guinea imposes a blanket obligation on mining companies to undertake social works in townships, communities, or municipalities where they conduct mining operations,[146] while in Ethiopia mining companies owe a general obligation to provide development assistance to their host communities.[147] In some instances, mining statutes oblige mining companies to make a formal commitment to provide development assistance to host communities prior to the grant of mineral rights. In fact, an applicant for a mineral right may be required to submit a community development plan as part of their application for a mineral right, as is the case in Central African Republic,[148] Mali,[149] and the DRC.[150] In these countries, mining companies often utilize CDAs to fulfil their community development obligations, and hence these statutory provisions create a platform for CDAs in the African mining industry.

However, CDAs are not statutorily mandated in most African countries; rather, mining companies voluntarily sign CDAs with host communities partly because

---

140 *Ibid.* s. 139.(1).
141 *Ibid.* s. 139.(1).
142 See, for example, *Proclamation No. 678/2010* (Ethiopia), art. 60.
143 Kendra E. Dupuy, "Community Development Requirements in Mining Laws" (2014) 1 *The Extractive Industries and Society* 200 at 201.
144 *Ibid.* at 201.
145 *Law No. 58/2018 of 13/08/2018 on Mining and Quarry Operations* (Rwanda), art. 66.
146 *Mining Law for the Republic of Equatorial Guinea, Law No. 9/2006*, art. 54.
147 *Proclamation 678/2010* (Ethiopia), art. 60.
148 *Mining Code of 2009, Law No. 09-005*, art. 33.
149 *Law 2012-015 (Mining Code 2012; Decree No. 2012-311/P-RM)*.
150 *Law No. 007/2002 of July 11, 2002 Relating to the Mining Code*, art. 69 (as amended by *Law No. 18/001 of 9 March, 2018*).

they desire to obtain a social licence from the host communities. For example, Ghana's mining statute does not require CDAs, but several mining companies in Ghana have voluntarily negotiated and signed CDAs with host communities,[151] including the Ahafo Social Responsibility Agreement[152] and the Newmont Ahafo Development Foundation Agreement.[153]

5.1.1 Benefit-sharing under CDAs

The contents of CDAs are statutorily prescribed in African countries where CDAs are mandatory. However, in countries where CDAs arise from an obligation to provide development assistance to host communities, the contents of CDAs may be mutually negotiated by mining companies and host communities. Generally speaking, CDAs in the African mining industry focus primarily on the provision of social amenities, such as schools, hospitals, roads, electricity, and portable water to local host communities, as well as the employment of local indigenes.[154] In this sense, CDAs in Africa's mining industry adopt the recommendations of the International Bar Association[155] and the World Bank[156] to the effect that CDAs should facilitate benefit-sharing with host communities. For example, the purpose of the CDA regime in Kenya is to ensure that

> (a) [the] benefits of mining are shared between the [mineral right] holder and the community; (b) mining is consistent with the continuing economic, social and cultural viability of the community; (c) mining significantly contributes to the improved economic, social and cultural wellbeing of the community; and (d) there is accountability and transparency in mining related community development.[157]

---

151 World Bank, *Mining Community Development Agreements – Practical Experiences and Field Studies* (June 2010) at 26, http://documents.worldbank.org/curated/en/697211468141279238/pdf/712990v30WP0P10IC00CDA0Report0FINAL.pdf [*Mining CDAs*].

152 *Ahafo Social Responsibility Agreement between the Ahafo Mine Local Community and Newmont Ghana Gold Limited*, https://static1.squarespace.com/static/5bb24d3c9b8fe8421e87bbb6/t/5c3bd93040ec9ab9b9f409dd/1547426116789/Ahafo-Social-Responsibility-Agreement.pdf [*Ahafo Social Responsibility Agreement*].

153 *Agreement between Newmont Ahafo Development Foundation and Newmont Ghana Gold Limited*, https://static1.squarespace.com/static/5bb24d3c9b8fe8421e87bbb6/t/5c3bd96521c67c086c302c14/1547426163768/Newmont-Ahafo-Development-Agreement.pdf [*Ahafo Development Foundation Agreement*]

154 See Chilenye Nwapi, "Legal and Institutional Frameworks for Community Development Agreements in the Mining Sector in Africa" (2017) 4 *The Extractive Industries and Society* 202 at 208.

155 International Bar Association, *Model Mine Development Agreement* (April 4, 2011) at 109–17, www.extractiveshub.org/servefile/getFile/id/1256

156 World Bank, *Community Development Agreement: Model Regulations and Example Guidelines* (June 2010), http://documents.worldbank.org/curated/en/278161468009022969/pdf/614820WP0P11781nal0Report0June02010.pdf

157 *Mining (Community Development Agreement) Regulations, 2017* (Kenya), reg. 3.

Thus, CDAs in Kenya must address

> (b) educational scholarship, apprenticeship, technical training and employment opportunities for the people of the community; (c) employment for members from the communities; (d) financial or other forms of support for infrastructural development and maintenance including education, health, roads, water and power; (e) assistance with the setting up of and support to small-scale and micro enterprises; (f) special programmes that benefit women, youth and persons with disabilities; (g) agricultural product marketing; (h) protection of the environment and natural resources; (i) support for cultural heritage and sports; (j) protection of ecological systems; (k) funding and control mechanisms to ensure funds are utilized as intended and accounting processes are transparent and audited; dispute resolution; and (l) any other areas as may be agreed between the parties.[158]

Likewise, CDAs in the Nigerian mining industry must "contain undertakings with respect to the social and economic contributions that the project will make to the sustainability of [the host] community".[159] In addition, mining-related CDAs in Nigeria must address

> (a) educational scholarship, apprenticeship, technical training and employment opportunities for indigenes of the communities; (b) financial or other forms of contributory support for infrastructural development and maintenance such as education, health or other community services, roads, water and power; (c) assistance with the creation, development and support to small scale and micro enterprises; (d) agricultural product marketing; and (e) methods and procedures of environment[al] and socio-economic management and local governance enhancement.[160]

In Sierra Leone, CDAs are required to specify the undertakings of the mining licence holder regarding the social and economic contributions that the project will make to the host community, including educational scholarships, apprenticeship, technical training, and employment opportunities; financial and material contribution towards infrastructural development; and the creation and development of small-scale and micro enterprises in the host community.[161] Likewise, Guinea requires Local Development Agreements to "establish conditions that are conducive to the efficient and transparent management" of the funds provided by the mineral right holder, including provisions covering the training of indigenes, environmental protection and health measures, and the processes for the

---

158 *Ibid.* reg. 8.(3).
159 *Nigerian Minerals and Mining Act, 2007*, s. 116.(2).
160 *Ibid.* s. 116.(3).
161 *Mines and Minerals Act, 2009* (Sierra Leone), s. 140.

development of social infrastructure, such as schools, hospitals, roads, water supply, and electricity.[162]

Some African countries with mandatory CDA regimes specify the value of development projects to be provided by mining companies under a CDA. For example, mining companies in Sierra Leone are required to "expend in every year that the [CDA] is in force no less than one percent of one percent of the gross revenue amount earned by the mining operations in the previous year to implement the agreement".[163] Guinea requires holders of Mining Operations Title to contribute "zero point five percent (0.5%) of the turnover of the company made on a Mining Title of a zone for category 1 mine substances and one percent (1%) for other mine substances".[164] Similarly, the minimum expenditure under a CDA in Kenya is "one per cent of the [mineral right holder's] gross revenue from the sale of minerals in every calendar year".[165]

Aside from the provision of infrastructural facilities, some mining companies operating in Africa voluntarily execute CDAs which provide for profit-sharing with local communities. In Ghana, for example, Newmont Ghana Gold Limited (Newmont) voluntarily signed CDAs with the Ahafo community, one of which obliges Newmont to provide a specified minimum amount for community projects. More specifically, the CDA requires Newmont to pay to the Foundation established under the CDA

> One US dollar (US$1) for every ounce of gold sold by Newmont in its operations under the Ahafo Mining Lease as reported to the government of Ghana [and] One per centum (1%) of Newmont's net pre-tax income after consideration of all inter-company transactions in each year derived from the Ahafo Mining Lease and computed pursuant to generally accepted accounting practice, any gains Newmont receives from the sale of assets when such gains are equal to or more than 100,000 United States Dollars in any such year.[166]

5.1.2 An appraisal of CDAs in the African mining industry

The CDA regimes in the African mining industry are nascent, and thus it is impracticable at this early stage to measure the utility of CDAs in relation to their ability to improve the economic conditions in host communities. Undoubtedly, CDAs have the potential to contribute significantly to the economic development of host communities, particularly where CDAs are properly managed and implemented. The provision of development assistance through CDAs could alleviate a common grievance held against mining companies by host communities regarding the economic neglect of these communities by governments and mining companies. The provision of social amenities through CDAs not only acknowledges that

---

162 *Amended 2011 Mining Code* (Guinea), art. 130.
163 *Mines and Minerals Act, 2009* (Sierra Leone), s. 139.(4).
164 *Amended 2011 Mining Code* (Guinea), art. 130.
165 *Mining (Community Development Agreement) Regulations, 2017* (Kenya), reg. 12.(1).
166 *Ahafo Development Foundation Agreement, supra* note 153, art. 11.(1).

such grievances are genuine but also recognizes that local communities ought to participate in sharing the enormous wealth generated from mining projects on their land. CDAs could thus help to pacify host communities who are often dissatisfied with resource projects, primarily due to the lack of economic benefits from resource projects and the adverse impacts of resource projects on the environmental wellbeing of host communities. The pacification of host communities through benefit-sharing schemes such as CDAs could reduce instances of violent resistance to resource projects by local indigenes in Africa.

However, as this author has discussed elsewhere, the utility of CDAs in Africa is hindered by several factors, including the non-disclosure of the contents of CDAs; the problems of identification of host communities and choice of community representatives; the power imbalance between mining companies and host communities; and the incapacity of host communities to negotiate CDAs.[167] Perhaps more significantly, there are lingering questions regarding whether CDAs are legally binding and enforceable in court.[168] A CDA may not be binding or legally enforceable if it does not satisfy the legal requirements for a binding contract, including, in the case of common law, the requirements of offer, acceptance, consideration, and intention to enter into legal relations. In addition, a CDA is unenforceable in court where the parties expressly provide that the CDA shall not attract legal consequences. For example, the two CDAs governing Newmont's goldmine in Ahafo, Ghana, provide that:

> The parties further agree, acknowledge and confirm that this document does not create any legally enforceable rights to the benefit of either of them and that all disputes or grievances of any kind arising out of or related to this document or policies described herein, shall be settled through mediation and conciliation making use of the Dispute Resolution Committee provided for in this Agreement. The parties hereby renounce their rights to enter into any form of litigation or arbitration on any disputes or grievances arising out of this Agreement.[169]

That said, in countries where CDAs are mandated by statutes, CDAs have the force of law. Thus, in Kenya, CDAs are legally binding and effective once they are signed by the parties.[170] Similarly, mining-related CDAs in Nigeria have legislative force and are binding on the parties,[171] while the CDA regime in Sierra Leone requires parties to a CDA to expressly stipulate in the CDA a commitment

---

167 Evaristus Oshionebo, "Community Development Agreements as Tools for Local Participation in Natural Resource Projects in Africa", in Isabel Feichtner, Markus Krajewski, & Ricarda Roesch, eds, *Human Rights in the Extractive Industries: Transparency, Participation, Resistance* (Cham, Switzerland: Springer, 2019) 77 at 97–106.
168 See Nwapi, *supra* note 154 at 210.
169 *Ahafo Social Responsibility Agreement*, *supra* note 152, s. 4.2.
170 *Mining (Community Development Agreement) Regulations, 2017* (Kenya), reg. 21.
171 *Nigerian Minerals and Mining Act, 2007*, s. 116.(5).

to be bound by the CDA.[172] Moreover, Sierra Leone imposes a positive obligation on mining companies to implement and fulfil their commitments under CDAs.[173] CDAs are equally binding in these countries because CDAs form part of the conditions attached to mineral rights. Hence, in countries such as South Sudan, non-compliance with the requirements of community development is grounds for suspending a mineral right.[174]

However, host communities in common law African countries may be unable to enforce CDAs against mining companies because of the doctrine of privity of contracts that says that only parties to a contract can sue to enforce the contract. The doctrine of privity of contract prevents third-party beneficiaries under a contract from suing to enforce the contract unless they come within any of the exceptions to the doctrine. While the persons that signed a CDA as representatives of the host community can enforce the CDA on behalf of the community, these representatives may refuse or fail to sue to enforce the CDA against the mining company even where there are obvious breaches of the CDA. Traditional rulers and other representatives of host communities may acquiesce in the breaches committed by mining companies, and therefore they may not want to enforce the relevant CDAs. In other instances, community representatives may be incentivized through bribery and corruption to not enforce CDAs against mining companies. In such instances, individual members of the host community may not be able to enforce the CDA in a court of law because they are not parties to the CDA. This problem can be resolved by African countries through the formal recognition of the right of third-party beneficiaries under a contract to sue to enforce the contract.[175] The United Kingdom did so in 1999 when it enacted a law that allows third-party beneficiaries to enforce a contract if the contract expressly grants them the right to enforce the contract.[176] To be effective, legislative provisions regarding the third-party beneficiary principle must expressly vest in the host communities a right to enforce CDAs even though they not parties to the CDAs.

5.1.3 Harmonization of the CDA regimes in Africa

Although a few African countries have enacted statutes requiring mining companies to sign CDAs with host communities, most African countries do not have such statutory enactments. Thus, for the most part, CDAs in Africa are negotiated voluntarily by mining companies and host communities. Even in those countries where CDAs are mandated by statutes, some of the statutes do not elaborate on

---

172 *Mines and Minerals Act, 2009* (Sierra Leone), s. 140.(1e).
173 *Ibid.* s. 139.(1).
174 *Mining Act, 2012* (South Sudan), s. 68.(2).
175 See James Thuo Gathii, "Incorporating the Third Party Beneficiary Principle in Natural Resource Contracts" (2014) 43(1) *Georgia Journal of International & Comparative Law* 93; Marissa Marco, "Accountability in International Project Finance: The Equator Principles and the Creation of Third-Party-Beneficiary Status for Project-Affected Communities" (2011) 34 *Fordham International Law Journal* 452.
176 See the United Kingdom's *Contracts (Rights of Third Parties) Act 1999*.

the contents of CDAs. A by-product of the voluntariness of CDAs in Africa is that the contents of CDAs vary depending on the interests of the contracting parties, as well as the capacity of host communities to negotiate favourable terms. However, because host communities in Africa lack the knowledge, skill, and resources to negotiate CDAs, they are often unable to extract favourable contractual commitments from mining companies, particularly commitments relating to the protection of community rights such as the right to a clean environment. The effect, then, is that the voluntariness of CDAs in Africa detracts from their utility as mechanisms for benefit-sharing with host communities. This begs the question whether African countries should enact legal regimes compelling and dictating the terms of CDAs.

This question is particularly pertinent in view of the AMV which aspires to fully integrate the African mining industry into a single market by creating "mutually beneficial partnerships between the state, the private sector, civil society, local communities and other stakeholders".[177] The AMV also seeks the creation of continent-wide linkages between the mining industry and other sectors, including down-stream linkages into mineral beneficiation and manufacturing; up-stream linkages into capital goods, consumables, and services; and side-stream linkages involving infrastructure, skills, and technology.[178] Such partnerships and linkages could be enhanced through the harmonization of legal regimes across Africa. The harmonization of legal regimes is in fact a condition precedent to the attainment of true economic integration in Africa.[179] If Africa's desire to foster mutually beneficial partnerships between the state, private sector, and local communities is to be realized, then Africa needs to harmonize the legal regimes governing CDAs, including the enactment of legislation mandating CDAs across the continent.

A mandatory CDA regime appears appropriate for African countries given the historical neglect of host communities by governments and extractive companies alike. Compelling extractive companies to share the economic benefits of mineral exploitation directly with host communities could assuage and pacify these communities, thus instilling a sense of ownership in the communities. The justification for a mandatory CDA regime in Africa is that the lands on which mining and other extractive activities take place historically belonged to the host communities. These communities were forcibly deprived of their lands and natural resources first by European colonizers and subsequently by African governments upon attainment of political independence. In Nigeria, for example, the *Land Use Act* forcibly divested local communities of ownership of lands and instead vested in the Governor of each state "all land comprised in the territory of each State".[180] Likewise, the statutes governing resource exploitation in Africa vest ownership of minerals in

---

177 *Africa Mining Vision, supra* note 108 at v.
178 *Ibid.*
179 Richard F. Oppong, *Legal Aspects of Economic Integration in Africa* (Cambridge: Cambridge University Press, 2011) at 10.
180 *Land Use Act*, CAP. L4, LFN 2004, s. 1.

the governments of African countries.[181] The divestment of ownership of land and mineral resources deprives local communities of the economic benefits accruing from mineral extraction, and hence many of these communities are poor. In addition, these communities suffer the adverse impacts of resource exploitation such as environmental pollution. This being the case, it is fair to require extractive companies to share the economic benefits of resource exploitation with local host communities.

Moreover, mandatory CDA regimes will lead to the widespread use of CDAs in Africa, which could then potentially ratchet up sustainable development standards across the continent. This is so because CDAs could spur competition among mining companies to outperform their peers. The World Bank alludes to this issue when it observes that "[i]f CDAs become mandated in legislation the mining sector would be provided with an opportunity to monitor and contrast different examples of CDAs increasing the best practice methodologies within the sector as experience develops and lessons are learnt".[182] Furthermore, the statutory prescription of the contents of CDAs could alleviate the power imbalance between mining companies and host communities, which prevents host communities from extracting favourable contractual terms from mining companies. A mandatory statutory regime could equally address the lack of transparency plaguing CDAs in Africa at this moment. To be effective, however, such mandatory CDA regimes should expressly prohibit confidentiality clauses in CDAs and prescribe a timeline within which parties to CDAs should disclose the contents of their CDAs.

That said, statutory provisions mandating CDAs must be flexible and adaptive to local circumstances, as well as the constantly changing circumstances surrounding mining projects.[183] This is particularly so given the diverse cultures, traditions, and customs within individual African countries. A mandatory CDA regime must provide broad but flexible guidelines for CDAs, thus ensuring the emergence of 'best practice' in the mining industry. In this regard, some of the CDA regimes in Africa contain a measure of flexibility, particularly with regard to the contents of CDAs. For example, mining-related CDAs in Nigeria need not address all of the issues prescribed in the *Nigerian Minerals and Mining Act, 2007*. Rather, CDAs in Nigeria may address only those issues that are "relevant to the host community".[184] Mandatory CDA regimes should equally allow parties to renegotiate their CDA if the circumstances surrounding a project have changed significantly. Such appears to be the case in Nigeria, Kenya, and Sierra Leone where CDAs are reviewable every five years, even though the CDAs are designed

---

181 See, for example, *Nigerian Minerals and Mining Act, 2007*, s. 1; *Minerals and Mining Act, 2006* (Ghana), s. 1; *Mining Act, 2016* (Kenya), s. 6; *Mining Law No. 20/2014*, dated 18th August 2014 (Mozambique), art. 4; *MPRDA*, s. 3; *Mines and Minerals Act, 2009* (Sierra Leone), s. 2; *Mining Act, 2012* (South Sudan), s. 6.
182 World Bank, *Mining CDAs, supra* note 151 at 64.
183 *Ibid.* at 65–6.
184 *Nigerian Minerals and Mining Act, 2007*, s. 116.(3).

to last for the duration of mining operations.[185] In essence, statutory regulation of CDAs must avoid a one-size-fits-all model by ensuring that the regime is flexible enough to cater to the constantly changing circumstances that often accompany mining projects.[186]

Although there is growing resort to CDAs in Africa and other parts of the developing world, and while CDAs foster resource benefit-sharing with host communities, some authors have cautioned against mandatory CDA regimes primarily because of the fear that the widespread adoption of CDAs in any country allows the government to abdicate and outsource its responsibility to provide for the welfare of its citizens. For example, it has been observed that in the past some government officials in Australia deprived certain Aboriginal communities of public funds on grounds that these communities could afford to provide their own facilities by utilizing the payments they received from mining companies.[187] Thus, where mining companies provide financial and material benefits to the host communities through CDAs, the government may wittingly reduce its financial support for these communities. In the words of another author, CDAs enable governments to transfer many of their "responsibilities to project proponents, leaving only to its judicial branch the option to adjudicate in case of disagreements" arising from the CDAs.[188] This could be particularly so in African countries with weak governance institutions because, in these countries, "CDAs can inadvertently create great dependence on the mining companies almost replacing governments".[189]

Mandatory CDA regimes could also create a culture of 'entitlement' among community members which could lead to more conflicts, particularly if community expectations are not met.[190] In addition, new investors might be dissuaded from investing in a community because they may not be able to match the benefits provided to that community by their more established competitors. Notwithstanding these concerns, CDAs are potential instruments for advancing the sustainable development of mineral resources in Africa because CDAs not only provide a compensation framework for extractive projects, but also constitute "part of an overall social policy to benefit" host communities, thus assuaging the adverse impacts suffered by these communities as a result of extractive projects.[191]

---

185 See *Nigerian Minerals and Mining Act, 2007*, s. 116.(5); *Mining (Community Development Agreement) Regulations, 2017* (Kenya), regs 14 & 15; and *Mines and Minerals Act, 2009* (Sierra Leone), s. 140.(1)(e).
186 World Bank, *Mining CDAs, supra* note 151 at 66.
187 Ciaran O'Faircheallaigh, "Denying Citizens their Rights? Indigenous People, Mining Payments and Service Provision" (2004) 63(2) *Australian Journal of Public Administration* 42 at 43–4.
188 Guillaume P. St-Laurent & Philippe Le Billon, "Staking Claims and Shaking Hands: Impact and Benefit Agreements as a Technology of Government in the Mining Sector" (2015) 2 *The Extractive Industries and Society* 590 at 596.
189 World Bank, *Mining CDAs, supra* note 151 at 66.
190 *Ibid*. at 66.
191 *Ibid*. at 18

5.2 Company-funded foundations, trusts, and funds

Company foundations, trusts, and funds (FTFs) are privately funded by mining companies and are established primarily for the benefit of host communities. The terms 'foundations', 'trusts', and 'funds' are often used interchangeably, but the legal nature of FTFs differ depending on the legal system of the host country. However, the distinction between foundations, trusts, and funds is moot in practice. FTFs are often referred to as 'trusts' in common law countries while 'foundations' is the preferred term in civil law countries.[192]

A trust is a legal relationship arising where a person transfers property to another (the trustee) for the specific purpose of managing the property on behalf, and for the benefit, of a third party (the beneficiary). A foundation is an entity registered or incorporated under relevant domestic law such that the incorporation confers a separate legal personality on the entity. In common law African countries such as Kenya, Nigeria, and Zambia, however, both foundations and trusts may be registered or incorporated as a charitable trust, thus blurring the distinction between the two.[193] The registration or incorporation of a trust or foundation requires an application for a certificate of incorporation under the relevant domestic statute such as Nigeria's *Companies and Allied Matters Act*,[194] Kenya's *Trustees (Perpetual Succession) Act*,[195] and Zambia's *The Land (Perpetual Succession) Act*.[196] The advantages of incorporation include the conferment of a separate legal personality with perpetual succession; capacity to acquire and own property; and the ability to sue and be sued in its name. Foundations and trusts are thus legal creatures with juridical personality and a separate legal status.[197] The term 'funds' is not a legal term but is generally used to describe legal trusts and foundations.[198] In reality 'funds' are informal entities that usually lack a separate legal status.[199]

Regardless of the difference in the legal nature of FTFs, FTFs are commonly adopted as vehicles to accomplish a multitude of objectives, including conservation of natural resources, environmental protection, and development of host

---

192 Wall & Pelon, *supra* note 130 at 20.
193 See *Trustee (Perpetual Succession) Act*, Chapter 164, Laws of Kenya, s. 3; *The Land (Perpetual Succession) Act*, Chapter 186, Laws of Zambia, ss. 2–4; and Nigeria's *Companies and Allied Matters Act*, CAP. C20, LFN 2004, ss. 673–9.
194 *Companies and Allied Matters Act*, CAP. C20, LFN 2004, ss. 673–9.
195 *Trustee (Perpetual Succession) Act*, Chapter 164, Laws of Kenya, s. 3.
196 *The Land (Perpetual Succession) Act*, Chapter 186, Laws of Zambia, ss. 2–4.
197 World Bank, *Mining Foundations, Trusts and Funds: A Sourcebook* (June 2010) at 20, http://documents.worldbank.org/curated/en/418481468158366874/pdf/828560WP0Sourc00Box379875B00PUBLIC0.pdf [*Mining Foundations Sourcebook*]
198 Ondotimi Songi, "Defining a Path for Benefit Sharing Arrangements for Local Communities in Resource Development in Nigeria: The Foundations, Trusts and Funds (FFTs) Model" (2015) 33(2) *Journal of Energy & Natural Resources Law* 147 at 157.
199 World Bank, *Mining Foundations Sourcebook*, *supra* note 197 at 13.

communities.[200] FTFs in the mining industry are established primarily for the latter purpose – that is, to provide development assistance to host communities.

5.2.1 Nature of foundations, trusts, and funds

Company-funded FTFs have been established in several African countries, including Botswana, Ghana, Mali, Namibia, Mozambique, South Africa, Zambia, and Zimbabwe.[201] These FTFs focus primarily on the provision of infrastructural facilities and social services in host communities, including education, healthcare, entrepreneurial skills development, and local economic development through the provision of loans to small-scale and medium-scale enterprises. For example, the Palabora Foundation in South Africa aims to ensure that "people from Ba-Phalaborwa are well-educated, have employable skills, are empowered to start their own businesses and are well enough to work".[202] Most FTFs engage in direct provision of infrastructure facilities and social services, while some FTFs are grant-making FTFs in the sense that they provide grants to NGOs who in turn utilize the grants to provide infrastructural facilities and other social amenities to mining-affected communities.

FTFs may be mandated by a mining lease or predicated on statutes requiring mining companies to provide development assistance to host communities. Some mining contracts and statutes impose 'community funding obligations' on mining companies in relation to the development of infrastructure in host communities. Funds provided by mining companies under this arrangement may be paid into a trust account and managed by a trust-like body which may or may not be formally registered. In Liberia, for example, some mining companies are contractually obliged to make 'annual social contribution' to the communities in which they operate. To that end, the contract between the government of Liberia and China-Union Mining Company provides that "the Concessionaire shall provide an annual social contribution of US$3.5 million which shall be managed and disbursed for the benefit of Liberian communities in the counties affected by its operations".[203] A more recent mining contract in Liberia provides that:

> Commencing on the Effective Date and thereafter on each anniversary of the Effective Date until the end of the year prior to the year in which the Start of

---

200 Emeka Duruigbo, "Managing Oil Revenues for Socio-Economic Development in Nigeria: The Case for Community-Based Trust Funds" (2004) 30 *North Carolina Journal of International Law and Commercial Regulation* 121 at 170–1.
201 Wall & Pelon, *supra* note 130 at 51–3.
202 Palabora Foundation, www.pafound.org
203 *Mineral Development Agreement between the Government of the Republic of Liberia, China-Union (Hong Kong) Mining Co. Ltd. and China-Union Investment (Liberia) Bong Mines Co., Ltd.* (January 19, 2009), s. 8.2, www.leiti.org.lr/uploads/2/1/5/6/21569928/152412379-mineral-development-agreement-between-the-government-of-the-republic-of-liberia-china-union-hong-kong-mining-co-ltd-and-china-union-investment.pdf

Commercial Production occurs with respect to the first Production Area, the Company shall pay Two Million Dollars (US$2,000,000) within ten (10) days of the Effective Date for the first year of the Term; Two and a Half Million Dollars (US$2,5000,000) on each anniversary of the Effective Date until the year in which the Start of Commercial Production occurs and on each anniversary of the Effective Date. Thereafter, the Company shall pay an Annual Social Contribution of Three Million and One Hundred Thousand Dollars (US$3,100,000) (adjusted annually for inflation ...).[204]

These funds are utilized for projects approved by a committee consisting of representatives of the company, the government of Liberia, and the host communities.[205] The committee disburses funds "only for direct delivery of services and community infrastructure improvements" and for the benefit of Liberian communities impacted by the mining project.[206] The committee does not make payments to individuals, except where it pays for goods supplied or services rendered by individuals in the execution of projects authorized by the committee.[207] In order to promote transparency, all projects executed by the committee as well the actual disbursements by the committee are publicly disclosed and audited.[208] The company has a right to conduct an independent audit of the activities, disbursements, and expenditure of the committee and, for this purpose, the committee is obliged to provide the company with all relevant documents and information regarding its activities.[209] While the disclosure of information promotes transparency, the problem is that the contractual scheme governing the activities of the committee provides neither oversight nor accountability for the committee. The operations and activities of the committee are thus susceptible to abuse by unscrupulous managers.

Aside from instances where FTFs arise from contractual or statutory obligations, mining companies voluntarily establish FTFs for the purpose of sharing mining revenues with host communities. The vast majority of FTFs in the mining industry are voluntarily initiated by mining companies, although governments facilitate the creation of FTFs through policy instruments that mandate or encourage the sharing of mining revenues with host communities.[210] Examples of such policy instruments include the CDA regimes discussed earlier and the Mining Charter of South Africa. As discussed above, mining companies usually adopt CDAs as mechanisms for discharging their obligation to provide development assistance to host communities. CDAs may establish FTFs for the purpose of executing development projects in, and engaging in benefit-sharing with, host communities.

---

204 *An Act to Ratify the Concession Agreement among the Government of the Republic of Liberia, Western Cluster Limited, Sesa Goa Limited, Bloom Fountain Limited and Elenilto Minerals and Mining LLC* (August 22, 2011), s. 8.2(a).
205 *Ibid*. s. 8.2(c).
206 *Ibid*. s. 8.2(c).
207 *Ibid*. at Exhibit 6 (Principles Relating to Community Funding).
208 *Ibid*. s. 8.2(c).
209 *Ibid*. s. 8.2(c).
210 World Bank, *Mining Foundations Sourcebook*, *supra* note 197 at 22.

Management of mining revenues 311

For example, the Newmont Ahafo Development Foundation in Ghana was established under the CDA between Newmont and Ahafo communities. In South Africa, mining companies are increasingly adopting FTFs in furtherance of the objective of the Mining Charter to transfer mineral wealth to historically disadvantaged persons and communities. Even in countries where development assistance obligations are not imposed on mining companies, FTFs may be voluntarily established by mining companies for the purpose of obtaining a 'social licence to operate' from host communities.

5.2.2 Funding and governance of foundations, trusts, and funds

Mining-industry FTFs are funded by mining companies, although some FTFs are funded jointly by mining companies and third-party donors. FTFs are funded either by way of an endowment or through annual financial contribution by mining companies or both.[211] Endowment funding involves the donation of funds and other assets to FTFs so that the funds and assets are invested for the perpetual benefit of the beneficiaries – that is, the host communities. Under the annual contribution model, the amount of funding may be based on a prescribed formula, such as a percentage of the net profits or gross sales of the company derived from its operations in the host community, or a percentage of the dividends distributed to the company's shareholders. For example, Newmont pays to the Newmont Ahafo Development Foundation in Ghana one US dollar (US$1) for every ounce of gold sold under the Ahafo Mining Lease, as well as 1% of its net pre-tax income in each year derived from the Ahafo Mining Lease.[212] As of 2017, the Newmont Ahafo Development Foundation had about US$10 million, out of which US$1.2 million was spent on development projects in Ahafo mine communities.[213] Similarly, the Rossing Foundation, established by Rio Tinto Rossing Uranium Limited in Namibia, receives annually 2% of all dividends distributed to the company's shareholders after tax or such greater amount as the company may determine.[214] MNCs that have several subsidiary companies in a host country may structure their FTFs in a manner that all companies within the group contribute to the funding of the FTFs. For example, the Anglo American Chairman's Fund in South Africa is funded annually by all companies in the Anglo American group, with the value of contribution linked directly to the profitability of each company.[215]

The endowment model is more appropriate for FTFs that seek to exist beyond the lifespan of a mining project, while the annual contribution model is better suited to FTFs designed to provide development assistance during the life of a

---

211 *Ibid.* at 31.
212 *Ahafo Development Foundation Agreement, supra* note 153, art. 11.(1).
213 Sam A. Kasimba & Paivi Lujala, "There is No One Amongst Us with Them! Transparency and Participation in Local Natural Resource Revenue Management" (2019) 6 *The Extractive Industries and Society* 198 at 202.
214 World Bank, *Mining Foundations Sourcebook, supra* note 197 at 146.
215 *Ibid.* at 114.

mining project.[216] The capital of an endowed FTF is invested in profit-making ventures and the proceeds of such investment can be used to fund the development activities of the FTF during the life of the mine and long after the ceasing of mining operations.[217] Moreover, while FTFs funded by annual contributions may be significantly impacted by commodity price fluctuations and other external influences, endowed FTFs are insulated from such factors, particularly where the capital of the FTF is invested in a diversified portfolio.[218]

FTFs are governed by a board of trustees consisting of representatives of mining companies, host communities, and independent third parties. However, some FTFs funded by mining companies exclude independent third parties from their governance. For example, the Palabora Foundation in South Africa, an endowment fund established by Rio Tinto Palabora Mining Company Ltd, is governed by a board of trustees consisting of representatives of the company and host communities.[219] In some instances, the board of trustees of FTFs may include representatives of the government even though the FTFs are funded by mining companies. For example, the Rossing Foundation in Namibia has a board of trustees consisting of representatives of Rio Tinto Rossing Uranium Limited, the Governor of the Bank of Namibia, the Chairman of the National Council, the Governor of Erongo Region, and independent third parties with requisite experience in business.[220] However, some FTFs are governed by representatives of the mining companies to the exclusion of host communities. For example, the Anglo American Chairman's Fund (South Africa), a grant-making fund, has a board of trustees consisting exclusively of representatives of companies in the Anglo American group.

5.2.3 Effectiveness of foundations, trusts, and funds

FFTs in the African mining industry have made modest positive impacts on host communities through the provision of infrastructural facilities. Schools, hospitals, roads, electricity, portable water, and other facilities have been provided by FTFs across Africa. However, there are concerns regarding the long-term viability and relevance of FTFs in Africa. The sustainability and continuity of some FTFs is suspect, particularly where FTFs are funded by mining companies through annual contributions. Given the volatility inherent in the commodities market, FTFs that rely solely on mining companies for funding may experience shortages in funding during times of low commodity prices. Moreover, while FTFs are governed by a board of trustees, some FTFs lack oversight and accountability, particularly where FTFs exclude host communities and independent third parties from their governance. Even where FTFs involve community participation in their governance, the

216 *Ibid.* at 31.
217 *Ibid.* at 31.
218 *Ibid.* at 32.
219 *Ibid.* at 131.
220 *Ibid.* at 145.

choice of community representation is flawed in the sense that these representatives are not chosen by the community members but by mining companies. In Ghana, for example, traditional rulers and chiefs represent their communities on the governing boards of FTFs but they do "not necessarily seek to advance the interests of community development through the trust funds but, rather, [seek] to exploit their own personal opportunities".[221] The illegitimacy of community representatives creates disaffection within the host community, which ultimately leads to the mistrust of FTFs by members of the community.

Grant-making FTFs are particularly problematic because host communities hardly participate in the decision-making processes of these FTFs. Rather, these FTFs award grants based on what the trustees think is good for the communities as opposed to what the communities think is good for them. It may be that some FTFs exclude host communities from their board of trustees due to incapacity and lack of expertise on the part of host communities. However, the incapacity of host communities ought not to be a predicate for their exclusion from the governance of FTFs. Rather, FTFs should strive to build the capacity of these communities through training and education of indigenes of the communities. The participation of host communities would ensure a buy-in from the communities, thus enhancing the legitimacy of FTFs. As the International Finance Corporation has rightly noted:

> Sustainable development is not possible without a local sense of ownership and empowerment. Many past development failures resulted from a top-down decision-making process. When project interventions are designed without local participation, communities are treated as passive recipients of outside assistance.[222]

Relatedly, the operations and activities of many FTFs in Africa are rarely monitored by independent third parties. Thus, it is difficult to ascertain the effectiveness of the FTFs, especially regarding the execution of policy objectives.

A further concern is that some FTFs in Africa focus on social programmes and services that ought to be provided by governments, thus appearing to usurp the role of governments. For example, the Anglo American Chairman's Fund focuses strongly on education, health, HIV/AIDS, and development programmes in South Africa.[223] While it is commendable for FTFs to provide these services, the downside is that the involvement of FTFs in the provision of social services could further alienate African governments from mining communities and, in extreme cases, exacerbate the unresponsiveness of African governments to the plight and concerns of mining communities.

221 Kasimba & Lujala, *supra* note 213 at 203.
222 International Finance Corporation, *Strategic Community Investment: A Good Practice Handbook for Companies Doing Business in Emerging Markets* (June 2010) at 38, http://documents1.worldbank.org/curated/en/230541468160771028/pdf/577870WP0FINAL101PUBLIC10BOX353774B.pdf
223 World Bank, *Mining Foundations Sourcebook, supra* note 197 at 112.

5.3 Equity participation by host communities

A more pragmatic approach to the sharing of mining revenues is to allow host communities to participate directly in mining operations. Host communities can participate in mining operations by owning equity interest in mining companies operating in the communities or by engaging in joint-venture or partnership arrangements with mining companies.[224] Community equity participation could be undertaken within the context of indigenization of the African mining industry. As discussed in Chapter 3, mining statutes in Africa actively encourage the indigenization of ownership of mineral rights. Relatedly, the government may grant a mining lease on the condition that the lessee shall assign a designated interest in the mining project to local host communities. While significant financial obstacles await host communities wishing to participate directly in mineral exploitation, these obstacles can be avoided by granting a free carried interest in mining operations to host communities.

In South Africa, the Minister of Minerals and Energy may, in granting mining rights, impose such conditions as they deem necessary to promote the rights and interests of host communities, including conditions requiring the participation of host communities in mining projects.[225] The Mining Charter of South Africa requires mineral extractors that acquired their mineral rights after the revised Charter came into effect in 2018 to allot "a minimum of 5% non-transferrable carried interest or a minimum 5% equity equivalent benefit as defined herein to host communities from the effective date of a mining right".[226] To attain the 'equity equivalent benefit', holders of new mining rights must assign 5% equivalent of their issued share capital to the host communities at no cost to the communities or to a trust or similar vehicle for the benefit of host communities.[227]

More significantly, South Africa allows host communities to participate in mining operations through what is referred to as "preferent right". The Minister is vested with power to grant to local communities a preferent right to prospect or mine any mineral and land.[228] To be entitled to a 'preferent right', however, the community must establish that the preferent right will contribute positively to the economic development and social upliftment of the community; submit a development plan indicating the manner in which the preferent right will be exercised; and establish that the envisaged benefits of the prospecting or mining project will accrue to the community.[229] A preferent right is registered in the name of the

---

224 See Emeka Duruigbo, "Community Equity Participation in African Petroleum Ventures: Path to Economic Growth?" (2013) 35 *North Carolina Central Law Review* 111 at 132–8 ["Community Equity Participation"].
225 See *MPRDA*, s. 23.(2A).
226 *Broad-Based Socio-Economic Empowerment Charter for the Mining and Minerals Industry*, 2018, s. 2.1.3.2(ii), www.gov.za/sites/default/files/gcis_document/201809/41934gon1002.pdf
227 *Ibid.* at s. 2.1.4.1.1.
228 *MPRDA*, s. 104.(1).
229 *Ibid.* s. 104.(2).

community and is valid for a period not exceeding five years, although it can be renewed for a further period of five years.[230]

Local communities in South Africa may not have access to the requisite technical and financial resources to exploit and produce minerals, but they can overcome these barriers by engaging in joint-venture and partnership agreements with mining companies. In reality, some host communities in South Africa own equity stake in mining operations within their communities. For example, the Royal Bafokeng Nation in South Africa owns equity in several mining companies, including a 12% equity stake in Impala Platinum (Implats).[231] The Royal Bafokeng Nation conducts its investment activities through a holding company, Royal Bafokeng Holdings (Pty) Limited.[232] As of 2017, the Royal Bafokeng Nation's equity participation in mining operations was reported to be worth R4.3 billion.[233]

The recognition by South Africa that local communities deserve to participate directly in the exploitation of mineral resources is laudable. Community equity participation in mining projects promotes redistribution of mining revenues to host communities. In addition, it reduces the amount of money available for looting by government officials, thus ultimately reducing the level of corruption within South Africa's mining industry.[234] More significantly, community equity participation provides direct financial and economic dividends for host communities. For example, profits realized by Royal Bafokeng Holdings (Pty) Limited are utilized in funding the socio-economic objectives of the Royal Bafokeng Nation, including the provision of social infrastructure such as schools, electricity, roads, and hospitals.[235] In fact, Royal Bafokeng Holdings (Pty) Limited has established the Royal Bafokeng Nation Development Trust, which is responsible for distributing and utilizing the profits arising from the Royal Bafokeng Nation's equity holdings in mining projects.[236]

6 Conclusion

Since the attainment of political independence, African citizens have desired and clamoured for economic development. This desire is predicated primarily on the idea that natural resource endowment could be used as catalyst for economic development in Africa. Unfortunately, Africa's natural resources have consistently been plundered by inept political elites through corruption and mismanagement.

---

230 *Ibid.* s. 104.(1) & (3).
231 Royal Bafokeng Holdings, *Shaped by Our Lineage, Growing Our Legacy: RBH Integrated Review 2018* at 42, www.bafokengholdings.com/images/pdf/rbh-integrated-review-2018.pdf
232 See Royal Bafokeng Holdings, www.bafokeng.com/organisation/entities/rbh
233 Royal Bafokeng Holdings, *Integrated Review 2017* at 19, www.bafokengholdings.com/media/images/reviews/pdf/2017.pdf
234 Duruigbo, "Community Equity Participation", *supra* note 224 at 147–8.
235 Royal Bafokeng Holdings, *Integrated Review 2017*, *supra* note 233 at 7.
236 *Ibid.* at 7.

As discussed in this chapter, the transformation of the economic fortunes of African citizens requires prudent management and utilization of resource revenues. If natural resource endowment is to spur Africa's economic development, resource-rich African countries should, at a minimum, enact statutory provisions on the prudent management and utilization of resource revenues; ensure accountability in the management of resource revenues by establishing oversight mechanisms; and make full disclosure of all resource revenue receipts, allocation, and utilization. In appropriate cases, African countries should afford resource-producing communities direct participation in mineral exploitation through a free carried interest in mining operations or by granting these communities licences and leases to exploit mineral resources. These measures are not only likely to promote social stability in African countries, but could also promote a conducive investment climate in Africa.

9 Mining and the environment

1 Introduction

This chapter analyzes the environmental and human rights impacts of mining operations in Africa and considers the degree to which African countries regulate the environmental activities of mining companies. Mining involves the physical disturbance of the earth's surface through excavation, drilling, and blasting. Thus, mining activities often lead to adverse environmental impacts, including pollution and degradation of the ecosystem. As some observers have noted, "mining activities accelerate the rate and degree of changes in the natural environment" and have long-term impacts on local communities "due to their physical degrading nature, as well as their use of chemicals and other harmful substances".[1] Although African countries have enacted specific rules governing the environmental practices of mining companies, these rules are ineffective because of the incapacity of regulatory agencies to enforce the rules. The ineffectiveness of domestic regulation has prompted some African citizens to resort to transnational tort litigation in the home countries of the MNCs operating in the African mining industry. These cases seek to hold parent companies based in Western countries liable for the environmental and human rights violations committed by their subsidiaries in Africa. As noted in this chapter, while these cases have yet to lead to direct liability on the part of parent companies, a recent decision by the United Kingdom Supreme Court raises the spectre of such liability.

2 Complicity of mining companies in environmental and human rights violations

Mining companies in Africa often engage in environmental degradation and pollution, and they are sometimes complicit in human rights and labour rights violations.[2] Mining activities such as blasting and crushing cause air pollution, which

---

1 UNECA & African Union, *Minerals and Africa's Development: The International Study Group Report on Africa's Mineral Regimes* (Addis Ababa: UNECA, 2011) at 46.
2 Evaristus Oshionebo, *Regulating Transnational Corporations in Domestic and International Regimes: An African Case Study* (Toronto: University of Toronto Press, 2009) at 19–25 [*Regulating TNCs*].

affects the respiratory health of local communities. In addition, mining operations sometimes lead to the contamination of rivers and streams on which local communities depend for drinking water. For example, in 2015 the Supreme Court of Zambia found that Konkola Copper Mines Plc illegally discharged effluent containing high acidic content into streams and rivers, including the Kafue river, as a result of which many residents of the area who drank water from the polluted streams and rivers suffered from diseases such as diarrhoea.[3] The contamination of rivers and streams is especially detrimental to the wellbeing of local communities not only because it causes ill-health, but also because it depletes fish stocks, thus depriving local communities of the opportunity to sustain themselves through fishing. Environmental degradation arising from mining operations is common in Africa because mining companies "frequently employ the cheapest option for exploration and processing" in order to curtail cost and enhance their profits.[4] Mining companies in Africa engage in waste disposal involving "releasing tailings and sludge directly into the sea or into rivers, with some companies failing to invest in the technologies needed to lower environmental impacts".[5]

Mining companies in Africa have also been alleged to engage in illegal labour practices such as compelling employees to work under unsafe conditions. For example, some Chinese mining companies in Zambia require their workers to work under hazardous "conditions for lengths of time that extend beyond what is permissible under Zambian law, or risk being fired".[6] Employees of some Chinese mining companies in Zambia "work 12-hour shifts, compared to the eight-hour shifts outlined in Zambian law and standard in every other copper mining and processing operation in the country".[7] In other cases, mining companies may be complicit in human rights violations by entering into joint-venture and partnership relationships with dictatorial governments in Africa. Some of these governments compel and conscript their citizens to work at joint-venture mines in deplorable and inhumane conditions. For example, the Eritrean government allegedly conscripted some Eritrean citizens to build and work a mine jointly owned by the government and Nevsun Resources Ltd, a Canadian mining company.[8] The private security personnel employed or hired by mining companies sometimes engage

---

3 *Konkola Copper Mines Plc v Nyasulu & others*, Appeal No. 1/2012; [2015] ZMSC 33, https://zambialii.org/zm/judgment/supreme-court-zambia/2015/33
4 Africa Progress Panel, *Equity in Extractives: Stewarding Africa's Natural Resources for All*, at 87, https://reliefweb.int/sites/reliefweb.int/files/resources/relatorio-africa-progress-report-2013-pdf-20130511-125153.pdf
5 *Ibid.* at 87
6 Human Rights Watch, "'You'll be Fired if You Refuse': Labour Abuses in Zambia's Chinese State-owned Copper Mines", at 4, www.hrw.org/sites/default/files/reports/zambia1111ForWebUpload.pdf
7 *Ibid.* at 4.
8 *Nevsun Resources Ltd v Araya*, 2020 S.C.C. 5.

in brutal repression of the human rights of local indigenes in apparent attempts to protect the physical assets of mining companies.[9]

In more extreme cases, mining operations lead to the involuntary displacement of local communities resulting in the disruption of lives, cultures, and traditions. For example, in Tanzania, pastoral and hunter-gatherer communities have been forcibly evicted from their ancestral lands in order to enable foreign companies to exploit minerals.[10] In Botswana, the Basarwa community was involuntarily displaced to make way for mining.[11] Likewise, in Zimbabwe, local communities in the Marange communal area have been forcibly displaced and relocated to enable diamond mining in the area.[12] Similarly, the Mbada diamond-mining company is projected to have displaced about 600 households in Chiadzwa, a ward located in the Mutare district of Zimbabwe.[13] Moreover, mining companies are sometimes involved in the illicit trade in conflict minerals, especially through the purchase of minerals from local warlords in Africa.[14] As the Africa Progress Panel has reported, "extractive industry investments are frequently associated with conflicts sparked by the displacement of local communities, or by local grievances".[15] In fact, many of Africa's most brutal civil wars, including the wars in Angola, the DRC, Liberia, and Sierra Leone, were sustained by mineral revenues.[16]

Instances of MNC complicity in environmental pollution and human rights violations are not confined to Africa and have been reported in other parts of the developing world. The Special Representative of the UN Secretary-General on

---

9 Cynthia Kwakyewah & Uwafiokun Idemudia, "Canada-Ghana Engagements in the Mining Sector: Protecting Human Rights or Business as Usual?" (2017) 4 *The Transnational Human Rights Review* 146 at 158.
10 Pacifique Manirakiza, "Asserting the Principle of Free, Prior and Informed Consent (FPIC) in Sub-Saharan Africa in the Extractive Industry Sector", in Isabel Feichtner, Markus Krajewski, & Ricarda Roesch, eds, *Human Rights in the Extractive Industries: Transparency, Participation, Resistance* (Cham, Switzerland: Springer, 2019) 219 at 224.
11 Motsomi Ndala Marobela, "The State, Mining and the Community: The Case of Basarwa of the Central Kalahari Game Reserve in Botswana" (2010) 43(1) *Labour, Capital and Society* 137 at 145–6.
12 Tumai Murombo, "Regulating Mining in South Africa and Zimbabwe: Communities, the Environment and Perpetual Exploitation" (2013) 9(1) *Law, Environment and Development Journal* 31 at 39.
13 Crescentia Madebwe, Victor Madebwe, & Sophia Mavusa, "Involuntary Displacement and Resettlement to Make Way for Diamond Mining: The Case of Chiadzwa Villagers in Marange, Zimbabwe" (2011) 1(10) *Journal of Research in Peace, Gender and Development* 292 at 293.
14 See United Nations, *Final Report of the Panel of Experts on the Illegal Exploitation of Natural Resources and Other Forms of Wealth of the Democratic Republic of the Congo*, UN Doc. No. S/2002/1146. See also *Final Statement by the UK National Contact Point for the OECD Guidelines for Multinational Enterprises: Afrimex (UK) Ltd*, www.oecd.org/investment/mne/43750590.pdf
15 Africa Progress Panel, *supra* note 4 at 87.
16 *Ibid.* at 87.

320 *Mining and the environment*

Human Rights and Transnational Corporations observed in 2006 that human rights abuse by MNCs were reported in 27 developing countries, with "the extractive sector – oil, gas, and mining – utterly [dominating] this sample of reported abuses, with two-thirds of the total".[17] Moreover, extractive industries accounted for

> most allegations of the worst abuses, up to and including complicity in crimes against humanity ... typically for acts committed by public and private security forces protecting company assets and property; large-scale corruption; violations of labor rights; and a broad array of abuses in relation to local communities, especially indigenous people.[18]

3 Environmental regulation of mining operations

3.1 Environmental standards

The environmental standards and rules governing mining operations in Africa are specified in mining statutes and generic environmental protection statutes, as well as mining contracts.[19] As discussed in Chapter 3, mineral rights in Africa are granted subject to certain standard conditions, including conditions regarding environmental protection. Generally speaking, mineral rights holders in Africa are obliged to conduct mining activities in a safe, efficient, and environmentally responsible manner. For example, in Botswana:

> The holder of a mineral concession shall, in accordance with the law in force from time to time in Botswana and in accordance with good mining industry practice, conduct his operations in such manner as to preserve in as far as is possible the natural environment, minimize and control waste or undue loss of or damage to natural and biological resources, to prevent and where unavoidable, promptly treat pollution and contamination of the environment and shall take no steps which may unnecessarily or unreasonably restrict or limit further development of the natural resources of the concession area or adjacent areas.[20]

---

17 United Nations, *Interim Report of the Special Representative of the Secretary-General on the Issue of Human Rights and Transnational Corporations and Other Business Enterprises*, U.N. Doc. E/CN.4/2006/97 (22 February 2006), paras 25 & 27, http://hrli brary.umn.edu/business/RuggieReport2006.html
18 *Ibid.* at para. 25.
19 See, for example, South Africa's *National Environmental Management Act 107 of 1998*; Ghana's *Environmental Protection Agency Act* (No. 490 of 1994); Nigeria's *National Environmental Standards and Regulations Enforcement Agency (Establishment) Act, 2007*; and Zambia's *Environmental Management Act, 2011*.
20 *Mines and Minerals Act, 1999* (Botswana), s. 65.(1).

Likewise, in South Africa mining operations

> must be conducted in accordance with generally accepted principles of sustainable development by integrating social, economic and environmental factors into the planning and implementation of prospecting and mining projects in order to ensure that exploitation of mineral resources serves present and future generations.[21]

The environmental protection regimes in Africa prohibit mineral right holders from polluting the environment, including contamination of rivers and streams. In Nigeria, "no person shall, in the course of mining or exploration for minerals, pollute or cause to be polluted any water or watercourse in the area within the mining lease or beyond that area".[22] In Zambia, a "person shall not, without a licence, discharge, cause or permit the discharge of, a contaminant or pollutant into the environment if that discharge causes, or is likely to cause, an adverse effect".[23] Similarly, in Kenya the holder of a mineral right shall ensure

> that the seepage of toxic waste into streams, rivers, lakes and wetlands is avoided and that disposal [of] any toxic waste is done in the approved areas only; that blasting and all works that cause massive vibration is properly carried out and muffled to keep such vibrations and blasts to reasonable and permissible levels in conformity with the Environmental Management and Coordination Act.[24]

Mineral rights holders are also obliged to mitigate the adverse environmental impacts of their activities and rehabilitate and reclaim mine sites. For example, South Africa requires:

> Every person who causes, has caused or may cause significant pollution or degradation of the environment must take reasonable measures to prevent such pollution or degradation from occurring, continuing or recurring, or, in so far as such harm to the environment is authorised by law or cannot reasonably be avoided or stopped, to minimise and rectify such pollution or degradation of the environment.[25]

In Nigeria, holders of mineral rights shall minimize, manage, and mitigate any environmental impact arising from mining operations; and rehabilitate and reclaim land disturbed, excavated, explored, mined, or covered with tailings to its natural

---

21 *MPRDA*, s. 37.(2).
22 *Nigerian Minerals and Mining Act, 2007*, s. 123.
23 *Environmental Management Act, 2011* (Zambia), s. 32.(1).
24 *Mining Act, 2016* (Kenya) s. 179.
25 *National Environmental Management Act 107 of 1998* (South Africa), s. 28.(1).

or predetermined state and in accordance with established best practices.[26] Botswana enjoins holders of mineral concessions to

> take such measures as are required from time to time to maintain and restore the top soil of affected areas and otherwise to restore the land substantially to the condition in which it was prior to the commencement of operations.[27]

Similarly, Kenya requires mineral right holders to ensure sustainable use of land, including restoration of mine sites.[28] Mineral right holders may be required to post a rehabilitation bond to finance the cost of rehabilitating mining-affected lands in the event that the mineral right holder fails to rehabilitate the lands on closure of their mines.[29] In other instances, mineral right holders are required to contribute to an environmental protection fund designed to ensure the reclamation and rehabilitation of mine sites.[30]

In some African countries, compliance with environmental laws is a condition precedent to the grant and renewal of mineral rights. For example, in Kenya a mining licence is not granted unless the applicant "has obtained an environmental impact assessment licence, social heritage assessment and the environmental management plan has been approved".[31] In addition, Kenya requires the submission of a mitigation and rehabilitation plan as a condition precedent to the grant of a mineral right.[32] In Zambia, mineral rights are granted or renewed subject to certain conditions intended to ensure that the holder observes their environmental and health and safety responsibilities, including conditions relating to the conservation and protection of the environment; the protection of human health; and the rehabilitation and reclamation of mining lands.[33] Mineral right holders are also required to conduct an environmental impact assessment and obtain requisite environmental approvals prior to the commencement of mining operations. In Ghana, holders of mineral rights must obtain the necessary approvals and permits from the Forestry Commission and the Environmental Protection Agency prior to commencing mining operations.[34] Similarly, both Nigeria and Zambia prohibit holders of mineral rights from commencing exploration, development work, or excavation of mineral resources without obtaining from the environmental management agency a written approval of the environmental impact assessment relating to the mining project.[35]

---

26 *Nigerian Minerals and Mining Act, 2007*, s. 118.
27 *Mines and Minerals Act*, (Botswana), s. 65.(4).
28 *Mining Act, 2016* (Kenya), s. 179.
29 See, for example, *Mining Act, 2010* (Tanzania), s. 47.(f).
30 See, for example, *Mines and Minerals Development Act, 2015* (Zambia), s. 86.
31 *Mining Act, 2016* (Kenya), s. 176.(2).
32 *Ibid.* s. 180.
33 *Mines and Minerals Development Act, 2015* (Zambia), s. 81.(1).
34 *Minerals and Mining Act, 2006* (Ghana), s. 18.
35 *Nigerian Minerals and Mining Act, 2007*, s. 71.(1); *Mines and Minerals Development Act, 2015* (Zambia), s. 12.(2).

Mining companies are obliged to pay compensation to persons and communities adversely impacted by their harmful environmental practices.[36] Unlike other common law African countries where environmental liability is based on a statutory duty of care or the tort of negligence, Zambia operates a strict liability regime for environmental damage caused by mining activities. Thus, holders of mineral rights in Zambia are "strictly liable for any harm or damage caused by mining operations or mineral-processing operations and shall compensate any person to whom the harm or damage is caused".[37] Under Zambia's strict liability regime, liability attaches primarily to the person who directly causes or contributes to the act or omission which results in the environmental harm or damage, but liability is joint and several where more than one person is responsible for the harm or damage.[38] The strict liability regime obviates the common law requirement of proof of a duty of care, causation, and loss. This is significant because in common law countries where environmental compensation is based on the tort of negligence, it is difficult for plaintiffs to prove that the loss or injury they sustained was caused by mining operations. Plaintiffs in Africa often lack the scientific evidence to prove not only the negligence of mining companies in causing environmental pollution, but that such negligence caused a loss or injury to the plaintiff. Hence, many negligence-based cases filed against mining companies in Africa are unsuccessful.

As a result of the difficulty in proving negligence in the context of environmental pollution, Tanzania recently introduced a 'no fault' liability regime under which mining companies are "liable for pollution damage without regard to fault".[39] The 'no fault' liability regime in Tanzania applies even in relation to events that would ordinarily constitute *force majeure*. More specifically,

> Where it is demonstrated that an inevitable event of nature, act of war, exercise of relevant Commission or a similar event of act of God has contributed to a considerable degree to the damage or its extent under circumstances, which are beyond the control of the licence holder or contractor, the liability may be reduced to the extent that is reasonable, with particular consideration to the-
> (a) scope of the activity;
> (b) situation of the party that has sustained the damage; and
> (c) opportunity for taking out insurance on both sides.[40]

In some African countries, the contravention of environmental standards is an offence punishable with a fine, imprisonment, or both. For example, in Nigeria the discharge of harmful quantities of any hazardous substance into the air or upon

---

36 See, for example, *Mines and Minerals Development Act, 2015* (Zambia), s. 87.
37 *Ibid.* s. 87(1).
38 *Ibid.* s. 87.(2) & (3).
39 *Mining Act, 2010* (Tanzania), s. 109.(1) (incorporated into the *Mining Act, 2010* by *The Written Laws (Miscellaneous Amendments) Act, 2017* (No. 7)).
40 *Ibid.* s. 109.(2).

the land and the waters of Nigeria or at the adjoining shorelines is an offence punishable with a fine or imprisonment for a term not exceeding five years.[41] Likewise, in Zambia the contravention of environmental standards, the unauthorized disturbance of the habitat of a biological resource, and the unauthorized discharge of hazardous waste materials are offences punishable with a fine or imprisonment or both.[42] Liability for environmental offences is not confined to mining companies but extends to the directors of such companies. For example, in South Africa company directors are personally liable for an environmental offence "if the offence in question resulted from the failure of the director to take all reasonable steps that were necessary under the circumstances to prevent the commission of the offence".[43] In fact, in South Africa proof that a mining company committed an environmental offence constitutes prima facie evidence that its directors are guilty of the offence.[44]

3.2 Environmental rights as human rights

The constitutions of some African countries expressly recognize environmental rights as human rights, thus providing broad protection for citizens in relation to the harmful environmental impacts of mining. For example, the Bill of Rights enshrined in the Constitution of South Africa encompasses the right to a clean and healthy environment, as follows:

> Everyone has the right – (a) to an environment that is not harmful to their health or wellbeing; and (b) to have the environment protected, for the benefit of present and future generations, through reasonable legislative and other measures that – (i) prevent pollution and ecological degradation; (ii) promote conservation; and (iii) secure ecologically sustainable development and use of natural resources while promoting justifiable economic and social development.[45]

Similarly, the Constitution of Kenya provides:

> Every person has the right to a clean and healthy environment, which includes the right – (a) to have the environment protected for the benefit of present and future generations through legislative and other measures, particularly those contemplated in Article 69; and (b) to have obligations relating to the environment fulfilled under Article 70.[46]

---

41 *National Environmental Standards and Regulations Enforcement Agency (Establishment) Act, 2007*, s. 27.
42 *Environmental Management Act, 2011* (Zambia), ss. 119–23; *Mines and Minerals Development Act, 2015* (Zambia), s. 111.
43 *National Environmental Management Act 107 of 1998* (South Africa), s. 34.(7).
44 *Ibid.* s. 34.(7).
45 *The Constitution of the Republic of South Africa, 1996*, s. 24.
46 *The Constitution of Kenya, 2010*, s. 42.

Among other things, Article 69 of the Constitution enjoins the government of Kenya to "ensure sustainable exploitation, utilization, management and conservation of the environment and natural resources, and ensure the equitable sharing of the accruing benefits", while Article 70 empowers Kenyan citizens to enforce environmental rights.

Even in countries such as Nigeria where the Constitution does not expressly provide for environmental rights, the undeniable link between the environment and human rights such as the right to life creates a platform for advocating for the constitutional recognition of environmental rights. Interestingly, unlike Western countries where constitutions apply exclusively to state actors such as governments and public agencies, the constitutions of some African countries apply to state actors (governments and public agencies) and non-state actors such as private individuals and companies. For example, the provisions of the Nigerian Constitution apply to governments, private persons, companies, and private organizations. Thus, where a mining company violates the fundamental human rights guaranteed under the Nigerian Constitution, the victim can sue the company to remedy the wrong.[47]

At the continental level, the *African Charter on Human and Peoples' Rights* [48] (African Charter) does not expressly apply to private corporate entities, but it establishes important standards for regulating the environmental practices of mining companies. The African Charter guarantees the right to life (Article 4); the right to "enjoy the best attainable state of physical and mental health" (Article 16); and the right to a clean and satisfactory environment (Article 24). These provisions can be enforced against African governments where they fail to take appropriate measures to regulate the environmental practices of mining companies. The African Commission on Human and Peoples' Rights (the Commission) has held that Article 24 of the African Charter imposes an obligation on African governments "to take reasonable and other measures to prevent pollution and ecological degradation, to promote conservation, and to secure an ecologically sustainable development and use of natural resources".[49] According to the Commission, governments desirous of complying with Article 24 of the African Charter must take several steps, including:

> ordering or at least permitting independent scientific monitoring of threatened environments, requiring and publicising environmental and social impact studies prior to any major industrial development, undertaking appropriate monitoring and providing information to those communities exposed to hazardous materials and activities and providing meaningful opportunities

---

47 See *Jonah Gbemre v Shell Petroleum Development Company Nigeria Ltd* (2005) AHRLR 151.
48 www.achpr.org/legalinstruments/detail?id=49
49 *Social and Economic Rights Action Center & Another v Nigeria* (155/96, October 27, 2001) at para. 52, www.achpr.org/sessions/descions?id=134

for individuals to be heard and to participate in the development decisions affecting their communities.[50]

The Commission has equally emphasized that African "[g]overnments have a duty to protect their citizens, not only through appropriate legislation and effective enforcement but also by protecting them from damaging acts that may be perpetrated by private parties".[51] Thus, any African country that fails to take reasonable regulatory steps to prevent mining companies from polluting and degrading the environment acts in violation of Articles 16 and 24 of the African Charter.[52]

3.3 Mine closure and rehabilitation standards

Mine closure may occur on the lapsing, abandonment, or termination of a mineral right; the cessation of prospecting or exploration activities; the relinquishment of a portion of the land covered by a mineral right; the cessation of mining operations; and on completion of mining operations.[53] A common feature of the African mining industry is the non-rehabilitation of mines upon closure of the mines. The non-rehabilitation of mine sites is prevalent in Africa even though domestic laws require mineral right holders to reclaim and rehabilitate lands upon cessation of mining operations. In Zambia, for example, mineral right holders are obliged, within six months of the cessation of mining operations,

> [to] cause to be removed from the land, on the surface or underground, any mining or mineral processing plant brought onto, or erected upon that land in the course of mining or mineral processing operations carried out under the mining right or mineral processing licence.[54]

Upon completion of mining operations, mineral right holders in Kenya must restore the land "to its original status or to an acceptable and reasonable condition as close as possible to its original state".[55] Nigeria requires mineral right holders to rehabilitate and reclaim any land disturbed, excavated, explored, or mined to its natural or predetermined state.[56] Hence, prior to the commencement of mining operations, mineral right holders must submit for approval an environmental protection and rehabilitation programme encompassing specific rehabilitation and reclamation actions, a reasonable estimate of the total cost of rehabilitation, and a timetable for rehabilitation and reclamation of the land to a safe and

---

50 *Ibid.* at para. 53.
51 *Ibid.* at para. 57.
52 *Ibid.* at paras 50–4.
53 See *MPRDA*, s. 43.(3).
54 *Mines and Minerals Development Act, 2015* (Zambia), s. 82.
55 *Mining Act, 2016* (Kenya), s. 179.
56 *Nigerian Minerals and Mining Act, 2007*, s. 118.

environmentally sound condition suitable for future economic development or recreational use.[57]

In some African countries, mineral rights are granted on specific environmental conditions, including conditions relating to mine rehabilitation. In Kenya, mineral rights are not granted unless the applicant submits "a site mitigation and rehabilitation or mine closure plans for approval".[58] In Zambia, the conditions under which a mineral right is granted or renewed include conditions regarding

> the rehabilitation, levelling, regrassing, reforesting or contouring of such part of the land over which the right or licence has effect as may have been damaged or adversely affected by exploration operations, mining operations or mineral processing operations; and the filling in, sealing or fencing of excavations, shafts and tunnels.[59]

Similarly, South Africa requires an environmental authorization as a condition precedent to the grant of mineral rights. The environmental authorization contains conditions regarding the rehabilitation of mine sites in accordance with the prescribed mine closure plan.[60] Moreover, African countries require the environmental impact assessment (EIA) of mining projects to include a detailed plan regarding mine closure and rehabilitation.

The mine closure process terminates when the government is satisfied that the mineral right holder has complied with requisite statutory and regulatory standards regarding rehabilitation of mine sites. The government may issue a formal approval or a certificate to the mineral right holder attesting that the holder has complied with statutory requirements regarding rehabilitation of mine sites. In South Africa, for example, a mineral right holder must apply for a closure certificate upon the lapsing, abandonment, cessation, or completion of mining operations. However, a closure certificate is not issued

> unless the Chief Inspector and each government department charged with the administration of any law which relates to any matter affecting the environment have confirmed in writing that the provisions pertaining to health and safety and management pollution to water resources, the pumping and treatment of extraneous water and compliance to the conditions of the environmental authorisation have been addressed.[61]

While the statutory obligations of mineral right holders to rehabilitate mine sites are clear and unambiguous, in reality many closed mines in Africa are unrehabilitated. Apart from the environmental degradation associated with unrehabilitated

---

57 *Ibid.* ss. 119 & 120.(1).
58 *Mining Act, 2016* (Kenya), s. 180.
59 *Mines and Minerals Development Act, 2015* (Zambia), s. 81.(1)(c) & (d).
60 MPRDA, ss. 38A & 43.
61 *Ibid.* s. 43.(5).

mines, these mines are hazardous to the health and wellbeing of host communities because they retain toxic and acidic liquid which, in some cases, spill on to the surface of land and into rivers.

4 Non-enforcement of environmental standards

The environmental standards prescribed by statutes in Africa are comparable to the standards in developed Western countries. However, environmental standards in Africa have failed to produce positive outcomes due to lack of enforcement. These environmental standards are not vigorously enforced; hence, mining companies in Africa seldom observe the environmental standards. Regulatory agencies in Africa rarely prosecute MNCs for environmental offences even when there is evidence that such offences have been committed. Moreover, environmental infractions are neither detected nor punished, while contaminated sites are hardly rehabilitated or cleaned up, even though mining statutes impose clear obligations on mineral right holders to rehabilitate contaminated sites. As discussed next, this regulatory failure is attributable to the incapacity of Africa's regulatory agencies to enforce environmental standards.

4.1 Incapacity of regulatory agencies

Public agencies responsible for enforcing environmental standards in Africa are vested with a plethora of powers as well as enforcement and compliance tools. Regulatory agencies have the power to inspect and search premises; request production of documents; arrest and seize materials and premises suspected to have contravened environmental standards; and issue enforcement orders such as cease and desist orders, compliance orders, and remedial orders. Some regulatory agencies in Africa possess quasi-judicial power such as the power to "request a person to attend at a time and place" to provide information to the agency.[62] Regulatory agencies can also impose sanctions on defaulting companies, including fines, withholding of regulatory approvals pending environmental compliance, and suspension and revocation of mineral rights. In Ghana, for example, the Environmental Protection Agency can order immediate cessation of mining operations where it determines that the operations contravene environmental standards.[63] Above all, regulatory agencies can institute criminal proceedings against companies and directors for environmental offences.

Although, as discussed above, African countries have clear provisions regarding environmental standards, and regulatory agencies in Africa have a wide variety of enforcement tools, environmental standards are rarely enforced by regulatory agencies. Hence, for the most part, mining companies seldom observe environmental standards in Africa. The lack of regulatory enforcement in Africa can be traced to a myriad of factors, including the incapacity and inefficiency of regulatory

62 See *Environmental Protection Agency Act* No. 490 (Ghana), s. 27.(1).
63 *Ibid.* s. 13.(2).

agencies; inadequate funding of regulatory agencies; corruption; lack of political will to enforce laws and regulations against mining companies; lack of independence on the part of regulatory agencies; and the complicity of African governments in mining-related environmental degradation through state participation in mineral extraction projects.[64]

Of these factors, the lack of regulatory capacity is the most profound. As some observers have noted, capacity is at the heart of the regulatory crisis in Africa as "many governments simply lack the technical capacity and information required to act".[65] The regulatory incapacity in Africa is compounded by the fact that mining and other extractive activities are technology-driven. Most African countries lack the equipment and expertise to effectively regulate the technically sophisticated operations of MNCs in the extractive industries.[66] Regulatory agencies seldom conduct environmental inspection and audit of mining operations. In most instances, mine inspectors lack financial and technical resources,[67] including equipment and laboratories to undertake proper scientific measurements, and therefore they are often unable to detect violation of environmental standards.[68] Furthermore, regulatory agencies in Africa are understaffed and overburdened due to the broad environmental mandate bestowed on these agencies, including enforcement of environmental standards, monitoring of environmental practices, and the setting of environmental policies. The Africa Progress Panel has reported:

> Sierra Leone's Environmental Protection Agency (EPA-SL) illustrates the capacity problem. Established as a self-standing agency reporting directly to the president's office, the EPA-SL had a 2010 budget of US$150,000 a year, with nine staff in three cramped rooms. Given such limited resources, carrying out its broad mandate – setting environmental standards, monitoring the impacts of all activities nationwide and mainstreaming environmental priorities across government – was barely possible. With just one technical expert responsible for reviewing all environmental impact assessments on a part-time basis, there was unsurprisingly a backlog of more than 200 EIAs pending review. While the agency's capacity has increased, its reach and effectiveness remain limited – and the challenges faced by the organization are common to those shared in many other countries across Africa.[69]

Moreover, regulatory agencies in Africa lack the political will to enforce laws and regulations against companies in the extractive industries because these companies

---

64 Oshionebo, *Regulating TNCs*, *supra* note 2 at 71–8.
65 Africa Progress Panel, *supra* note 4 at 96.
66 Jedrzej G. Frynas, *Beyond Corporate Social Responsibility: Oil Multinationals and Social Challenges* (Cambridge: Cambridge University Press, 2009) at 88.
67 *Ibid*. at 61.
68 Kwame A. Domfeh, "Compliance and Enforcement in Environmental Management: A Case of Mining in Ghana" (2003) 5(2) *Environmental Practice* 154 at 160.
69 Africa Progress Panel, *supra* note 4 at 87.

are considered by African governments to be vital to their national economies.[70] As discussed in Chapter 7, African countries are increasingly participating in mineral extraction through equity participation, joint ventures, and partnerships with mining companies. State participation in mineral extraction hinders the ability of regulatory agencies to regulate mining companies because it makes these agencies susceptible to interference from the government. African governments may interfere with regulatory agencies where they perceive that regulation will adversely affect joint-venture projects.[71] In essence, because the economic interests of mining companies align with those of African governments that engage in equity participation with mining companies, African governments are incentivized to ignore environmental violations by these companies. While a few African countries, such as South Africa, can be said to possess a limited degree of regulatory expertise, these countries are fearful that stringent regulation could dissuade MNCs from investing in their countries, and therefore they generally refrain from regulating the activities of MNCs.

With regard to mine closure and rehabilitation, the non-rehabilitation of mine sites can be situated within the context of the structural and institutional deficiencies in Africa. Africa lacks the resources and personnel to successfully implement mine closure standards and regulations. Even South Africa, a country that is more resourced than most African countries, is said to have a "shortage of relevant mine closure skills and knowledge within" the regulatory agencies.[72] In addition, there is apparent role conflict among regulatory agencies with regard to enforcement of mine closure standards. Mine closure involves complex issues that are regulated by a multitude of statutes, including mining statutes, environmental statutes, water statutes, and nuclear energy statutes. Given that these statutes are enforced by different regulatory agencies, the mine closure process usually involves a multitude of government departments and agencies "with overlapping requirements and different interpretations of the law".[73] The conflation of regulatory responsibilities not only creates conflicts between agencies, but unnecessarily renders the mine closure process cumbersome, thus adding to the cost of mine closures. Moreover, because mine closures involve considerable costs, mining companies in Africa are not motivated to successfully rehabilitate and close their mines.[74]

---

70 See Engobo Emeseh, "Limitations of Law in Promoting Synergy between Environment and Development Policies in Developing Countries: A Case Study of the Petroleum Industry in Nigeria" (2006) 24(4) *Journal of Energy & Natural Resources Law* 574 at 606.
71 Oshionebo, *Regulating TNCs, supra* note 2 at 75.
72 E.S. van Druten & M.C. Bekker, "Towards an Inclusive Model to Address Unsuccessful Mine Closures in South Africa" (2017) 117(5) *Journal of the South African Institute of Mining and Metallurgy* 485 at 489.
73 I. Watson & M. Olalde, "The State of Mine Closure in South Africa – What the Numbers Say" (2019) 119 *Journal of the South African Institute of Mining and Metallurgy* 639.
74 *Ibid.* at 639.

There is also the issue of 'outsourcing' of mine closure obligations by some mining companies. Large mining companies in Africa sometimes sell their mines to lower-cost producers in order to avoid the cost of mine rehabilitation. Mining companies do so when they determine that a mine is only marginally profitable and hence a low-cost producer may be better able to profit from the mine. While such sales are legal, the problem is that the purchasers are often "less well-resourced companies" and hence are less able to rehabilitate the land upon closure of the mine. Moreover, such sales enable mining companies to abdicate their responsibilities to rehabilitate the mine sites. That said, some African countries appear to have partially resolved this issue by providing that the sale or transfer of a mineral right shall not extinguish the obligations of the mineral right holder that accrued while they held the mineral right. Overall, the inability of regulatory agencies to ensure that mining companies rehabilitate mine sites means that African governments are ultimately saddled with responsibility to rehabilitate these sites, including, of course, the financial cost of rehabilitation and reclamation of closed mines.

To forestall mine closure problems, Africa must adopt responsive mine closure policies that incorporate international best practices, including the planning for mine closure from the inception of mining projects and "throughout the life of the mine until final closure and relinquishment".[75] Mine closure plans must be required as part of the criteria for the grant of mineral rights. In addition, applicants for mineral rights in Africa should be required to submit evidence regarding the successful implementation of mine closure plans in the past. An applicant with a poor record regarding past mine closure obligations should be denied a new mineral right unless they successfully reclaim closed mines. Furthermore, African countries should rigorously scrutinize transferees and assignees of mineral rights to determine whether they have the technological and financial "ability to fulfil [mine] closure commitments already provided by the transferor".[76] Finally, the incorporation of mine closure obligations into mining contracts could encourage mining companies to rehabilitate contaminated mine sites in Africa. As the International Bar Association has recommended, mining contracts should contain express provisions regarding mine closure obligations, mine closure plans, as well as the mineral right holder's guarantees regarding mine closure expenses.[77]

5 Citizen enforcement of environmental standards

The incapacity of regulatory agencies in Africa makes a compelling case for the diversification of regulatory enforcement, including private enforcement of environmental standards. Citizen enforcement of environmental standards

---

75 Nwaka Nakazwe, "Life beyond the Glitz and Glamour of Mining: Strengthening the Mine Closure Regime in Zambia" (2017) 35(3) *Journal of Energy & Natural Resources Law* 325 at 333.
76 *Ibid.* at 334.
77 International Bar Association, *Model Mineral Development Agreement*, at 119–20, www.extractiveshub.org/servefile/getFile/id/1256

332 *Mining and the environment*

complements the work of public regulatory agencies and helps to fill the regulatory void arising from the non-enforcement of regulatory standards in Africa. Private citizens sometimes possess the expertise that regulatory agencies in Africa lack. Moreover, the involvement of citizens in the enforcement of environmental standards could lessen the burden on regulatory agencies in Africa and free up scarce resources which could then be devoted to other areas of need.[78] In this regard, South Africa, Kenya, and Zambia have enacted statutory and constitutional provisions enabling private enforcement of environmental standards. The Constitution of the Republic of South Africa grants citizens "the right to approach a competent court, alleging that a right in the Bill of Rights has been infringed or threatened".[79] As mentioned previously, one of the rights guaranteed under the Bill of Rights in both South Africa and Kenya is the right to a clean and healthy environment, and hence citizens of both countries can enforce environmental laws provided they can establish that the alleged infringement directly impacts their right to a clean and healthy environment. The constitutional provision empowering South African citizens to enforce environmental laws is reinforced by the *National Environmental Management Act, 1998*, which empowers any person or group of persons to seek appropriate relief in respect of any breach or threatened breach of any provision of the Act, or of any provision of a specific environmental management Act, or of any other statutory provision concerned with the protection of the environment or the use of natural resources.[80] In addition, South Africa allows private prosecution of environmental offences.[81] Thus, South African citizens can institute proceedings against mining companies for environmental offences where public regulatory agencies fail to institute such proceedings against the companies. In the case of Kenya, the Constitution provides:

> If a person alleges that a right to a clean and healthy environment recognised and protected under Article 42 has been, is being or is likely to be, denied, violated, infringed or threatened, the person may apply to a court for redress in addition to any other legal remedies that are available in respect to the same matter.[82]

Similarly, Zambia accords its citizens the right to enforce environmental standards and obligations.[83] The *Mines and Minerals Development Act, 2015* provides that:

> A person, group of persons or a private or State organisation may bring a claim and seek redress in respect of the breach or threatened breach of any

---

78 Oshionebo, *Regulating TNCs, supra* note 2 at 217.
79 *The Constitution of the Republic of South Africa, 1996*, s. 38.
80 *National Environmental Management Act, 1998*, s. 32.(1).
81 *Ibid*. s. 33.
82 *The Constitution of Kenya, 2010*, s. 70.(1).
83 See *Mines and Minerals Development Act, 2015* (Zambia), s. 87.(7); *Environmental Management Act, 2011* (Zambia), ss. 108–10.

provision relating to damage to the environment, biological diversity, human and animal health or to socio-economic conditions –
(a) in that person's or group of persons' interest;
(b) in the interest of or on behalf of, a person who is, for practical reasons, unable to institute such proceedings;
(c) in the interest of, or on behalf of, a group or class of persons whose interests are affected;
(d) in the public interest; and
(e) in the interest of protecting the environment or biological diversity.[84]

The constitutional and statutory regimes in South Africa, Kenya, and Zambia could potentially enhance the regulation of mining companies, given that they cast a wide net with regard to the persons who may institute public interest litigation. Public interest litigants in these countries include persons acting in their own interest; persons acting on behalf of another person who cannot act in their own name; persons acting as a member of, or in the interest of, a group or class of persons; and persons acting in the public interest.[85] However, the high cost of litigation is a significant barrier to the private enforcement of environmental standards in Africa given the high rate of poverty in the continent. The cost of litigation could become prohibitive because of the inordinate delays in the litigation process in Africa.[86] Moreover, rural communities adversely impacted by the environmental practices of mining companies may not be aware of their right to enforce environmental standards against mining companies due to rampant illiteracy in these communities.

6 Liability of parent companies for the wrongful actions of subsidiaries

A plethora of international initiatives have been deployed to address the problem of corporate complicity in environmental and human rights abuse in developing countries, including the OECD Guidelines for Multinational Enterprises, the International Finance Corporation's Environmental and Social Performance Standards, the UN Guiding Principles on Business and Human Rights, the UN Global Compact, and the Voluntary Principles on Security and Human Rights. Although these initiatives have enhanced the profile of corporate social responsibility in international discourse, they are yet to produce appreciable changes in corporate behaviour. Despite the application of these initiatives in the last few decades, instances of MNC complicity in environmental and human rights abuses persist in

---

84 *Mines and Minerals Development Act, 2015* (Zambia), s. 87.(7).
85 *The Constitution of the Republic of South Africa, 1996*, s. 38; *Mines and Minerals Development Act, 2015* (Zambia), s. 87.(7).
86 Emeka P. Amechi, "Strengthening Environmental Public Interest Litigation through Citizen Suits in Nigeria: Learning from the South African Environmental Jurisprudential Development" (2015) 23(3) *African Journal of International and Comparative Law* 383 at 387.

the developing world. The ineffectiveness of these international initiatives is partly attributable to the voluntariness of the initiatives. While the initiatives seek to influence positive changes in corporate behaviour, they do not impose mandatory obligations on companies to observe environmental and human rights in developing countries.

Moreover, international law does not impose human rights obligations on companies because they are not 'subjects' of international law. Companies are not 'subjects' of international law since they do not meet the criteria for 'subjects' of international law, which are the capacity to make claims for breach of international law; the capacity to enter into valid international agreements; and the capacity to enjoy privileges and immunities from national jurisdictions – that is, sovereignty.[87] International law does not hold companies liable for violating international law standards, including human rights, because companies are not its subjects. However, as this author has long argued, MNCs and other companies ought to be bearers of international human rights obligations because they enjoy certain rights under that realm of law, including the human rights guaranteed under international conventions such as the European Convention on Human Rights.[88] In fact, the day may not be far when MNCs are held liable for violation of international law. As the Supreme Court of Canada held recently:

> [I]nternational law has so fully expanded beyond its Grotian origins that there is no longer any tenable basis for restricting the application of customary international law to relations between states. The past 70 years have seen a proliferation of human rights law that transformed international law and made the individual an integral part of this legal domain, reflected in the creation of a complex network of conventions and normative instruments intended to protect human rights and ensure compliance with those rights.[89]

Hence, in the absence of any contrary domestic law, customary international law norms such as prohibition of forced labour, slavery, cruel, inhumane, or degrading treatment, and crimes against humanity form part of the Canadian common law and potentially apply to corporations.[90]

While the Supreme Court of Canada has moved the needle forward with regard to the potential application of customary international law norms to companies, the position remains that extant international law does not expressly impose human rights obligations on companies. The non-imposition of international obligations on companies has led to the creative use of the domestic laws of Western countries (home countries of MNCs) by foreign plaintiffs who seek to hold parent companies accountable for the wrongful actions of their subsidiaries in

---

87 Ian Brownlie, *Principles of Public International Law*, 6th Edition (Oxford: Oxford University Press, 2003) at 57.
88 Oshionebo, *Regulating TNCs*, supra note 2 at 146.
89 *Nevsun Resources Ltd v Araya*, 2020 S.C.C. 5 at para. 107.
90 *Ibid*. at para. 116.

developing countries. In recent years, a number of cases were filed in developed Western countries seeking to hold parent companies liable for the wrongful actions of their subsidiaries in developing countries. These cases attempt to impute liability to parent companies through the piercing of the corporate veil, or based on domestic tort law in Western countries. In the United States, for example, transnational tort litigation is premised primarily on the *Alien Tort Statute*,[91] which vests original jurisdiction in District Courts in relation to "any action by an alien for a tort, committed in violation of the law of nations or a treaty of the United States". Most of these cases have been dismissed on procedural grounds such as the *forum non conveniens* doctrine which allows a court to dismiss a case where a more convenient forum exists for adjudicating the dispute. More significantly, the US Supreme Court held in *Kiobel v Royal Dutch Petroleum Co.*[92] that the *Alien Tort Statute* does not apply extraterritorially to facts that occurred in foreign countries because "nothing in the text of the statute suggests that Congress intended causes of action recognized under it to have extraterritorial reach".[93] The *Kiobel* decision appears to have effectively dealt a fatal blow to lawsuits by foreign plaintiffs seeking to impute liability to US parent companies for the wrongful actions of their subsidiaries in developing countries.

6.1 Piercing the corporate veil

A subsidiary company possesses a legal personality that is separate and distinct from its parent company even though the parent company owns all of the shares of the subsidiary company.[94] Thus, while the parent company owns shares in the subsidiary company, in law the assets and liabilities of the subsidiary belong to the subsidiary, not the parent company. In effect, the parent company is not liable for the wrongful conduct of its subsidiary. In appropriate cases, however, the corporate veil can be pierced or lifted in order to impute liability to parent companies for the wrongful actions of their subsidiaries. Where a court pierces the corporate veil of a subsidiary company, it is in effect disregarding the separate legal personality of the company by attributing the wrongful actions of the subsidiary company to the parent company, its controlling shareholder. However, the veil is pierced only in specific instances where the court finds that the subsidiary company is used by the parent company as a shield for fraudulent, illegal, or other wrongful conduct. In Canada, for example, the corporate veil is pierced in three circumstances – namely, where the corporation is "completely dominated and controlled and being used as a shield for fraudulent or improper conduct"; where the corporation acted as the authorized agent of its controlling shareholders in engaging in fraudulent or improper conduct; and where a statute or contract requires that the

---

91 28 U.S.C. § 1350.
92 569 U.S. 108 (2013).
93 *Ibid.*
94 *Salomon v Salomon & Co.* [1897] A.C. 22 (H.L.).

corporate veil be pierced.[95] The Court of Appeal for Ontario, Canada, held recently that, in the context of a parent–subsidiary relationship, the veil of incorporation is disregarded where the court is

> satisfied that: (i) there is complete control of the subsidiary, such that the subsidiary is the "mere puppet" of the parent corporation; and (ii) the subsidiary was incorporated for a fraudulent or improper purpose or used by the parent as a shell for improper activity.[96]

In the United Kingdom, the circumstances under which the veil of incorporation is pierced appear to have been circumscribed, thus making it harder to impute liability to parent companies. The United Kingdom Supreme Court (UKSC) has held that while "the court may be justified in piercing the corporate veil if a company's separate legal personality is being abused for the purpose of some relevant wrongdoing", the veil may be pierced only where the wrongdoing involves the use of the separate legal personality of a company by the controlling shareholder to evade liability.[97] Thus, in the United Kingdom the piercing of the corporate veil is based on the 'evasion principle' which requires the corporate veil to be disregarded only where a company is used by its controlling shareholders to evade a legal liability. Expatiating on the 'evasion principle', the UKSC observed in *Prest v Petrodel Resources Limited* that

> the court may disregard the corporate veil if there is a legal right against the person in control of it which exists independently of the company's involvement, and a company is interposed so that the separate legal personality of the company will defeat the right or frustrate its enforcement.[98]

Even where the evidence shows that the separate legal personality of a company is being used to evade liability, the piercing of the corporate veil is not automatic, as the corporate veil will be pierced only where the court finds that there is no alternative legal remedy. Thus, in the UK the court will pierce the corporate veil as a last resort where no other realm of law provides a remedy for the wrongdoing for which the corporate veil is being pierced.

In the United States, the criteria for piercing the corporate appear to be in flux as courts adopt differing and inconsistent approaches to the piercing of the corporate veil.[99] Professor Phillip Blumberg weaves these differing approaches together when he states that "traditional 'piercing' jurisprudence rests on a

---

95 *Transamerica Life Insurance v Canada Life Assurance Co.* (1996), 28 O.R. (3d) 423 at 433–4 (Gen. Div.), *affirmed* (1997) 74 A.C.W.S. (3d) 207 (Ont. C.A.).
96 *Yaiguaje v Chevron Corporation*, 2018 ONCA 472 at para. 66.
97 *Prest v Petrodel Resources Limited & others* [2013] UKSC 34 at paras 27–8.
98 *Ibid.* at para. 28.
99 John H. Matheson, "The Modern Law of Corporate Groups: An Empirical Study of Piercing the Corporate Veil in the Parent-Subsidiary Context" (2009) 87 *North Carolina Law Review* 1091 at 1099.

demonstration of three fundamental elements: the subsidiary's lack of independent existence; the fraudulent, inequitable, or wrongful use of the corporate form; and a causal relationship to the plaintiff's loss".[100] The factors regularly considered by US courts in determining whether to pierce the corporate veil include whether the transaction is fraudulent or induced by misrepresentation; whether the parent company commingled its funds with the funds of the subsidiary; whether the subsidiary is undercapitalized; whether the parent company assumed the risk of the transaction; and whether the parent company exercised control or dominance over the subsidiary company.[101] As in other countries, US courts are reluctant to pierce the corporate veil due to their reverence for the separate legal personality principle. However, it appears that US courts are more willing to pierce the corporate veil in contract-based cases than in tort-based cases.[102] Given that foreign plaintiffs seeking to impute liability to US-based parent companies usually rely on tort law, these plaintiffs face seemingly insurmountable hurdles in piercing the corporate veil. Even then, it is unlikely that foreign plaintiffs would be able to invoke the jurisdiction of US courts given the US Supreme Court decision in *Kiobel*.

While in theory the corporate veil can be pierced in order to make a parent company liable for the wrongful actions of its subsidiary company, piercing of the corporate veil rarely occurs in practice. The piercing of the corporate veil seldom occurs in relation to parent companies in common law countries because, as the cornerstone of Western company law, the separate legal personality principle is zealously guarded by the courts. For example, the Supreme Court of Canada has held that "unless there is a legal basis for ignoring the separate corporate personality of separate corporate entities, those separate corporate existences must be respected".[103] In the words of the England and Wales Court of Appeal, courts are reluctant to pierce the corporate veil because a "subsidiary and its [parent] company are separate entities [and] [t]here is no imposition or assumption of responsibility by reason only that a company is the parent company of another company".[104] Thus, the separate legal personality principle "applies even where the evidence demonstrates that the corporation has been involved in impropriety".[105] In effect, the mere fact that a parent company or its subsidiary has engaged in some impropriety is not a ground for piercing the corporate veil. As one court has observed, "it is not permissible to lift the veil simply because a company has been involved in wrong-doing, in particular simply because it is in

---

100 Phillip I. Blumberg, "The Transformation of Modern Corporation Law: The Law of Corporate Groups" (2005) 37 *Connecticut Law Review* 605 at 612.
101 Matheson, *supra* note 99 at 1113.
102 *Ibid.* at 1122; Robert B. Thompson, "Piercing the Corporate Veil: An Empirical Study" (1991) 76 *Cornell Law Review* 1036 at 1058.
103 *Sun Indalex Finance v United Steelworkers* [2013] 1 S.C.R. 271 at para. 238.
104 *Chandler v Cape PLC* [2012] EWCA Civ. 525 at para. 69.
105 *Shoppers Drug Mart Inc. v 6470360 Canada Inc.* [2012] O.J. No. 4320, 2012 ONSC 5167 at para. 72.

breach of contract".[106] In this sense, the separate legal personality principle is a significant barrier to the imposition of liability on parent companies for the wrongful actions of their subsidiaries.

Moreover, in the context of MNCs, the numerous levels of ownership separating the parent company from the subsidiary often makes it impracticable to pierce the corporate veil, thus insulating parent companies from tortious liability. The MNCs operating in Africa's mining industry often have an intricate web of ownership involving parent companies, holding companies, and subsidiary companies spread across multiple countries and continents. The UNCTAD has reported that the top 100 multinational enterprises in its Transparency Index "have 7 hierarchical levels in their ownership structure (i.e. ownership links to affiliates could potentially cross 6 borders), they have about 20 holding companies owning affiliates across multiple jurisdictions, and they have almost 70 entities in offshore investment hubs".[107] This complex ownership arrangement is widespread in Africa where "hundreds of offshore-registered companies" operate in the mining industry.[108] For example, Sierra Leone Hard Rock (SL) Limited, a mining company, operates through "three separate offshore holding companies (two registered in Guernsey and one in Bermuda) with a primary owner registered in Bermuda, owned in turn by three separate holding companies (two of which were registered in London and one in China)".[109] This complex ownership structure inhibits attempts through litigation to hold parent companies liable for the wrongful actions of their subsidiaries in Africa.

The multi-layered ownership structure of MNCs is deliberately adopted in order for the parent company to take full advantage of the doctrine of separate legal personality to the effect that the subsidiary company is separate from its parent company. A Canadian court has held that

> a subsidiary, even a wholly owned subsidiary, will not be found to be the alter ego of its parent unless the subsidiary is under the complete control of the parent and is nothing more than a conduit used by the parent to avoid liability.[110]

The task of piercing the corporate veil is extraordinarily difficult in circumstances where MNCs are structured in such a manner that the parent companies based in the developed countries do not, at least on paper, have direct control over the affairs of the subsidiaries in developing countries. In order not to be seen as controlling and dominating their subsidiary companies, parent companies usually assign their shares in the subsidiary companies to a holding company. By doing so,

106 *Dadourian Group International Inc. v Simms & others* [2006] EWHC 2973, at para. 683 (Ch), *affirmed* [2009] EWCA Civ. 169 (CA).
107 UNCTAD, *World Investment Report 2016 – Investor Nationality: Policy Challenges* (Geneva: United Nations Publications, 2016) at xiii.
108 African Progress Panel, *supra* note 4 at 60.
109 *Ibid.* at 61.
110 *Gregorio v Intrans-Corp. et al.* (1994), 18 O.R. (3d) 527 (C.A.).

parent companies avoid exercising any direct control over the subsidiary companies, thus shielding the parent companies from the wrongful actions of their subsidiary companies in developing countries.

6.2 Parent company's duty of care

In recent years, Western courts have been asked to impute liability to parent companies for the wrongs committed by their subsidiaries in developing countries based on the common law duty of care.[111] While tort-based cases have yet to be litigated successfully on the merits, some of the cases raise the possibility that such a duty of care could be imposed on parent companies. For example, a Canadian court has held that, in appropriate cases, a parent company may owe a duty of care to third parties that are adversely affected by the operations of its subsidiary companies in developing countries. In *Choc v Hudbay Minerals Inc.*,[112] the plaintiffs, who are indigenous Mayan Q'eqchi' from El Estor, Guatemala, brought three related actions against Hudbay Minerals, and its wholly owned subsidiary companies. The plaintiffs alleged that the security personnel working for Hudbay's subsidiaries in Guatemala committed human rights abuses, such as shooting, killing, and gang-rapes of women, including some of the plaintiffs. The plaintiffs attempted to hold Hudbay Minerals liable for these violations because, in their view, the two subsidiaries were under the control and supervision of Hudbay Minerals. Predictably, Hudbay Minerals filed motions for dismissal of the cases primarily on grounds that the plaintiffs failed to disclose a reasonable cause of action. While noting that "the plaintiffs are not claiming that Hudbay is responsible for the torts of the security personnel, but that Hudbay was, itself, negligent in failing to prevent the harms that they committed",[113] the Superior Court of Justice for the province of Ontario held that "the plaintiffs have pled all material facts required to establish the constituent elements of their claim of direct negligence as against Hudbay, separate and distinct from any claims framed in vicarious liability as against it".[114] The court observed that, in appropriate cases, a parent company may owe a duty of care regarding the wrongful actions of its subsidiary companies. Such would be the case where it is established by evidence

> that the harm complained of is a reasonably foreseeable consequence of the alleged breach; that there is sufficient proximity between the parties that it would not be unjust or unfair to impose a duty of care on the defendants; and, that there exist no policy reasons to negative or otherwise restrict that duty.[115]

---

111 See, for example, *Garcia v Tahoe Resources Inc.*, 2017 BCCA 39 (CanLII); *Nevsun Resources Ltd v Araya*, 2020 S.C.C. 5; *Chandler v Cape Plc* [2012] EWCA Civ. 525; *Connelly v RTZ Corporation* (1999) CLC 533; *Lubbe v Cape Plc* [2000] 1 WLR 1545; *Okpabi v Royal Dutch Shell Plc* [2018] EWCA Civ. 191; [2018] Bus LR 1022.
112 2013 ONSC 1414, 116 O.R. (3d) 674.
113 *Ibid.* at para. 52.
114 *Ibid.* at para. 54.
115 *Ibid.* at para. 57, citing *Odhavji Estate v Woodhouse*, 2003 S.C.C. 69 [2003] 3 S.C.R. 263.

The court concluded thus:

> It is possible that, based on the foregoing, the defendants have brought themselves into proximity with the plaintiffs. The pleadings disclose a sufficient basis to suggest that a relationship of proximity between the plaintiffs and defendants exists, such that it would not be unjust or unfair to impose a duty of care on the defendants. Based on the foregoing, I find that it is not plain and obvious that no duty of care can be recognized. A *prima facie* duty of care may be found to exist for the purposes of this motion.[116]

The *Choc v Hudbay Minerals* case was not decided on the merits, but its significance lies in the fact that it is the first case in Canada to open the door to the possibility of imposing a duty of care on parent companies regarding the wrongful conduct of their subsidiaries in developing countries.[117]

While Canadian courts are yet to formally recognize a parent company's duty of care toward third parties, UK courts have held that, in appropriate cases, parent companies owe a duty of care to the employees of their subsidiaries operating in foreign developing countries.[118] This duty of care is imposed on parent companies where the facts and circumstances satisfy the general principles of tort law regarding imposition of a duty of care. As Sales LJ observed in *AAA v Unilever Plc*:

> [A] parent company will only be found to be subject to a duty of care in relation to an activity of its subsidiary if ordinary, general principles of the law of tort regarding the imposition of a duty of care on the part of the parent in favour of a claimant are satisfied in the particular case.[119]

The House of Lords has long established a three-part test for determining whether a duty of care arises, as follows:

> [I]n addition to the foreseeability of damage, necessary ingredients in any situation giving rise to a duty of care are that there should exist between the party owing the duty and the party to whom it is owed a relationship characterised by the law as one of "proximity" or "neighbourhood" and that the situation should be one in which the court considers it fair, just and reasonable that the law should impose a duty of a given scope upon the one party for the benefit of the other.[120]

---

116 *Ibid.* at para. 70.
117 See Chilenye Nwapi, "Resource Extraction in the Courtroom: The Significance of *Choc v. Hudbay Minerals Inc.* for Transnational Justice in Canada" (2014) 14 *Asper Review of International Business and Trade Law* 121.
118 *Chandler v Cape PLC* [2012] EWCA Civ. 525.
119 *AAA v Unilever Plc* [2018] EWCA Civ. 1532 at para. 36.
120 *Caparo Industries Plc v Dickman & Others* [1990] 2 A.C. 605 at 617–8.

In essence, UK courts require claimants to prove foreseeability of harm or damage, proximity between the claimant and the defendant, and fairness and reasonableness of the imposition of a duty of care on the defendant. Thus, in *Chandler v Cape PLC*,[121] the England and Wales Court of Appeal held that in situations where the policy of a parent company "on subsidiaries was that there were certain matters in respect of which they were subject to parent company direction", the parent company owes a direct duty of care to the employees of its subsidiary company in relation to those matters.[122] The Court of Appeal elaborated in *Chandler* thus:

> [I]n appropriate circumstances the law may impose on a parent company responsibility for the health and safety of its subsidiary's employees. Those circumstances include a situation where, as in the present case, (1) the businesses of the parent and subsidiary are in a relevant respect the same; (2) the parent has, or ought to have, superior knowledge on some relevant aspect of health and safety in the particular industry; (3) the subsidiary's system of work is unsafe as the parent company knew, or ought to have known; and (4) the parent knew or ought to have foreseen that the subsidiary or its employees would rely on its using that superior knowledge for the employees' protection.[123]

The parent–subsidiary relationship does not, in and of itself, create a duty of care. Rather, whether a parent company owes this duty of care depends on the degree to which the parent company exercises supervision and control over its subsidiaries. Where there is evidence that the parent company subjects its subsidiaries to its rules and policies, the duty of care can be imposed on the parent company, provided the requirements of foreseeability, proximity, and fairness are satisfied.

This duty of care has been expressly affirmed by the UKSC in the more recent case of *Vedanta Resources PLC and another v Lungowe and others*,[124] where the court held that the parent company's duty of care "depends on the extent to which, and the way in which, the parent availed itself of the opportunity to take over, intervene in, control, supervise or advise the management of the relevant operations (including land use) of the subsidiary".[125] In order to put this seminal decision in proper context, it is necessary to discuss the facts of the case, albeit briefly. Vedanta Resources PLC (Vedanta) is the parent company of Zambia-based Konkola Copper Mines Plc (KCM). Vedanta owns 79.42% interest in KCM through its subsidiary company, Vedanta Resources Holdings Limited (VRHL). The remaining 20.58% of the shares in KCM are owned by the Zambian government through the state-owned mining company. KCM owns and operates the

---

121 [2012] EWCA Civ. 525.
122 *Ibid.* at paras 73, 78–9.
123 *Ibid.* at para. 80.
124 [2019] UKSC 20.
125 *Ibid.* at para. 49.

Nchanga mine located in an area that encompasses waterways which flow into the Kafue river. The plaintiffs rely on the waterways as the primary source of clean water for drinking and other domestic purposes, as well as irrigation of crops, sustenance of livestock, and fishing. The plaintiffs sued the defendants, Vedanta and KCM, in negligence, alleging that the defendants, through the Nchanga mine, knowingly discharged harmful effluent into the waterways and the local environment, thus adversely impacting their livelihoods as well as their physical, economic, and social wellbeing. The plaintiffs' allegations appear to be corroborated by the Zambian Government's Auditor General who reported in 2014 that the Nchanga mine discharged effluent containing high levels of toxic metals and other substances into surface water. The plaintiffs claimed that, as a parent company that exercised a very high level of control and direction over the mining operations of its subsidiary (KCM), Vedanta assumed responsibility for ensuring that the mining operations of KCM do not cause harm to the environment or local communities. The plaintiffs alleged that Vedanta breached this duty, and hence they are entitled to compensation for the injury and loss arising from the breach of duty.

The primary issue in this case was whether English courts had the jurisdiction to adjudicate the dispute. At the trial court, the defendants brought a motion to strike the action based on the doctrine of *forum non conveniens*. They argued that English courts lacked jurisdiction to adjudicate the dispute because the facts arose in Zambia and the Plaintiffs are Zambian citizens resident in Zambia. Thus, Zambia is a more appropriate forum to adjudicate the dispute. The trial judge dismissed the jurisdictional challenge and the defendants appealed unsuccessfully to both the Court of Appeal and the UKSC. While dismissing the jurisdictional challenge, the UKSC held that, in appropriate cases, a parent company owes a duty of care to third parties (such as host communities in developing countries) that are adversely impacted by the operations of its subsidiary company.[126] The UKSC observed that the circumstances giving rise to the duty of care in *Chandler v Cape PLC* would surely result in a duty of care to other third parties "if the dust had escaped to neighbouring land where third parties worked, lived or enjoyed recreation".[127] The UKSC articulated other instances where a parent company may owe a duty of care to third parties thus:

> Even where group-wide policies do not of themselves give rise to such a duty of care to third parties, they may do so if the parent does not merely proclaim them, but takes active steps, by training, supervision and enforcement, to see that they are implemented by relevant subsidiaries. Similarly, it seems to me that the parent may incur the relevant responsibility to third parties if, in published materials, it holds itself out as exercising that degree of supervision and control of its subsidiaries, even if it does not in fact do so. In such

---

126 *Ibid.* at para. 52.
127 *Ibid.* at para. 52.

circumstances its very omission may constitute the abdication of a responsibility which it has publicly undertaken.[128]

The duty of care could thus arise from the policy documents adopted by a parent company or, in the absence of such policy documents, a parent company's conduct in holding itself out as exercising a degree of supervision and control over its subsidiaries that is sufficient to support the duty of care.

At the time of writing, a parent company had yet to be found to have breached this duty of care. Should this duty of care become the norm, parent companies would be obliged to ensure proper supervision of their subsidiaries operating in foreign developing countries. However, parent companies could potentially evade the duty of care through risk-mitigation strategies. They could avoid adopting policies regarding control or supervision of their subsidiaries, or avoid holding themselves out as exercising such control or supervision. The complex ownership structure of MNCs could enable parent companies to evade the duty of care, particularly where the parent company is separated from the subsidiary by multiple layers of intermediary owners.

7 Conclusion

A seemingly intractable crisis facing the African mining industry is the complicity of mining companies in environmental and human rights abuses. Although most African countries have enacted clear rules for the sustainable exploitation of mineral resources, these countries have failed to effectively manage the environmental and social impacts of mineral exploitation. In many African countries, mining induces environmental degradation, the displacement of local communities, and, in some instances, the violation of human rights. The environmental crisis in the mining industry is exacerbated by Africa's lack of regulatory capacity. Regulatory agencies in Africa lack the expertise, funding, and equipment to enforce environmental standards. Hence, as a last resort, African citizens adversely impacted by mining operations are increasingly relying on litigation in the home countries of MNCs. While the UKSC's decision in *Vedanta Resources PLC and another v Lungowe* has the potential to produce positive changes in the environmental practices of MNCs in Africa, this decision alone will not resolve the environmental crisis in the African mining industry since most Africans do not possess the financial resources required to litigate cases in Western countries. Rather, the resolution of the environmental crisis in the mining industry rests primarily on the effective enforcement of environmental standards, which itself depends on the capacity of regulatory agencies in Africa. Thus, African countries must enhance the capacity of their regulatory agencies by ensuring adequate funding for the agencies, as well as the boosting of the number of staff, training of personnel, and provision of equipment, tools, and laboratories.

---

128 *Ibid.* at para. 53.

10 Conclusion

1 Towards pragmatic mining regimes in Africa

This book has undertaken a comparative analysis of the legal and fiscal regimes for hard-rock mining in Africa, focusing on core themes such as the types of mineral rights, the methods for acquiring mineral rights, the security of mineral tenure, the terms and conditions governing mineral rights, the royalty and tax regimes applicable to mineral exploitation, and the management and utilization of mineral rents. As observed in previous chapters, the legal and fiscal regimes for mineral mining in Africa have been deliberately liberalized in order to attract FDI to the mining industry. The liberalization of the legal and fiscal regimes appears to be yielding the desired outcome in the sense that until a few years ago, when commodity prices plummeted on the international market, there was an upsurge in FDI inflow to the African mining industry. Correspondingly, the volume of minerals produced in Africa has also increased. Surprisingly, however, mineral rents accruing to African governments have not increased proportionately with the increase in mineral production. This is due to a myriad of factors, including the generous nature of the financial incentives granted to investors in the mining industry, the incapacity of revenue administrators to collect and remit revenues to the government, tax avoidance by MNCs, corruption, and mismanagement of revenues.

Perhaps more significantly, although Africa has long desired to harness her enormous mineral wealth as a catalyst for economic development, the reality is that mineral resource exploitation has failed to catalyze the economic transformation of Africa due to a multitude of structural and institutional obstacles highlighted in earlier chapters of this book. These obstacles include rampant corruption among Africa's ruling elites, the mismanagement of mineral revenues, the exploitative relationship between host African governments and MNCs that has resulted in the disproportionate accumulation of profits by mining MNCs to the detriment of African countries, and the deliberate and intentional avoidance of taxes by MNCs.

What, then, should Africa do in order to catalyze her economic transformation through mineral wealth? This chapter offers recommendations to African countries regarding the optimization of mining benefits. The overarching recommendation is that Africa must design pragmatic mining regimes that cater to the interests of both host countries and mining companies, as opposed to the current regimes that offer inordinately generous incentives to investors. There is no question that foreign investors are essential to the growth of the African mining industry given the dearth of domestic capital and expertise in Africa. However, Africa's mineral wealth ought not to be exploited at the expense of African citizens.

1.1 Inventory of mineral resources

At this moment, the quantity and quality of minerals in individual African countries are unknown due to lack of geological mapping of the continent. Very little geological data exists on mineral deposits in Africa. Africa would be better able to plan for the economic development of the continent through mineral wealth if she took an inventory of her mineral resource endowment. The Africa Progress Panel recommends that Africa should "[d]evelop a regional inventory of natural resources through geological mapping, building on the foundations created by the African Minerals Geoscience Initiative".[1] The question is, how can Africa achieve this objective? One strategy that could incentivize mineral exploration activities in the continent is the liberalization of the mineral right acquisition process. For example, companies should be allowed to prospect for minerals in designated areas without prior acquisition of mineral rights over the area, provided that the land is not the subject of a subsisting mineral right or a pending application for a mineral right. The designated areas, which could be referred to as 'open prospecting areas', should be clearly delineated into blocks, grids, and cells. Mining companies should then be invited by the government to express and register their interest in any of these blocks, grids, and cells on a first-come, first-served basis. However, a registered interest should have a specified duration, the expiry of which would lead to the extinguishment of the company's interest in the blocks, grids, or cells. The duration should be relatively short in order to encourage companies to prospect the area in an expeditious manner, thus preventing warehousing of lands.

As an additional incentive, companies prospecting a particular block, grid, or cell should be entitled to priority regarding acquisition of mineral rights over the land. To avoid any confusion, African countries should keep an interactive and searchable electronic (online) register of the 'open prospecting areas', including the blocks, grids, and cells being prospected at any given time and by whom. This register should be accessible to the public, thus serving as a source of information for prospective investors in the mining industry. That said, a limit should be imposed on the size of land or the number of blocks, grids, and cells that can be prospected by one prospector at a given time. This would prevent mining

---

1 Africa Progress Panel, *Equity in Extractives: Stewarding Africa's Natural Resources for All*, at 96, https://static1.squarespace.com/static/5728c7b18259b5e0087689a6/t/57ab29519de4bb90f53f9fff/1470835029000/2013_African+Progress+Panel+APR_Equity_in_Extractives_25062013_ENG_HR.pdf

346   *Conclusion*

companies from making inordinate claims regarding the size of the land they are prospecting, thus also preventing land speculation on the part of mining companies.

1.2 Capacity-building

Mining benefits in Africa can be optimized only if African countries consciously nurture the institutional capacity to manage their mineral wealth. Thus, Africa must begin to build the institutional capacity to manage her mineral wealth, including the capacity to negotiate mining contracts in a manner that enhances the benefits accruing to African countries. More specifically, Africa needs to build the capacity of local companies to exploit mineral resources; the capacity to administer and enforce laws and regulations, including the ability to undertake social and environmental impact assessment of mining projects; and the capacity to utilize mineral revenues in an efficient and equitable manner for the benefit of citizens. The building of capacity must also encompass the teaching of mining law as a distinct course at African universities and law schools. It is rather surprising that, despite Africa's abundant mineral wealth, very few African universities teach courses relating to mining. In addition, capacity-building in relation to the exploitation of minerals must entail statutory and contractual provisions compelling mining MNCs to help African-citizen employees to acquire exploration and production skills.

The building of domestic capacity to manage mineral resources should not be restricted to public institutions but must encompass host communities. As noted in Chapter 8, the capacity to negotiate and implement CDAs is crucial to the effectiveness of CDAs. Host communities in Africa lack the capacity to negotiate and manage CDAs, and therefore mining MNCs are often able to insert contractual terms that are inimical to the long-term interests of these communities. Host communities lack appropriate expertise and competence to understand the complex nature of mineral extraction projects. Thus, they are unable to articulate a coherent position regarding proposed projects in their communities. African governments and civil society groups must aid host communities to build requisite capacity, including the provision of funds for capacity-building programmes and the convening of capacity-building workshops and seminars. Such "capacity building programs should strategically target specific groups, and should aim to develop skills in areas that will support the functioning of a CDA and the longer term sustainability of community development projects".[2] Capacity-building must involve the training of local communities on various aspects of CDAs such as the management and utilization of funds; project execution and monitoring, including environmental monitoring; accountability of managers of CDAs; and conflict resolution. Civil society organizations can assist host communities by organizing workshops and training on the legal, financial, and managerial aspects of CDAs

---

2   World Bank, *Mining Community Development Agreements – Practical Experience and Field Studies* (June 2010) at 77, http://documents.worldbank.org/curated/en/697211468141279238/pdf/712990v30WP0P10IC00CDA0Report0FINAL.pdf

and by representing these communities in the course of negotiating CDAs. However, NGOs representing host communities must ensure that representatives of these communities are included in the negotiation team so that, in the course of their participation in CDA negotiation, community representatives can gain first-hand knowledge and insight regarding the negotiation process. In addition, host communities must cooperate and share experiences among themselves regarding CDAs. Host communities in individual African countries must harmonize their negotiation strategies and draw appropriate lessons from the experiences of other communities that have successfully negotiated CDAs in the past.

1.3 Transparency and accountability

Mining regimes in Africa must incorporate minimum levels of transparency and accountability such that African citizens are able to hold their governments to account for the mismanagement of mineral resources. African governments should "provide civil society groups with the political space to, for example, monitor contracts, concessions and licensing agreements in the extractive sector, and to remove restrictions on legitimate scrutiny".[3] There must be statutory provisions mandating and compelling the disclosure of all mining receipts by African governments, including taxes, royalties, and other fees paid by mining companies to African governments. In addition, all mining contracts should be disclosed to the public in a timely manner so that citizens are kept abreast of the terms on which their mineral wealth is exploited. The disclosure of such information must be instantaneous through an online portal that enables citizens to see in real time the amount of money paid to the government by mining companies. Statutory disclosure provisions should be complemented by access to information legislation that vests a right in citizens to compel the government to disclose mining-related information through established channels, including the courts. Moreover, African countries must provide their citizens with a credible and transparent process for monitoring and assessing the utilization of mining revenues by the government, including mechanisms that facilitate public scrutiny of government business.[4] The public scrutiny of mining revenues must involve independent third parties such as a parliamentary committee or, better still, an independent judicial commission with power to conduct investigations and compel production of documents.

A vital component of transparency is the disclosure of information regarding proposed mining projects to host communities, including the potential harm and the economic benefits arising from mining projects. As some observers have noted, Africa's mining regimes "should protect societies, communities and the environment by assessing the potential impacts of extractive industry activities, through research, consultation and information sharing, with an emphasis on public disclosure and public engagement".[5] An ideal transparency regime must ensure that

3 Africa Progress Panel, *supra* note 1 at 93.
4 *Ibid.* at 94.
5 *Ibid.* at 93.

host communities are consulted prior to the grant of mineral rights in Africa. Such prior consultation would enable communities to express their concerns regarding proposed mineral operations, as well as afford governments and project proponents an opportunity to adjust projects to accommodate the concerns and interests of local communities. However, the consultation being advocated here should not be seen as conferring on host communities a veto power over mining projects; rather, consultation should be a deliberative and dialogic process designed to obtain a buy-in from communities.

1.4 Responsive fiscal and tax reforms

The fiscal and tax reforms undertaken by African countries in the last few years are justifiable based on the lopsidedness of the mining regimes in Africa, which has helped mining companies to reap enormous profits, while the mineral rents accruing to African countries stagnated and in some cases dwindled. However, fiscal and tax reforms must be pragmatic in the sense that while they seek to increase the mineral rents accruing to the host country, they must take into account the interest of investors as dictated by prevailing circumstances. In this regard, the Africa Progress Panel recommends that African countries should:

> Avoid generalized use of extensive tax concessions – such as tax holidays, reduced royalty fees and the waiving of corporation tax – but when projects demand extra capital because they involve high levels of commercial risk or technical difficulties, provide tax relief in the early years on a transparent basis and with full public disclosure.[6]

The restriction of fiscal reliefs to the early years of a project is particularly appropriate in circumstances where initial projections show significant deposits of mineral resources. However, fiscal incentives may be extended based on prevailing market and economic conditions.

Africa must plug the tax loopholes inherent in extant tax policies by enacting and enforcing anti-tax-avoidance rules, including the express prohibition of tax-avoidance practices. In particular, transactions between parent companies and subsidiary companies must comply with the arm's-length standard. Non-arm's-length transactions between parent and subsidiary companies must be discounted for tax purposes and fines should be imposed on companies for engaging in such transactions. In addition, Africa must streamline and reduce the financial incentives granted to mining companies. In particular, the number of items in regard to which companies can deduct their expenses for tax purposes should be reduced to a minimum level. These incentive schemes not only impose serious financial costs on host countries, but also encourage rent-seeking on the part of tax officials, thus promoting corruption within tax agencies.

---

6 *Ibid.* at 94.

More significantly, fiscal reforms in the African mining industry must be responsive to the developmental aspirations of African peoples. For example, fiscal incentives schemes must encourage the optimization of economic benefits through the creation of linkages between the mining industry and other sectors of the economy, thus ensuring value addition to the domestic economy. Africa must encourage and incentivize domestic beneficiation of mineral ores by providing fiscal and other "incentives to favour foreign investors who build links with domestic suppliers, undertake local processing and support skills development".[7] Domestic beneficiation of mineral ores within Africa can create spill-over effects for other industries such as the manufacturing industry. Thus, statutory and contractual provisions should require a specified minimum level of beneficiation of mineral ores within Africa.

1.5 A clear contractual framework for mineral exploitation

African countries must establish a contract template that balances the economic interest of the contracting parties, as opposed to the current situation where mining contracts are negotiated on a case-by-case basis. The negotiation of contracts on a case-by-case basis is disadvantageous to African countries, given the power disparity between these countries and mining MNCs. While MNCs are able to retain the services of top lawyers, host African countries negotiate mining contracts through lawyers employed by the Ministry of Mines or the Ministry of Justice. Quite often, government lawyers are not experts in mining law and may be unfamiliar with the intricacies of the mining business. Hence, lawyers negotiating contracts on behalf of MNCs have the upper hand and are often able to extract terms that are inordinately favourable to MNCs.

As a core part of the pragmatic mining regimes suggested in this book, African countries must adopt legislative provisions requiring periodic renegotiation of mining contracts. Mining contracts should be renegotiated at specified intervals or based on specified triggers, such as a significant change in circumstances. For example, mining contracts should be renegotiated when the tax arrangements under the contracts "are out of line with international practice or generate windfall profits as a result of higher-than-expected export prices".[8]

Moreover, the variable nature of mining operations makes the case for the renegotiation of mining contracts. As argued in Chapter 7, mineral extraction contracts are negotiated based on speculative projections regarding mineral deposit and the profitability of mining projects. Such projections may underestimate or overestimate the extent of mineralization in the area covered by the mining contract. Where the mineral right holder subsequently discovers heavy mineralization in the area, the host state may justifiably feel that the mining contract should be reviewed because, under this scenario, the projections underestimated the profitability of the project. Conversely, where exploration activities

---

7 *Ibid.* at 93.
8 *Ibid.* at 94.

reveal that the area is less mineralized than originally projected, the mineral right holder may justifiably demand that the contract be reviewed to cater to the obvious overestimation of mineralization in the area.

The recommendation that mining contracts should be renegotiated in specified circumstances is not far-fetched and is, in fact, backed by international practice. International organizations have advocated for a duty to renegotiate investment contracts where the contractual equilibrium existing at the time of execution of the contract has been materially and significantly altered to the detriment of one party or both parties.[9] Several decades ago, the defunct United Nations Commission on Transnational Corporations recommended that even in the absence of renegotiation clauses in investment contracts between governments and MNCs, such contracts should be renegotiated in good faith "where there has been a fundamental change of the circumstances on which the contract or agreement was based".[10] In fact, international law recognizes the doctrine of *rebus sic stantibus*, which states that contracting parties have an obligation to renegotiate the terms of their contract where there is a change in the circumstances underlying the contract such that the economic viability of the contract is significantly jeopardized.[11] In this regard, the Vienna Convention adopts the principle of 'changed circumstances' by allowing parties to a treaty to terminate, withdraw from, or suspend the operation of a treaty where a fundamental change of circumstances has occurred with regard to the circumstances existing at the time of the conclusion of the treaty, and which was not foreseen by the parties, provided that

> (a) the existence of those circumstances constituted an essential basis of the consent of the parties to be bound by the treaty; and (b) the effect of the change is radically to transform the extent of obligations still to be performed under the treaty.[12]

Likewise, the UNIDROIT Principles of International Commercial Contracts, a non-binding set of principles designed to aid international business transactions, recognize the principle of 'hardship', which is that contracting parties may renegotiate a contract in exceptional situations where "supervening circumstances are such that they lead to a fundamental alteration of the equilibrium of the contract".[13]

---

9   See, for example, the United Nations Commission on Transnational Corporations, *Draft UN Code of Conduct on Transnational Corporations*, art. 11 (1983) 22 I.L.M. 192 at 194.
10 *Ibid.* art. 11.
11 Sangwani P. Ng'ambi, "Efficient and Flexible: The Case for Renegotiation Clauses in Concession Agreements" (2014) 45 *Zambian Law Journal* 1 at 15.
12 *Vienna Convention on the Law of Treaties between States and International Organizations or between International Organizations 1986*, art. 62.(1), https://legal.un.org/ilc/texts/instruments/english/conventions/1_2_1986.pdf
13 International Institute for the Unification of Private Law, *UNIDROIT Principles of International Commercial Contracts 2016*, art. 6.2.1 (read with the Commentary), www.unidroit.org/instruments/commercial-contracts/unidroit-principles-2016

A corollary observation is that African countries should desist from the indiscriminate grant of stabilization clauses or stability agreements. At the very least, freezing stabilization clauses should be prohibited and outlawed. Where it is imperative that stabilization clauses be granted, the scope and duration of the clauses should be narrowed to a minimum. For example, stabilization clauses should cover a few specified items as opposed to the current clauses that cover virtually all aspects of the legal and fiscal regimes for mineral exploitation. Human rights and environmental rights should be excluded from the scope of stabilization clauses. In addition, stabilization clauses should aim solely at maintaining the economic equilibrium struck by the contracting parties at the time of execution of the contract. Even then, stabilization clauses should have a short duration, preferably five years, with a provision for mutual renegotiation and renewal of the clauses at prescribed intervals. The advantage in renegotiating stabilization clauses after a prescribed period is that the contracting parties would then be privy to geological and other information regarding the mining project, information that may not have been available at the time of execution of the original contract. In addition, stabilization clauses should be granted subject to the condition that the investor shall comply with the domestic laws of the host country in order to be able to enforce the stabilization clause against the government.

2 A mutually beneficial relationship

Finally, a word for foreign investors in the African mining industry. Africa desires the inflow of FDI – hence the generous fiscal incentives offered to investors in Africa. Although these incentives appear to be attracting FDI to Africa, the problem is that MNCs in the African mining industry focus almost exclusively on wealth accumulation, sometimes to the detriment of African countries. Given the current state of affairs, particularly the enormous profits made by mining companies from operations in Africa, the time has come for companies and other investors to pursue a mutually beneficial and sustainable relationship with African countries. Such a relationship must entail the negotiation of contracts in good faith; the exploitation of mineral resources in a manner that respects the rights of host communities; the payment of all rents, royalties, and taxes due to African governments without circumscribing such financial obligations through tax-avoidance schemes; and the disclosure of all payments made to governments. Such a relationship will create a conducive atmosphere for investments to thrive, thus benefiting both investors and host African countries.

Index

Acacia Mining Plc, 246–7
Acquisition of mineral rights, 15, 72–113, 344
Africa Mining Vision, 6, 10–11, 17, 98, 254, 305
Africa Progress Panel, 319, 329, 345, 348
African Charter on Human and Peoples' Rights, 107, 109, 111–12, 226–7, 325–6
African Commission on Human and Peoples' Rights, 109–110, 227, 325–6
African Union, 6, 10–11, 101–2, 232, 238, 292
Ahafo Social Responsibility Agreement, 300
Algeria, 284
Alien Tort Statute, 335
AngloGold Ashanti, 177, 195,
Angola, 21 (**tab. 2.1**), 84, 91, 107, 136, 152 (**tab. 5.1**), 284
  Anti-tax avoidance rules, 185–91, 348
  arm's length standard, 187–8, 348
  exclusivity standard, 188
  ring-fencing rules, 188–9
  suspicious and fictitious transactions, 185–6
Anvil Corporation, 246
Application for mineral rights, 81–83
Artisanal and small-scale mining, 11–12

Badenhorst, P.J. & Mostert, Hanri, 126
Barrick Gold, 235
Beneficiation, 8–9, 49, 53–4, 98, 247–50, 261, 263, 349
Benefit-sharing schemes, 296–315
Benin, 197, 202, 203
Beny Steinmetz Group Resources, 276
BHP Billiton, 267
Bilateral investment treaties, 202–210, 212, 219, 265–6

Blomstrom, Magnus, 174
Botchway, Francis, 176
Botswana, 2, 12–13, 21 (**tab. 2.1**), 33–6, 45–7, 49, 52, 55–6, 58–62, 66, 69, 72–3, 75–7, 79–80, 83, 87–8, 95, 133, 137, 142, 146, 151–2 (**tab. 5.1**), 155–6, 164, 175, 185–91, 225, 239, 241, 250, 252, 262, 265, 273, 284–6, 309, 319–20
  Bank of Botswana Act, 285–6
  Debswana Diamond Company Limited, 239
  Income Tax Act, 18, 155
  Mines and Minerals Act, 1999, 75
  Ministry of Minerals, Energy and Water Resources, 81
BSG Resources, 235
Burkina Faso, 14 (**tab. 1.1**), 152 (**tab. 5.1**), 202, 287–8
  Community Development Fund, 288–9
  Mining Code, 2015, 288
Burundi, 203

Cameron, Peter, 213,
Cameroon, 202–3, 287–8, 291
Camps and temporary buildings, 28
Capacity-building, 346–7
Cases:
  AAA v. Unilever Plc, 340
  Agri South Africa v. Minister for Minerals and Energy, 126
  Baleni and Others v. Minister of Mineral Resources and Others, 106, 126
  Chandler v. Cape PLC, 17, 341–2
  Choc v. Hudbay Minerals Inc., 339–340
  Cortec Mining Kenya Limited v. Republic of Kenya, 210
  Kiobel v. Royal Dutch Petroleum Co., 335, 337

Index   353

Maledu and Others v. Itereleng Bakgatla Mineral Resources (Pty) Limited and Another, 126
Meepo v. Kotze, 116
Minister of Mineral Resources v. Mawetse (SA) Mining Corporation (Pty) Ltd., 115–6, 121–2
Ondombo Beleggings (EDMS) Beperk v. Minister of Mineral and Energy Affairs, 117
Vedanta Resources PLC and another v. Lungowe and others, 17–18, 341–343
Central African Republic, 14 (**tab. 1.1**), 54, 299
Chad, 284
China, 226, 231, 243, 260, 265
  Chinese mining companies, 260–1, 318
  'off-take agreements', 260
Citizen enforcement of environmental standards, 331–3
Community development agreements, 9, 13, 17, 54, 68, 88, 108, 167, 258, 271, 273, 297–307, 346–7. *See also* local development agreements
Community Property Associations, 100
Compulsory acquisition of land, 70
Congo, 202–3
Consultation with host communities, 15, 102–112
Contractual framework for mineral exploitation, 349–351
Contractual guarantees regarding payment of royalties and taxes, 189–90. *See also* performance bond
Controlled relationship, 158
Convertibility and transition of mineral rights, 136–7
Corporate income tax, 151–163, 232
Cortec Mining Kenya Limited, 233
Cote d'Ivoire, 202–3

De Beers, 240, 250
Democratic Republic of Congo, 1–2, 13–14, 21 (**tab. 2.1**), 54, 78–9, 91–2, 104, 127, 144, 152 (**tab. 5.1**), 194, 202–3, 206, 221–2, 232, 236, 239, 241–3, 245–6, 248–9, 256, 266, 274–6, 287, 299
  La Générale des Carrieres et des Mines, 239, 274–5
  Mining Contract Review Commission, 236, 245
Development agreement, 134, 183, 215, 223–4, 255. *See also* mineral development agreement

Economic Community of West African States, 102, 110, 224
Economic significance of mining, 12–14
Egypt, 197, 202–3
Environmental and human rights impacts of mining, 317–20, 324
Environmental authorization, 44, 58, 67, 87, 327
Environmental degradation and pollution, 317–9, 329
Environmental impact assessment, 322, 327, 346
Environmental liability, 323
Environmental Permit, 29
Environmental regulation of mining operations, 320–328
Environmental rehabilitation fund, 173
Equatorial Guinea, 19, 54, 284, 299
Equity participation by host communities, 314–315
Equity participation, 9, 17, 47. *See also* state participation in mining projects
Eritrea, 14 (**tab. 1.1**)
Eswatini, 203
Ethiopia, 5, 13–14, 22 (**tab. 2.1**), 28, 30, 42–3, 54, 60–1, 67, 84, 91, 137, 299
Eurasian Natural Resources Corporation PLC, 235
Exploration licence, 37–43, 82, 87, 90, 101, 115
  duration and renewal, 42
  exploration area, 39
  exploration operations, 37–8
  record-keeping and reporting obligations, 41–2
  relinquishment of land upon renewal of exploration licence, 42–3
  rights conferred by an exploration licence, 38–40
  terms and conditions, 40–41
Expropriation of mineral rights, 137, 140–1, 143, 192, 214, 234, 267–8. *See also* nationalization of mining assets
Extractive Industries Transparency Initiative, 292–3

First Quantum, 235
Fiscal and tax reforms, 231, 233, 348
Fiscal incentives, 8, 13–14, 163–73
  allowances regarding environmental rehabilitation, 173
  exemption from import and export duties, 164, 172, 179
  expatriate quotas, 172

354 Index

investment / capital allowances, 164–8
loss deductions, 164
resident permits, 172
tax credits, 164–8
tax relief / holidays, 168–71
Fiscal regimes, 146–191. *See also* royalties & taxes
Flow–through share arrangement / agreement, 98–100, 112
Force majeure, 323
Foreign direct investment, 7–8, 16, 94, 146, 148, 173–6, 191, 226, 230–1, 253, 259, 267, 269–272, 344, 351
Forum non conveniens, 335, 342
Foundations, Trusts and Funds, 17, 308–313
  Anglo American Chairman's Fund (South Africa), 311–13
  Newmont Ahafo Development Foundation (Ghana), 311
  Palabora Foundation (South Africa), 309, 312
  Rossing Foundation (Namibia), 311–12
Free, prior and informed consent, 111

Gabon, 284
Generations of mining statutes, 5–12
Ghana, 3–5, 12–14, 19–20, 22 (**tab. 2.1**), 29–30, 33, 35–6, 51–4, 57, 59, 61–2, 66, 69, 72–4, 79, 82, 85, 90, 95, 107, 109, 120, 128–9, 133, 135, 137, 146, 151, 153 (**tab. 5.1**), 156–8, 165, 171–3, 175–8, 184–9, 191, 195, 197, 199, 203, 216, 222, 224–5, 232, 238–40, 242, 251–2, 264–5, 270, 280, 284, 286–92, 299–300, 302–3, 309, 311, 313, 322, 328
  Ahafo Social Responsibility Agreement, 300
  District Assembly, 107, 290
  Environmental Protection Agency, 322, 328
  Ghana Chamber of Mines, 153 (**tab. 5.1**), 157, 290
  Ghana Revenue Authority, 274
  Income Tax Act, 157–8, 165, 189
  Minerals and Mining (Amendment) Act, 2015 (Act 900), 157
  Minerals and Mining Act (Act 703), 2006, 18, 20, 54, 199, 224
  Minerals Commission, 29, 53, 81, 90, 290
  Minerals Development Fund Act, 2016, 289–290
  Mining Community Development Scheme, 289

  Newmont Ahafo Development Foundation Agreement, 300
  Newmont Ahafo Development Foundation, 311
  Newmont Ghana Gold Limited, 302–3, 311
  Office of the Administrator of Stool Lands, 289, 291
Glencore Plc, 235,
Grant of mineral rights, 74–81
  criteria for grant of mineral rights, 85–8
  eligibility requirements, 74–7
  financial and technical competency, 79–81
  free entry method, 74
  indigenization and domestic incorporation, 77–9
  order of processing of applications, 84–5
  sovereign discretion method, 74
  timeframe for determining applications, 90–91
Gross domestic product, 13, 175,
Gross market value, 152–163
Guinea, 12, 14 (**tab. 1.1**), 22 (**tab. 2.1**), 54, 108, 117, 119, 127, 138–9, 153 (**tab. 5.1**), 194, 202, 204, 222, 232, 239–40, 242–3, 257, 262, 267, 276, 297–8, 301–2
  local development agreements, 297–8, 301
  Mining Code 2011, 127, 240

Haddow, Ken, 93
High-grading practices, 170
Historically disadvantaged persons, 10, 115, 236, 251, 257, 311
History of mining in Africa, 2–3
Human rights and labour rights violations, 317–9

Incapacity of regulatory agencies, 328–31
India, 231, 243
Indigenization, 9, 15, 17, 64, 236–8, 261, 263, 314. *See also* indigenous participation
Indigenous participation, 15, 97–100, 112
Institutional incapacity, 276, 296
Inventory of mineral resources, 345–6
Involuntary displacement, 319, 343

Junior mining companies, 94–5,

Kabanga nickel project, 235
Kenya, 11, 14, 19, 23 (**tab. 2.1**), 35–7, 43, 45–6, 54, 56, 59, 61, 64, 66, 72, 75,

91–2, 95, 98, 103–4, 109, 118–120, 127, 137, 142, 144, 153 (**tab. 5.1**), 203, 210, 232–4, 239–42, 252, 257, 284, 287–8, 297, 300–3, 306, 308, 321–2, 324–7, 332–3
  Community Assembly, 103
  Community Land Management Committee, 103
  Cortec Mining Kenya Limited, 233
  Mineral Rights Board, 103
  Mining Act, 2016, 18, 98, 103
  National Mining Corporation, 242
  Trustees (Perpetual Succession) Act, 308
Konkola Copper Mines Plc, 318, 341–2

Lacy, Willard, 37–8
Land available for mining operations, 73–4
Large scale mining, 11, 54, 56, 79, 84, 86, 92, 97, 120, 134
Legal nature of mineral rights, 15, 114, 122–130
Lesotho, 14 (**tab. 1.1**)
Liability of parent companies, 333–43
Liberia, 14, 17, 23 (**tab. 2.1**), 80, 117, 120, 138–9, 195, 218, 222, 242–3, 252, 256, 266, 291, 293–4, 296, 309–310
  China-Union Mining Company, 309
  Liberia Extractive Industries Transparency Initiative, 294
Libya, 198, 214, 284
Limited real right, 96
Local content requirements / provisions, 9, 17, 47, 53, 231, 250–2, 261, 264–5
Local development agreements, 297–8, 301
London Metal Exchange, 181

Madagascar, 14 (**tab. 1.1**), 288
Major mining companies, 94
Malawi, 242
Mali, 14 (**tab. 1.1**), 169, 202, 242, 299, 309
  Randgold Resources Limited, 169
Management of mining revenues, 273–316. *See also* mismanagement of mining revenues
Mauritania, 14 (**tab. 1.1**), 284
Mauritius, 203
Methods of acquisition of mineral rights, 81–95
  discretionary grant, 81–83
  mergers and take-over transactions, 93–95
  public auction of mineral rights, 91–93

Index 355

Mine closure and rehabilitation, 173, 326–8, 330
Mineral development agreement, 195, 218, 245, 266. *See also* development agreement
Mineral development funds, 17, 249, 273, 287–92
Minerals as 'strategic national assets', 241–2
Minimum investment requirements, 80
Minimum work and minimum expenditure requirements, 20, 33, 40, 42, 47, 49–50
Mining and the environment, 317–343
Mining lease / licence / right, 19, 47–58, 77, 83, 86–90, 115
  duration, renewal and amendment, 56–8
  programme of mining operations, 57–8, 60
  record-keeping and reporting obligations, 52–3
  rights conferred by a mining lease, 47–8
  standard terms and conditions, 49–56
Ministerial discretion, 87, 132–6
Mismanagement of mining revenues, 256
Mittal Steel, 195, 218, 245, 266,
Morocco, 202–3
Mozambique, 2, 14, 23 (**tab. 2.1**), 54, 91–2, 108, 118, 138, 153 (**tab. 5.1**), 202–3, 257, 297, 309
  Mining Law No. 20/2014, 118, 138
Multilateral Investment Guarantee Agency, 225
Multilateral investment treaties, 202, 212, 265
Multinational corporations, 16, 18, 98, 125, 148, 176–7, 179–184, 192–3, 196, 202, 211–12, 215, 218, 222–3, 226, 229, 243, 245–7, 252–3, 255–6, 258–9, 261, 263–6, 272, 276, 317, 319–20, 328–30, 333–4, 338, 343–4, 346, 349, 351

Namibia, 14, 19, 23 (**tab. 2.1**), 43–6, 98, 137, 153 (**tab. 5.1**), 225, 236–7, 239–42, 249–50, 288, 309, 311–12
  Diamond Act 13 of 1999, 249
  Epangelo Mining Company (Pty) Limited, 239, 241
  Namdeb, 240, 250
  New Equitable Economic Empowerment Framework, 237
  Rio Tinto Rossing Uranium Limited, 311–12
  Rossing Foundation, 311–12

Transformation of Economic and Social Empowerment Framework, 98
Nationalization of mining assets, 6, 234, 267–8. *See also* Expropriation of mineral rights
Newmont Ahafo Development Foundation Agreement, 300
Newmont Ghana Gold Limited, 302–3, 311
Niger, 54, 299
Nigeria, 4, 14, 17, 19–20, 24 (**tab. 2.1**), 28, 30, 37, 39, 42–3, 48, 50, 52–4, 59, 66, 68, 70, 72, 75–6, 79, 81, 83–4, 89–92, 95, 108–9, 128–9, 133, 135, 142, 144, 146, 151, 154 (**tab. 5.1**), 158–9, 165–7, 169, 172–3, 175, 182, 188, 191, 197, 202–3, 216, 225, 258, 278, 284, 286–7, 293–4, 296–8, 301, 303, 305–6, 308, 321–6
  Companies and Allied Matters Act, 308
  Companies Income Tax Act, 151, 166, 169
  Federal Inland Revenue Service, 274
  Land Use Act, 305
  Mines Inspectorate Department, 172
  Mining Cadastre Office, 52, 81, 84, 89, 95
  Nigeria Extractive Industries Transparency Initiative Act, 2007, 293
  Nigeria Extractive Industries Transparency Initiative, 295
  Nigeria LNG (Fiscal Incentives, Guarantees and Assurances) Act, 199, 202, 216
  Nigerian Geological Survey Agency, 52
  Nigerian Investment Promotion Commission Act, 138
  Nigerian Investment Promotion Commission, 78
  Nigerian Minerals and Mining Act, 2007, 76, 128–9, 306
  Nigerian Minerals and Mining Regulations, 2011, 76
  'pioneer status' tax holiday, 169
  rural investment allowance, 166–7
Non-enforcement of environmental standards, 328–331
Non-governmental Organizations, 111
Norm price, 162
Norm value, 154 (**tab. 5.1**), 162

OECD Guidelines for Multinational Enterprises, 333
Organisation for Economic Co-operation and Development, 170

Otto, James, 130, 149
Oversight and audit mechanisms, 278–280, 290–1,
Ownership of minerals, 3–5, 15, 114

Parent company's duty of care, 339–43
Parliamentary oversight, 290
Parliamentary ratification of mining contracts, 120, 224–5
Performance bond, 189
Permanent sovereignty over natural resources, 229
Piercing the corporate veil, 335–9
Power and influence of MNCs, 211
Preferent right, 314
Priority of mining operations over other uses of land, 70–71
Privatization of state-owned mining companies, 8
Profit à prendre, 47, 122–4, 128, 130,
Programme of exploration operations, 92
Programme of mining / mineral operations, 83, 86, 90
Programme of prospecting operations, 82, 92, 133. *See also* prospecting work programme
Prospecting licence / right, 30–37, 82, 85, 88–9, 115
  duration and renewal, 35–6
  prospecting operations, 33–4
  relinquishment of land upon renewal of prospecting licence, 36–7
  rights conferred by a prospecting licence, 31–32
  terms and conditions, 33–5
Prospecting operations, 31, 79
Prospecting work programme, 86, 89. *See also* programme of prospecting operations
Public auction of mineral rights, 91–3, 100–101
Public-private partnerships, 101
Publish What You Pay, 292

Randgold Resources Limited, 169
Reclamation and rehabilitation of mine sites, 322, 326–8
Reconnaissance licence / permit, 20, 22–5, 28–30, 90
  duration and renewal, 30
  rights conferred by reconnaissance licence, 28–29
  terms and conditions, 29
Register of Mineral Rights Applications, 85

Registration of mineral rights, 15, 95–97, 121
Renegotiation of mining contracts, 17, 231, 242–8, 266, 349–350
Renewal of mineral rights, 21–27 (**tab. 2.1**), 30, 35–6, 42, 45–6, 56–8, 61, 81, 132, 135
Resettlement of host community, 69
Resource nationalism, 9, 15–17, 228–72
Retention licence / permit, 19, 21–3, 25–7 (**tab. 2.1**), 43–6
  duration and renewal, 45–6
  retention area, 44–6
  rights conferred by a retention licence, 43–4
  terms and conditions, 44–5
Revenue management agencies, 274–296
Revocation / termination of mineral rights, 141–2, 231, 233–6
Right to a clean and healthy environment, 324–5, 332
Royal Bafokeng Holdings (Pty) Limited, 315
Royal Bafokeng Nation, 315
Royalties, 4, 9, 16, 119, 146–191, 231–3, 344
  ad valorem value-based royalty, 147–163, 175, 181
  profit-based royalty, 148–9, 175
  sliding-scale royalty, 150, 152–3 *(tab. 5.1)*, 156–7, 255
  types of royalties, 147–150
  unit-based royalty, 148–9, 175
  valuation methods for royalties, 150
Rwanda, 14 (**tab. 1.1**), 24 (**tab. 2.1**), 202, 299

Santiago Principles, 293
Sao Tome and Principe, 284, 295
Security of mineral tenure, 8, 11, 15, 32, 61–2, 96, 114, 130–45, 212, 344
Senegal, 202–3
Sierra Leone, 11, 14, 19, 25 (**tab. 2.1**), 28–30, 39, 42, 54, 59, 61, 66, 75, 84, 91–2, 96, 108, 121, 154 (**tab. 5.1**), 176, 203, 222, 242, 258, 287–9, 291, 297–8, 301–4, 306, 319
  Diamond Area Community Development Fund, 289
Simandou iron ore project, 235
Small-scale mining lease / minerals permit / mining permit, 19, 59–61, 79
  duration and renewal, 61

rights conferred by a minerals/mining permit, 59–60
  terms and conditions, 60
Small-scale mining, 59–61, 79, 97–8, 134
Social licence to operate, 271, 297, 300, 311
Sornarajah, M., 244
Sources of mineral rights, 15, 114–120
South Africa, 1–2, 4–5, 9, 12–14, 19, 25 (**tab. 2.1**), 29–35, 37, 43–50, 52–4, 57, 59, 61, 63–5, 67, 69–70, 72–3, 84–5, 87–90, 95–6, 100, 104–105, 109, 112, 115–17, 121, 126, 132–3, 136–7, 144, 146, 151, 154 (**tab. 5.1**), 159, 161, 167, 171, 175, 180, 185–8, 191, 198–201, 216, 219, 224–5, 232, 235–6, 239, 241, 249, 251–5, 257, 259, 261–2, 265–6, 270, 273, 287, 299, 309–15, 321, 324, 327, 330, 332–3
  African Exploration Mining and Finance Corporation (Proprietary) Limited, 239, 261
  African National Congress, 235, 259
  Anglo American Chairman's Fund, 311–3
  Broad-Based Socio-Economic Empowerment Charter for the Mining and Minerals Industry, 98, 115, 236, 251, 259, 263, 266, 310–311, 314
  Community Property Association, 100
  Income Tax Act, 1962, 160–1, 167
  Industrial Development Corporation, 239
  Interim Protection of Informal Land Rights Act 31 of 1996, 105–106
  Mine Health and Safety Act, 1996, 86–7
  Mineral and Petroleum Resources Development Act, 2002, 4, 9, 32, 37, 45, 52, 64, 70, 85, 88, 105, 115–7, 121, 126, 249, 251, 253–4, 257, 263
  Mineral and Petroleum Resources Royalty Act, 2008, 159, 198, 200–202, 261–2
  Mineral and Petroleum Titles Registration Office, 34, 49, 95, 126
  Mining Titles Registration Act, 1967, 65, 126
  MPRRA,
  National Environmental Management Act, 1998, 91, 332
  Palabora Foundation, 312

Regional Manager, Department of Minerals and Energy, 34, 44, 57, 81, 89–91, 104–105, 121
Regional Mining Development and Environment Committee, 104
Rio Tinto Palabora Mining Company Ltd., 312
Royal Bafokeng Holdings (Pty) Limited, 315
Royal Bafokeng Nation, 315
social and labour plan, 50, 58
South Sudan, 19, 25 (**tab. 2.1**), 39, 43, 46, 54, 67, 91–2, 108, 119, 137, 258, 284, 297–8, 304
Mining Act, 2012, 298
Sovereign wealth funds, 17, 273–4, 284–7, 293
  Minerals Income Investment Fund of Ghana, 284
  Pula Fund (Botswana), 284–6
  Sovereign Wealth Fund of Zimbabwe, 284–6
Special Representative of the UN Secretary-General on Human Rights and Transnational Corporations, 319
Stability / stabilization agreement, 16, 183, 195, 199–200, 216, 219, 224, 351
Stabilization clauses, 16, 137, 192–202, 211–27, 245, 265–6, 351
  economic equilibrium clauses, 193, 196–8
  freezing stabilization clauses, 193–6, 351
  hybrid stabilization clauses, 193, 198
State participation in mining projects, 54–6, 58, 87, 118, 238–40, 261, 329–30
State-owned mining companies, 6–8, 94, 261, 274, 276–7
  African Exploration Mining and Finance Corporation (Proprietary) Limited, 239, 261
  Debswana Diamond Company Limited, 239
  Epangelo Mining Company (Pty) Limited, 239, 241
  Industrial Development Corporation, 239
  La Génerale des Carrieres et des Mines, 239, 274–5
  Namdeb, 240, 250
  National Mining Corporation, 242
  Zambia Consolidated Copper Mines Limited, 7–8

Zimbabwe Consolidated Diamond Company, 237
Zimbabwe Mining Development Corporation, 7, 234, 237, 239
Surface rights, 67–9

Tanzania, 3, 14, 17, 26 (**tab. 2.1**), 35–6, 43, 45, 56, 64, 72, 75, 77–8, 84, 90–2, 95, 109, 128–9, 133–7, 140, 142–4, 149–50, 154 (**tab. 5.1**), 174, 177, 179, 190, 197, 202–3, 219, 222–3, 225, 232, 238–40, 242–3, 246–7, 249, 251–2, 254, 260, 262, 264, 279, 284, 293–4, 296, 319, 323
  Acacia Mining Plc, 179
  Barrick Gold Corp., 179
  Mining Advisory Board, 78
  Natural Wealth and Resources Contracts (Review and Renegotiation of Unconscionable Terms) Act, 2017, 243, 246–7
  Tanzania Minerals Audit Agency, 279
  The Written Laws (Miscellaneous Amendments) Act, 2017, 222
Tax avoidance practices / schemes, 8, 16, 177–82, 262, 344, 351
  round-tripping, 180
  transfer pricing, 8, 177–8, 180, 187, 262, 276
Taxes, 4, 9, 16, 146–7, 150–91
Togo, 197
Transferability of mineral rights, 62–7
Transnational tort litigation, 17, 335
Transparency and accountability, 9, 17–18, 274, 282–4, 286–7, 290–6, 347
Treaty–based stabilization regimes, 202–210
  fair and equitable treatment clause, 203–5, 209
  full protection and security clause, 203–5, 209
  umbrella clause, 206–9
Tunisia, 203
Types of mineral rights, 15, 19–61, 344

Uganda, 27 (**tab. 2.1**), 203, 278, 287–8
UN Global Compact, 333
UN Guiding Principles on Business and Human Rights, 333
UNIDROIT Principles of International Commercial Contracts, 350
United Nations Commission on Transnational Corporations, 350

United Nations Conference on Trade and Development, 173-4, 270, 338
United Nations Declaration on the Rights of Indigenous Peoples, 102
United Nations Economic Commission for Africa, 277, 282
United Nations, 229-30

Vesting of mineral rights, 120-2

Windfall tax, 151, 232
Withholding tax, 156, 176
World Bank, 6-8, 130-1, 180, 212, 225, 255, 300, 306

Zambia, 4, 13-14, 27 (**tab. 2.1**), 37, 39-40, 42-3, 47, 50-6, 62-3, 68-9, 72, 75-7, 78-9, 82, 86, 89-90, 99-100, 104, 127, 133, 137, 142, 144, 146, 150-1, 154 (**tab. 5.1**), 161-3, 168, 170, 175-8, 181, 183, 188-9, 191, 196, 215-6, 223-5, 232, 234, 236, 239, 242, 249, 252, 254-5, 262, 267, 308-9, 318, 321-4, 326-7, 332-3, 341-2
  'citizen-empowered company', 78, 236
  'citizen-influenced company', 78, 236
  'citizen-owned company', 78, 236
  Director of Mines Safety, 41
  Director of Mining Cadastre, 41, 50-1
  Income Tax Act (Chapter 323), 168
  Konkola Copper Mines Plc, 318, 341

Mines and Minerals Development Act, 2015, 10, 18, 134, 162, 332
  Mining Appeals Tribunal, 69
  Mining Cadastre Office, 41
  Mining Licensing Committee, 82, 86, 89-90, 104
  Mopani Copper Mines Plc, 178
  The Land (Perpetual Succession) Act, 308
  Zambia Revenue Authority Act, 163
  Zambia Revenue Authority, 274
  Zambia Consolidated Copper Mines Limited, 7-8
  Zambian Environmental Management Agency, 39-40, 86
Zimbabwe, 13-14, 27 (**tab. 2.1**), 54, 82, 117, 154 (**tab. 5.1**), 234, 236-7, 239, 241, 249, 257, 284-5, 291, 299, 309, 319
  'appropriate designated entity', 237
  Indigenization and Economic Empowerment Act, 237
  Indigenization and Economic Empowerment Regulations, 98
  National Indigenization and Economic Empowerment Fund, 237
  Zimbabwe Consolidated Diamond Company, 237
  Zimbabwe Iron and Steel Company, 234
  Zimbabwe Mining Development Corporation, 7, 234, 237, 239

Printed in the United States
By Bookmasters